T0122984

Papal Bull

Singleton Center Books in Premodern Europe

LAWRENCE PRINCIPE, SERIES EDITOR

PAPAL BULL

PRINT, POLITICS, AND PROPAGANDA IN
RENAISSANCE ROME

Margaret Meserve

Johns Hopkins University Press
Baltimore

© 2021 Johns Hopkins University Press
All rights reserved. Published 2021
Printed in the United States of America on acid-free paper
2   4   6   8   9   7   5   3   1

Johns Hopkins University Press
2715 North Charles Street
Baltimore, Maryland 21218-4363
www.press.jhu.edu

Library of Congress Cataloging-in-Publication Data
Names: Meserve, Margaret, author.
Title: Papal bull : print, politics, and propaganda in Renaissance Rome /
Margaret Meserve.
Description: Baltimore : Johns Hopkins University Press, 2021. | Series:
Singleton Center books in premodern Europe | Includes bibliographical
references and index.
Identifiers: LCCN 2020020200 | ISBN 9781421440446 (hardcover ; acid-free
paper) | ISBN 9781421440453 (ebook)
Subjects: LCSH: Printing—Italy—Rome—History—Origin and antecedents. |
Printing—Italy—Rome—History—16th century. | Book industries and
trade—Political aspects—Italy—Rome—History—To 1500. | Book
industries and trade—Political aspects—Italy—Rome—History—16th
century. | Catholic Church—Publishing—Italy—Rome—History—To 1500. |
Catholic Church—Publishing—Italy—Rome—History—16th century. |
Bulls, Papal.
Classification: LCC Z156.R7 M47 2021 | DDC 686.209456/32—dc23
LC record available at https://lccn.loc.gov/2020020200

A catalog record for this book is available from the British Library.

*Special discounts are available for bulk purchases of this book. For more
information, please contact Special Sales at specialsales@jh.edu.*

Johns Hopkins University Press uses environmentally friendly book materials,
including recycled text paper that is composed of at least 30 percent
post-consumer waste, whenever possible.

For Robert

CONTENTS

I am grateful to Lawrence Principe and the Charles Singleton Center for the Study of Premodern Europe at Johns Hopkins University for the invitation to deliver the Singleton Lectures in September 2016. Although this book covers more than the original lectures, the talks I gave at Hopkins—on rituals of publication, inventions of tradition, and the branding of papal bulls in early modern Rome—form the backbone of the present study. For their hospitality during my stay at Hopkins, I also want to thank Shane Butler, Stephen Campbell, Earle Havens, Eugenio Refini, and Joaneath Spicer.

I'm also grateful for the support of many agencies and institutions. The American Academy in Rome, the National Endowment for the Humanities, and the American Council of Learned Societies provided time in Roman collections. A fellowship year at the Newberry Library in Chicago provided time to research and write. Three separate stays at Civitella Ranieri in Umbertide provided inspiration and solitude, and I am grateful to Dana Prescott and Diego Mencaroni for their kindness. I also thank the Institute for Scholarship in the Liberal Arts in the College of Arts and Letters at Notre Dame, the Notre Dame Center for Italian Studies, Notre Dame Research, and Notre Dame's Rome Global Gateway for grants supporting research and travel. Thanks, too, to the American Philosophical Association for a travel grant.

Librarians and archivists facilitated visits to collections across Europe and North America, and I thank the staff at the British Library; the Biblioteca Apostolica Vaticana and Archivio Apostolico Vaticano; the Archivio di Stato, Biblioteca Casanatense, Biblioteca Corsiniana, and Biblioteca Nazionale in Rome; the Staatsarchiv Basel; the Pierpont Morgan Library and New York Public Library in New York City; the Newberry Library in Chicago; and the library at Bryn Mawr College. Librarians in Tübingen, Munich, Berlin, Novara, the Library of Congress, the Huntington, and the Boston Public Library answered queries from afar. At Notre Dame, David Gura, Julie Tanaka, and Tracy Bergstrom tracked down references, solved bibliographical problems, and located sources. Special thanks to Julia Schneider, librarian of Notre Dame's Medieval Institute, for making the MI such a superb place to work. Katie Labor took photos at the Library of Congress. At Notre Dame, Erica Hastings and Mihow McKenny provided research assistance. Mousa Mohammadian organized a morass of images and permissions.

For expert advice, helpful references, and illuminating discussions, I also thank Monica Azzolini, Carrie Beneś, the late Joe Buttigieg, Ted Cachey, Dorigen Caldwell, Mary Baine Campbell, Tom Cohen, Jon Coleman, Anthony D'Elia, Simon Ditchfield, Cristina Dondi, Patricia Edbury, Falk Eisermann, Margot Fassler, Catherine Fletcher, Irene Fosi, Danielle Fosler-Lussier, Paul Gehl, Ken Gouwens, Daniel Greene, Brad Gregory, Patrick Griffin, Julia Hairston, James Hankins, Dan Hobbins, Peter Holland, Alizah Holstein, Philippa Jackson, Kate Jansen, Katie Jarvis, Peter Jeffery, Claire Taylor Jones, Richard Kieckhefer, Joëlle Rollo Koster, Tom Kselman, Claudia La Malfa, Susanne Meurer, Emily Michelson, Lisa Mignone, Steve Milner, Vernon Minor, Anna Modigliani, Christian Moevs, Emily O'Brien, Anthony Ossa-Richardson, Alessandra Petrina, Andrew Pettegree, Manu Radakrishnan, Diana Robin, Massimo Rospocher, Ingrid Rowland, Paolo Sachet, Chris Schabel, Michaela Schedl, Michael Schreffler, Marcello Simonetta, Luka Špoljarić, Francesca Trivellato, David Whitford, Lila Yawn, and Carla Zecher.

Certain friends lent special help at just the right moment. Annie Giletti shared anecdotes of legal process serving on a memorable day in the Musei Capitolini as we sat by the *Lion Attacking a Horse*. Mary Laven listened to talk about Loreto on several continents and sent photos of Loreto editions in Cambridge when catalogues failed. Sheila Pierce Ortona sent a photo of a Roman inscription when I badly needed it. Caroline Goodson read and offered encouragement on chapter one when I nearly lost faith in it. Heather Hyde Minor shook photographs loose from a Roman library. CJ Jones helped me think about German Dominicans and German bibliography. Sarah McKibben was ruthless with writing encouragement. Scott Meserve brought to my attention what became the closing anecdote of the book. Anthony Grafton came through with reassuring advice at a critical point. Jill Kraye continues to be the most exacting mentor and critic, whose standards inspire me daily.

Martin Davies, Mary Laven, Tom Noble, John van Engen, Tommy Goulding, Brian Krostenko, Marcello Simonetta, and Denis Robichaud read individual chapters and provided helpful critiques. Tom Burman, Brad Gregory, Dan Hobbins, Ted Cachey, and Tom Kselman read a draft of the entire manuscript and helped me think through revisions. Liza McCahill read both the proposal and the manuscript and offered generous suggestions for improvement. To her, and to the press's other, anonymous reviewer, I am immensely grateful. Of course, the faults and errors remain my own.

A portion of chapter two appeared in an earlier version in the article "News from Negroponte," published in *Renaissance Quarterly* 59 (2006), which generously gave permission for me to use the material again here.

Two periods of institutional service bracketed this work and in different ways helped to shape it. I began my academic career at the Incunabula Short-Title Catalogue in the British Library. My colleagues Martin Davies, John Goldfinch, and Lotte Hellinga provided an education in the study of incunabula I could not have found anywhere else, and it was there I began thinking about how to integrate the study of bibliography with cultural and political history. This book is not a work of formal bibliography, by any means, but I could never have written it without the training and experience my colleagues at the BL provided.

Two decades later, John McGreevy asked me to serve as the associate dean for the humanities in Notre Dame's College of Arts and Letters. I did not expect to find parallels between university administration and the Renaissance papal court, but there they were. From using formularies in order to draft official correspondence, to the problems of administrative publication (If we email it, is it binding? What if we "post" it online?), to the politics of sticking up posters in public spaces, the dynamics of communication in the modern university have informed the arguments of this book in myriad ways. After leading an effort to renovate a campus coffee shop, I even found hostile placards stuck to my office door. For help navigating these experiences and much more, I'm grateful to Megan Snyder, Jenny Petersen, Chloe Leach, Kathy Fischer, Karin Dale, Mo Marnocha, Rob Becht, Dawn Foster, Michelle LaCourt, Allison Collins, Matthew Fulcher, Kathy Knoll, Matt Zyniewicz, Kate Garry, Tom Merluzzi, Alison Rice, Mark Schurr, Peter Holland, Jim Brockmole, Elisabeth Köll, John McGreevy, and Sarah Mustillo.

Emily Osborn, Sara Maurer, Maria di Pasquale, Liz Zapf, Joanie Downs Krostenko, and Sophie White have shared the peculiar joys of life as a parent, in the academy, in northern Indiana. I couldn't ask for better friends. For as long as I can remember, Helen and Hamilton Meserve have been thinking and talking about history, news, print, and politics, and they've inspired me with their passions. My children have contributed too: Alice, with her fascination with politics and political communication; Tommy, with his love of Italy and Italian history; and Lily, with her meticulous eye for the art of the material text. They slowed this project down terribly but made life so much richer. Without Robert Goulding, none of it would have been possible. Thank you.

Papal Bull

Introduction

In March 1455, Aeneas Sylvius Piccolomini sent a letter from the imperial court in Austria to a friend in Rome to report some intriguing news. The previous October, a "marvelous man" had turned up in Frankfurt carrying sheets of a mechanically produced Bible, whose words were so remarkably clear they could even be read without spectacles. The traveling salesman was taking orders for a complete book produced in this same uniform style, and by the time Piccolomini wrote, the print run had sold out. The printed papers were almost certainly unbound sheets of the forty-two-line Bible. Whether the marvelous man was Johannes Gutenberg himself or one of his commercial partners, sent upriver from Mainz to Frankfurt to find customers for their venture, Piccolomini's letter provides important evidence for the Gutenberg Bible on the cusp of completion and a preview of the advances in European communication that printing with movable type would soon usher in.[1]

The forty-two-line Bible is traditionally considered Europe's first printed book. By the time Piccolomini caught sight of it in early 1455, however, Gutenberg had already printed a number of smaller works, test pieces that he used to refine and perfect his technique.[2] These ephemeral texts include two single-sheet papal indulgences, certificates guaranteeing a reduction of the time a sinner must spend in Purgatory, printed in the autumn of 1454 for distribution to the faithful who contributed to a new crusade against the Turks. Also dating to around this time is the so-called *Türkenkalendar*, a printed pamphlet containing a German poem encouraging the princes of Europe to take action against the Turkish threat. A little later, in 1456, Gutenberg's press would issue a bull of Pope Calixtus III encouraging yet more contributions to the crusade, first in Latin and then in a German translation.[3]

Scholars of printing history have scrutinized these short texts for what they reveal about the technical development of Gutenberg's art.[4] They have also been studied as examples of the urgency of the Turkish question—and the lingering persistence of the crusade as an ideal—in fifteenth-century Europe.[5] Constantinople fell to the Ottoman Turks in May 1453; the next year saw Frederick III convene two imperial diets to discuss a European response to the Ottoman threat. The mass reproduction of crusade indulgences, anti-Turkish poetry, and a crusading bull suggests how deeply concern for the Turkish problem resonated throughout German society in the mid-1450s.

Another angle from which to view these earliest printed texts is that of the history of the Catholic Church. They signal the extraordinarily early adoption of print by agents of the Church in one of its key operational areas, raising funds from the laity. Indulgences themselves were not new, but the industry of selling them grew rapidly with the spread of printing; over the next decades, printers across the Holy Roman Empire and beyond would turn out hundreds of thousands of copies of indulgences and confessional certificates, advertisements of new indulgences, notices announcing the renewal of older offerings or the credentials of new commissioners, and papal bulls and briefs encouraging the faithful to take advantage of the spiritual benefits indulgences could offer.[6] Entrepreneurial churchmen seized on printing as a way to supply and expand the market for indulgences on an industrial scale. The flood of printed matter would eventually provoke complaints from Martin Luther, among others, who challenged the theology of indulgence preaching and the Church's aggressive marketing of pardons, and these complaints were themselves soon printed and sold across the same popular markets. Luther's protest against indulgences was a phenomenon both prompted and spread by the new technology of print; a through line runs from Gutenberg's earliest printed indulgences to the Ninety-Five Theses Luther promulgated in Wittenberg in October 1517.

Following the pioneering work of Robert Scribner, much recent scholarship on the Reformation has focused on its character as a "media event," exploring how print enabled the spread of Luther's challenges to indulgences, the theology of penance, and papal authority.[7] In this view, an enterprising group of theologians and scholars, working in collaboration with equally enterprising printers, spread a revolutionary message to the European reading public. Using cheap formats like the broadside or the quarto pamphlet, publishing texts both in the high scholastic Latin of academic theology and in local vernaculars, and using images as well as text to convey their key points, reformers and printers exploited a new technology in order to launch a movement of momentous social change.

But what of the papacy? In this scheme, the Roman Church, the conservative and authoritarian institution on the receiving end of the Reformers' printed critiques, is supposed to have stood by, unsure how to respond or unwilling to seize the same communication tools or engage in the same popularizing discourse.[8] This historical model, which pits innovation against stagnation, and German technological ingenuity against Roman reliance on tradition, overlooks the fact that long before the advent of print the Catholic Church had engineered a communications revolution of its own. The bishop of Rome presided over the largest and most elaborate communications apparatus in the late medieval world. The papal chancery produced thousands of papal bulls every year and could reproduce these documents in significant numbers when required. Three hundred handwritten copies of the 1439 bull *Laetentur caeli*, announcing the union between the Orthodox and Latin churches after the Council of Florence, were distributed across Europe according to the account books of the Apostolic Camera.[9] From within the secretariat, papal letters were routinely copied to "the Christian princes," those heads of state whom the pope wished to keep informed of his actions. Papal secretaries thought nothing of drawing up dozens of copies of a single papal letter and sending them to the farthest corners of the continent. Nor were the pope's communications restricted to the powerful. The pope claimed spiritual jurisdiction over every Latin-rite Christian from Iceland to the Middle East and beyond, and since the Gregorian reforms of the late eleventh century, popes and their functionaries had been issuing more pedestrian documents—indulgences, dispensations, pardons, and more—to these far-flung communities.[10] Arguably, it was the ubiquity of papal documents, and the vast economy of grace and pardon that papal documentary practices kept afloat, that prompted Luther's new theology in the first place.

The papacy added print to its communications apparatus early on. When Piccolomini heard of the Gutenberg Bible in the spring of 1455, Luther's call to reform lay more than sixty years in the future. It took a few years for printing to reach Rome and for the pope (as opposed to his agents abroad) to commission the printing of official documents. But once the popes began to print, they did so with enthusiasm. They printed bulls and briefs, treaties and declarations of war, announcements of new doctrines, canonizations, and spiritual initiatives like the Jubilee, the proceedings of their councils, and denunciations of their enemies. By 1517, six successive popes had been using print to broadcast their vision of the way the world was and ought to be, in many dozens of editions, many thousands of copies. Just how they did so—how the Renaissance popes adopted print, how their chancery officials put it to use, how they deployed printed texts in a multitude of political, ecclesiastical, and cultural contexts in order to support and defend the Roman Church and their supremacy over it—is a story known only in part, and the subject of this book.

In the six decades between Gutenberg and Luther, Popes Paul II, Sixtus IV, Innocent VIII, Alexander VI, Julius II, and Leo X and their administrations exploited print for a wide array of purposes: to prosecute diplomatic and military campaigns against political opponents; to condemn and silence ecclesiastical critics and rivals; to promote the sale of indulgences and to advertise new devotional initiatives; and to project images of their authority as ancient and unassailable and of Rome as their capital, the newly resplendent metropolis of Christendom from where the pope dispensed grace to the faithful and harsh justice to those who dared oppose him. Papal use of the press varied over time, and in this book I pay particular attention to the different ways that different popes deployed the medium. Other figures in the ambit of the papal curia—scholars, poets, and diplomats, authors of devotional texts and pilgrim guides—published texts that popes may or may not have endorsed but that added to the universe of Roman print and burnished the image of the Renaissance papacy the new technology was helping to spread. Critics of the papacy also used the press to spread more negative views of Rome and its policies, provoking the popes into some of Europe's first printed pamphlet wars. From printed bulls of excommunication and declarations of war to Jubilee announcements and new indulgences, from pilgrim guides and souvenir broadsides to diplomatic orations, humanist treatises, and panegyric verse, Renaissance Rome was a city awash in print, much of it papal and much of it political. The question, then, is not whether the Roman Church had as much interest in print as the Reformers, but what kind of print culture it fostered. How did a deeply traditional and autocratic institution adopt a new technology? What did the printing "revolution" look like, and just how revolutionary was it, when it came to the Eternal City?

Three years after filing his report on the Gutenberg Bible, Aeneas Sylvius Piccolomini was elected pope, taking the name of Pius II. Already a renowned humanist author, the new pontiff continued to write after his election: histories, geographical works, even his own autobiography, the only memoir ever composed by a reigning pope.[11] Even so, and despite his early encounter with the press in Germany, Piccolomini never saw his literary works appear in print. The works of his that were printed in his lifetime were official documents produced by his chancery and printed far away to the north, with most concerning a bitter ecclesiastical controversy that broke out in Gutenberg's city during Pius's reign.

In 1459, the archbishopric of Mainz fell vacant. Two candidates vied for the position, one backed by Pius and the other by the cathedral chapter and city council.[12] Sometime after August 8, 1461, a letter from Emperor Frederick III appeared in print from the press of Johann Fust and Peter Schoeffer, Gutenberg's former partners and now his successors in the printing trade in Mainz. The imperial letter declared that Pius had deposed the city's candidate, Diether von Isenburg, in favor of his own choice, Adolph von Nassau, the city's true bishop.[13] Soon after, three briefs and a bull from Pius himself, all supporting Adolph, appeared from the same printshop.[14] In the spring of 1462, Diether responded with three letters of his own, also printed by Fust and Schoeffer.[15] All were printed as broadsides, to be posted on walls and doors around town for all to see. In each, the author pleaded his case by framing it as part of a larger contest between the principles of papal supremacy (surely the pope had the right to appoint his own bishops) and German self-rule (surely the cathedral chapter of an imperial free city had the right to fill its own see). Soon Adolph weighed in with a manifesto of his own, another broadside printed by Fust and Schoeffer.[16]

The printers, Fust and Schoeffer, must have profited as they printed texts for both sides of the dispute; one has to admire their business sense. The episode also reveals how quickly leading prelates of the Church in Germany moved to adopt the new technology of print, still less than a decade old, not just to exhort the faithful to support a crusade, but also to attack and defend in the course of intramural controversies. The affair of the Mainz archbishopric has been called the first "pamphlet war" in the history of the European press.[17]

The printed papal texts mentioned thus far—the 1454 crusade indulgences, the Calixtus bull, and Pius's bull and briefs against Diether von Isenburg—were issued in the name of, and with explicit reference to, the reigning pope. This hardly means that Nicholas V, Calixtus III, or Pius II ordered their printing. Enterprising local clerics surely commissioned their publication. But the papal administration in Rome would soon see the potential of the press at closer hand. In October 1462, Adolph von Nassau occupied Mainz with troops and sent hundreds of citizens into exile, including many apprentice and journeyman printers whom Gutenberg, Fust, and Schoeffer had trained. Some made their way south over the Alps into Italy, bringing with them the art of printing.

By late 1464, Conrad Sweynheym and Arnold Pannartz had set up a press at the remote monastery of Subiaco, high in the Sabine hills east of Rome.[18] By November 1467, the printers had moved to Rome, where the Bavarian Ulrich Han had already opened a printshop, possibly with another German, Sixtus Riessinger.[19] By the early 1470s, at least half a dozen printers, almost all German craftsmen with connections to Mainz, were at work in the papal city.[20]

They were not brought to Rome by the pope, or indeed by any churchman acting in an official capacity.[21] Rather, they gravitated toward the vast crowd of educated clerics, jurists, scholars, curial officials, secretaries, and scribes who populated the overlapping worlds of the papal chancery, the Roman legal system, the university, and the papal court, many of them enthusiastic participants in the Renaissance revival of classical learning both on the job and in their spare hours. Guided by local humanist editors and correctors, the early printers of Rome tried to cater to this market with a stream of well-known and more obscure classical texts.

Their efforts were certainly heroic: Sweynheym and Pannartz, working with the humanist bishop Giovanni Andrea Bussi, turned out substantial editions of classical authors (Cicero, Livy, Virgil, Caesar) and early Church fathers (Augustine, Jerome, Lactantius) that they imagined would be attractive to humanistically educated Roman readers. Ulrich Han started with editions of Cicero, Juvenal, Livy, and Plutarch, though he added texts of civil and canon law to his catalogue and theological works by living luminaries like Juan de Torquemada and Rodrigo Sánchez d'Arévalo. Sixtus Riessinger turned out editions of Terence and Sextus Aurelius Victor before decamping to Naples; Georg Lauer started yet more ambitiously with two volumes of Chrysostom. The race was on to publish ever more difficult and arcane classical texts: Apuleius, Aulus Gellius, Persius, and Martial; Statius and Silius Italicus; Polybius and Columella. Before long, this strategy proved unsustainable: the market for classical texts was smaller than anticipated, and most of Rome's early printers went bust.[22]

In 1471, Bussi wrote to the newly elected Sixtus IV, seeking a subvention for his desperate printers: they sat in their shop, the bishop said, surrounded by stacks of unsold quires but with nothing to eat.[23] Soon after, Sweynheym disappeared from the scene, and Pannartz turned to printing less ambitious and more lucrative legal, theological, and vernacular texts. Among his last works were a guidebook cataloguing the city's indulgences, printed for pilgrims to the 1475 Jubilee, and two editions of Thomas Aquinas.

It is at this point that many histories of early printing take their leave of Rome. Given the abiding concerns of much traditional book history—tracing the appearance of first editions of canonical texts, on the one hand, and assigning credit for innovations in typography and design, on the other—bibliographical attention has tended to wander away from Rome after the early 1470s—and thus it has tended to overlook the print culture that developed in and around the papal court in the following decades. The conventional narrative has it that German technicians, fleeing civil war in Mainz, took bad advice from local enthusiasts of humanism, invested heavily in large folio volumes of arcane literary texts, and by their failure proved that early Renaissance Rome could not sustain a printing program of avant-garde classical scholarship. Those printers who remained (or arrived after the crash) set off in new but uninteresting directions. In a recent survey, Concetta Bianca describes how, after their overly optimistic early experiments, Roman printers confined themselves to one or two safe paths. One led to the university, where scholarly and humanistic printing continued, as professors ordered editions of books they intended to teach, and the other led to the papal curia, where the printing of papal bulls and orations delivered before the pope soon prevailed.[24]

Histories of printing have rarely paid much attention to the kind of official, or quasi-official, publication that Bianca describes.[25] Having surveyed early attempts to print the classics in Rome, most surveys of Renaissance printing shift focus fairly quickly to Venice,[26] where Nicholas Jenson cut a beautiful roman type that he used to print elegant editions of the classics (that sold!) and laid a foundation on which Aldus Manutius would build his even more successful program of classical publishing.[27] From Venice, print history moves on to Florence, where the Ripoli press and Niccolò Tedesco turned out popular ballads, religious verse, and elegant editions of Medici court poetry, while other Florentines carried scholarly publishing into the sixteenth century.[28] Historians of early print may glance back to Rome at certain points to note the proliferation of pilgrim texts, indulgence catalogues, and guides to the monuments of the city, which multiplied with the buildings they described as the Roman Renaissance reached its peak. But the standard narrative of printing history (the Anglophone narrative, at least) usually leaves Rome in the mid-1470s and rarely returns till the 1520s or even later. It's then that the papacy began to contract more regularly with particular printers to publish its decrees, eventually appointing *tipografici apostolici*, and later, dismayed by the spread of Lutheran ideas, establishing the Roman Inquisition (1542) and the Index of Prohibited Books (1559).[29] In this narrative, Rome seems to have responded to the revolutionary spread of print by doubling down on a strategy of regulation, centralization, and authoritarian control. The general picture is of a city and an institution behind the times: retrograde and resistant to print in its early years and quite unprepared for the coming shocks of religious reform.[30]

This book challenges these assumptions about the nature of Roman printing, about Renaissance Rome itself, and about the Roman Church that animated both, by offering a more complete picture of print culture in the city and the papacy's use of the press as a political, ecclesiastical, and cultural tool. As my opening examples suggest, the new technology was adopted by agents of the Church for operational purposes from the very start: raising money, issuing pardons, announcing appointments, and promulgating decrees. This in itself is not a controversial point,[31] and the proliferation of such basic and functional publications fits well with a recent argument that much early European printing amounted to little more than the production of stationery.[32]

Here, though, I am not so concerned with the printing of forms (not even the printing of indulgences) or the paperwork of ecclesiastical bureaucracy as with how the Renaissance popes used print as a political tool. The late fifteenth century saw the papacy involved in an unrelenting series of wars, diplomatic disputes, and ecclesiastical controversies. How the popes turned to the press for help in these conflicts has been little studied, despite the extraordinary role the Renaissance papacy played in the development of political iconography and political discourse in the early modern period.[33]

Those dozens of bulls and orations that poured off of Roman presses between 1470 and 1520—what were they about, and what circumstances prompted their publication? What did the popes and their chancery officials hope to achieve by commissioning their mass reproduction? And what can their efforts tell us about the history of the Renaissance papacy, its character and its concerns?

In Rome—as in Mainz—raising support for a crusade against the Turks and asserting papal authority over princes and prelates were urgent issues and ones the papacy used the press to pursue. The popes used the new technology for other purposes too: asserting Rome's temporal jurisdiction over the papal states; protesting the actions of foreign princes and republican governments; prosecuting political rivals with decrees of interdict and excommunication; summoning the faithful to Rome for successive Jubilees; and convening a major church council (Lateran V), communicating its decrees, and condemning rival gatherings as illegitimate. Popes from Paul II to Leo X, their administrations in Rome, and their diplomatic representatives abroad all used the press to prosecute these political goals. They published briefs to other princes, the texts of treaties and decrees, and many condemnations of their rivals. Though these communiqués could take various documentary forms, by far the most prevalent was the *bulla* or papal bull, the most formal type of papal letter produced by the papal chancery, named for the *bulla* or lead seal that was attached to the parchment and authenticated its contents.

Printed papal bulls thus feature in every chapter of this book, from Calixtus III's *Bulla Thurcorum* of 1456, to Sixtus IV's bulls excommunicating the Medici in the late 1470s, to the "thunderbolt" bulls Julius II lobbed against the Venetians and the French in the first two decades of the sixteenth century, to *Exsurge Domine*, Leo X's bull condemning Luther. Crafted by learned scholars in the papal chancery, fortified with ancient diplomatic formulas but also enlivened by flashes of the newly fashionable humanist rhetoric, papal bulls were Rome's most formal but also most common medium of political communication. How the physical and textual form of the manuscript bull was translated into print is an important part of my story, as various material practices of the late medieval chancery were preserved, but also transformed, in the new medium of print.

Here my work intersects with recent scholarship on the techniques of the early modern archive.[34] While European states continued to collect, store, and manage information in manuscript for many centuries after the invention of printing, Renaissance chanceries also adopted print, especially when the promulgation of decrees could serve a political or propagandistic function. We know less about this process of making legal records public. How did government bureaux that operated primarily in "manuscript space" compose, prepare, and distribute decrees in print? What was kept and what was lost in the process?

Papal bulls usually conveyed some kind of decree or legal sentence, but they could also deliver a good deal else: exhortation, character assassination, bluster, special pleading, vituperation, wishful thinking, appeals for sympathy, imprecations to God. Print allowed these documents to travel more widely and in greater numbers; increasingly, the rhetoric of papal bulls betrayed awareness of a broader audience whose opinions needed to be courted. The idea that the pope should persuade as well as command the faithful was not entirely new in the fifteenth century, but the papacy's political situation was increasingly precarious. The pope was at war almost constantly in these decades, sometimes with Italian powers, sometimes with a northern kingdom, sometimes with both. The French invaded Italy in 1494, and Rome itself would be sacked by imperial troops in 1527.

As humanist secretaries poured into the papal chancery, eager to display their command of Latin prose and their rhetorical skills, they added high color to traditional documents issued to defend the pope and prosecute his policies. The dense legal language of the medieval papal bull was never dispensed with; Renaissance bulls, like their medieval predecessors, contain vast thickets of legalistic prose: assertions of the authority of the pope to issue them, dire warnings against any who dared impede them, and legal hedges against future contingencies conveyed in long stretches of pleonastic fussing. Amidst these, secretaries and chancery staff began to add new passages intended to persuade, justify, defend, or spin the actions of the ruling pope in the best possible light, and to cast the greatest possible doubt and denigration on his enemies. The Renaissance bull was an appeal not just to divine law but also to public opinion; it was a documentary form evolving in just the right ways, and at just the right moment, to take off as it did via publication in print.

The papacy's adoption of print is one chapter in the larger story of the institution's late medieval reinvention as a territorial state and a bureaucratizing institution. After a century of "captivity" in Avignon and the tumult of the Great Schism, the papal court returned to a dilapidated Rome in 1420. Popes like Eugene IV, Nicholas V, Pius II, and Sixtus IV reimagined the role of pontiff as the absolute monarch of the Church and a territorial, perhaps even dynastic prince of Italy, with a splendid capital city to match. The Renais-

sance popes would stake their claims to authority (over Church and Christendom alike) not just on their apostolic succession from Peter but also on
their possession of the holy places of the Eternal City. Deploying innovations
in ceremony, diplomacy, ritual, art, architecture, and city planning, the popes
grounded their claims to supremacy in the topography of their ancient and
Christian metropolis and then sought to project that authority out from Rome
to the rest of Christendom.[35]

Beyond the walls of Rome, papal power was more spiritual than political,
but the popes were deeply invested in temporal affairs. Their finances and
their liberty depended on them. The popes had to contend with a host of
other powers across western Europe, from the empire and rising independent kingdoms like France, Spain, and Bohemia to republics like Florence
and Venice and the regimes of subject states and cities like Ferrara, Bologna,
and Urbino. Not least, the popes faced pressure from within their own walls,
from aristocratic Roman families and the city's communal administration,
which from the twelfth century had claimed certain rights of self-governance.

Different challenges arose from each quarter: from northern Europe, attempts to localize church income and appointments (as Diether von Isenburg
argued the city of Mainz could do); from the Italian states, incursions into
papal territories and disputes over taxes and trade; from within Rome, conspiracies by disaffected barons. All of these periodically erupted into violent
confrontations. One thing these challengers had in common was the tendency to invoke conciliar ideas—the late medieval idea that it was not the
pope but the "universal church in council" that held the highest authority
in the Church. Since the Council of Constance resolved the Great Schism in
1417 by deposing three rival popes and appointing Martin V, the question of
where supreme power rested in the Church lay open. Did the pope preside
over the council, or vice versa? Could a new council depose another pope? In
1439, the Council of Basel tried to, declaring Eugene IV deposed and electing
the antipope Felix V in his stead; for a decade, a rival papal court presided
from the shores of Lake Geneva. A Renaissance pope who overreached, insulted, or harmed the interests of a rival could find himself called to appear
before a new council and threatened with deposition or worse. Much papal
print sought to combat and condemn conciliar challenges from near and far.

Another challenge to Catholic authority came from outside the bounds
of Christendom: the Ottoman Turks, conquerors of Constantinople, loomed
over eastern Europe and the coasts of Italy itself. While conciliarism threatened to break the Latin Church apart from within, the idea of crusade stood

poised to unite it. Successive popes—and the humanists and diplomats who sought to flatter them—promoted the defense of Christendom against the infidel as an urgent crisis and one that only the pope, the head of the Church militant, could resolve.[36] Between them, the ideas of council and crusade animate many if not most of the printed texts I examine in this book.

Beyond documenting the simple fact that the popes used print, quite a lot, in the late fifteenth and early sixteenth centuries, this book offers further contributions to several conversations ongoing among historians of early modern Europe: debates over the printing "revolution," its character and extent; questions about the nature and achievements of the early modern papacy; and investigations into the origins of the Reformation.

The debate over how revolutionary the advent of print in Europe really was has been underway for decades.[37] How should we interpret the papacy's distinctive use of print amidst the larger body of evidence for and against a revolution in European communications? Arguments for print's revolutionary impact tend to stress its ability to popularize ideas, publicize discoveries, and decentralize authority. Print enabled a reforming preacher like Savonarola to sway the hearts of thousands of readers with striking images and texts in the vernacular. Print upended older truths, as scientific discoveries and theological debates proliferated across the European republic of letters. And print gave the early Reformers a platform from which to assail the prevailing orthodoxies of late medieval Catholicism. What I explore here is the other side of these revolutionary moments: print could also be used by an autocratic institution to assert its privileges and suppress dissent, print could publicize ancient religious devotions and traditions and present new ones as older than they were, it could strengthen ties between early modern monarchs, it could celebrate victories over rebellions; it could even be used to publish regulations limiting the ability of printers to print.[38] In short, print could perpetuate—and expand—an institution's existing documentary regime, its ability to reach into and regulate the lives of its subjects through the delivery of texts. Revolutionary technology could reinforce old authorities as well as explode them.

Second, the episodes I trace in this book shed light on the complex nature of the early modern papal monarchy. Since Paolo Prodi, historians of the papacy have debated whether and to what extent the papacy secularized as the early modern era dawned.[39] As popes contended with the absolute monarchs

of the new nation-states, they expanded their state bureaucracy, dug themselves deep into their domains in central Italy, pursued dynastic agendas for
their extended families, and reimagined themselves as territorial princes. The
papacy's use of print illustrates this story well but also complicates it in some
unexpected ways. Even as popes were mired in grubby disputes over territory and treasure, papal print asserted an extraordinarily elevated model of
papal power, one little changed from the hieratic claims of medieval popes
like Gregory VII, Innocent III, or Boniface VIII. The pope, at once priest and
prince, held the keys to heaven in one hand and, in the other, the secular
sword. He stood at the head of European Christendom, bestowing grace on
the weak, humbling the proud, correcting heresy and error, leading the fight
against the infidel. Papal print amplified this sacral model of papal power
even as the reality of papal politics veered far from this ideal. The Renaissance popes were weaker than their medieval predecessors, but print helped
to screen that fact for a long while.

Third, there is the prospective question of how prepared the Renaissance
papacy was for the coming shocks of Lutheran Reform. While the Roman
response to Luther was, initially, quite inadequate, still I argue that the papacy in 1517 had a greater awareness of the potential of print, and a greater
command of its rhetorical possibilities, than previously recognized. Papal
print could inflict real damage on the papacy's opponents, as contemporary
complaints about printed bulls of excommunication and interdict that flooded
local markets and embarrassed regional powers suggest. Moreover, papal use
of print was not just reactionary, or negative. The vast Roman publishing
project surrounding the Fifth Lateran Council reveals theologians, jurists, cu-
rialists, and popes at work constructing a new model of the "church in council," whose deliberations were presided over by the pope and whose decrees
spread across Europe with the stamp of papal approval.

To explore these issues, I employ a number of different approaches. Each
of the following chapters began as an exercise in bibliography, locating every
known publication on or related to a particular event or topic. Some extraordinary scholarly resources made this possible: the Incunabula Short-Title
Catalogue, the comprehensive census of European publications produced
with movable type before 1501, supplemented by the still incomplete but
essential *Gesamtkatalog der Wiegendrucke* (including recently digitized
manuscript files for volumes of the catalogue not yet published); EDIT-16,
the Italian census of sixteenth-century books; USTC, the Universal Short-
Title Catalogue of European handpress books; and the Bayerische Staatsbib-

liothek's digitized facsimiles of incunabula and cinquecentini. The availability of these resources online has made bibliographical research vastly easier to pursue; it is possible now, as never before, to identify all or close to all known editions by a particular author or all editions published in a particular city in a particular year.

In addition to enumerating the earliest examples of papal print, I have tried to contextualize their publication using insights gleaned from the study of law and diplomacy, court ceremonial, popular devotion, urban development, and material culture. As test cases for this methodology, some years ago I published two studies on printed responses to the fall of Negroponte to the Turks in July 1470.[40] By examining, in one case, every extant copy of a printed text that discussed the capture of the Venetian city and, in the other, every extant edition having to do with that event, no matter how humble its author's social standing or how modest its literary value, it became clear that bibliographical description could be used as a tool to illuminate more than the history of printing.[41] The contents of popular laments and the arguments proposed by scholars of middling status could be set against debates conducted at the highest levels of state in order to compose a more complete picture of a political event and the courtly, scholarly, diplomatic, and popular circumstances in which it was debated and understood.

Here, I have expanded my investigations to a longer period (1470–1520), an entire polity (the Renaissance papacy), and practically the entire corpus of printed works produced by the popes, the papal chancery, and its agents during this time. For a work of bibliography, this is an ambitious expansion of scale. But as an attempt at cultural history, limiting the focus to just one European state and its use of print over a period of just fifty years allows me to set a body of diplomatic, political, and devotional texts in far wider social and cultural contexts. What I am aiming at is a richer approach to book history, one that pays close attention to bibliographical detail while still integrating the study of individual editions into larger stories of political, social, and cultural activity.[42] Ultimately, I am most interested in the very basic questions of why a papal text got printed in Renaissance Rome, why it read and looked as it did, whose interests it served, and whether it was successful. The answers to these questions can open up new perspectives on the entire history of the Renaissance papacy, its relations with the European powers, and its means of communicating with the faithful.

Papal print broadcast the diplomatic and political interests of the bishop of Rome to new audiences while deliberately preserving traces of earlier manuscript practices, ritual actions, and topographical resonances in ways that underscored the emphatically Roman character of the post-Avignon papacy. Precisely because of the arcane, almost entirely Latin, and overwhelmingly legal nature of papal print, it is difficult to determine much about the reception of these texts. Wherever possible, I call on diplomatic, archival, or chronicle sources that reveal how papal publications were received, perceived, posted or purchased, ripped up, eaten (!), burnt, embargoed, denounced, or otherwise rejected. But we cannot have a history of reception without first knowing what it was that was received: what messages were transmitted, how they were rhetorically framed and materially presented; where they went; whether they were circulated by hand, posted on doors in the night, or sold on the open market; and whether they were reprinted. My contribution here is to show what kinds of messages the Renaissance papacy made public in print, how they were presented, and why.

Print was an integral part of the Renaissance papacy's cultural and political armamentarium. But papal use of the "revolutionary" new technology was hardly disruptive. Papal publications broadcast the institution's claims to authority derived from supposedly unchanging tradition. Here a technical innovation was used to shore up and protect, not to disrupt or overthrow. The popes used the press to publicize the antiquity of their monarchy. They mass-produced cheap bulls and briefs packed with difficult Latin. They printed texts full of references to the manuscripts from which they were made and the supposedly ancient rituals by which they were promulgated in Rome. They printed texts for a European audience that insisted on Rome as their point of origin, and they used the press to promote new traditions as though they had been around since time immemorial. They used print to cement their own grip on power, to stifle dissent, and to prosecute their foes.

The chapters that follow trace these themes across the archive of printed texts from and about Renaissance Rome: the physical projection of papal authority both *urbi et orbi*, the publication of legal decrees as a form of political ritual, the invention of new devotional and diplomatic traditions that elevated the position of the pope over his church, the development of new graphic and visual vocabularies that linked the pope to Rome and vice versa, the weaponization of spiritual penalties in the arena of early modern diplomacy, and the litigation of disputes in an expanding court of public opinion.

Chapter one examines precedents for political communication in late medieval Rome by surveying the specific sites in the city where political messages were conveyed and contested. This chapter strays some way from the question of print as it explores the legal assumptions that underpinned political communication in the medieval Italian commune and the communications practices that Renaissance popes co-opted for their own political ends. Publication in the papal city was a ritual event, with the promulgation of documents tied to certain places, times, and actions that endowed papal decrees with legal authority, efficacy, and even sanctity. Publication was also a tool of justice, as the dunning of debtors and the punishment of criminals offered ritualized opportunities for civic authorities to articulate their power. Critics of the papacy could mimic these practices for their own purposes. Thus, legal citations were slapped on the facades of private houses; papal bulls were posted on church doors; amputated hands were nailed to city gates; denunciations were draped across criminal bodies, alive and dead; placards critical of the papacy appeared on church doors and palace walls in the night; and ancient statues like Pasquino became sites for the publication of bawdy lampoons and hostile critiques of the pope. The arrival of print offered a new and different mechanism of "publication," and printed papal documents from the 1470s to the 1520s both participated in and departed from these older modes of urban communication.

The next three chapters survey the arrival of print in Italy and its development as a medium for political discourse. The protagonists of this story are not only the Renaissance popes but also their political opponents and the ambitious literary men who wrote in the service of each. When German printers arrived in Rome in the late 1460s, the papacy did not immediately adopt the new technology. It was figures of lower rank—humanist scholars, jurists, poets, and other pens for hire—who first recognized and exploited the potential of print as a medium for political commentary and self-promotion. After the fall of Negroponte to the Turks in 1470, humanist authors in Rome and Venice used their connections with early printers to publish tracts that called on the pope to move against the Turks while also advertising their own talents and expertise. In the unsavory case of the ritual murder trial of Simon of Trent (1475), an ambitious humanist bishop commissioned some of the same poets and scholars to publish tracts demonizing the Jews of Trent and demanding sainthood for the dead boy, expressly *against* the instructions of Pope Sixtus IV.

Within a few years, the story would change. In chapter three, Sixtus IV emerges as the first pope to exploit the technology of print and a key player in Italy's first printed pamphlet wars. In the Pazzi War (1478–79) and the War of Ferrara (1482–83), Sixtus issued political briefs and bulls of excommunication (which prompted printed responses from their targets in turn) in a fierce contest for diplomatic advantage and public opinion that ranged across Europe from Naples to Westminster. Chapter four continues this story with the little-known episode of Andrija Jamometić, a disaffected Croatian prelate who in 1482 attempted to reconvene the Council of Basel. Jamometić used the press to denounce Sixtus as a heretic and to call for a church assembly to depose him. Papal legates responded with printed attacks of their own, ultimately publishing a crusade against Basel before cooler heads prevailed and defused the conflict through diplomacy. The episode reveals the limits of print as a tool for managing conflict in the late fifteenth century.

Chapter five returns to Rome to examine how the popes used ritual devotions, invented traditions, and urban topography to articulate their claims to supremacy in the Church, and the centrality of Rome to Catholic Christendom, both before and after the arrival of print. The chapter traces the fortunes of the Veronica across the medieval city, from its emergence in the thirteenth century as an object of pilgrim devotion to its appropriation by the fifteenth-century popes as a political emblem and its triumph as the focal point of the 1475 and 1500 Jubilees, imprinted onto pilgrim badges and memorialized in woodcut images and printed pilgrim guides. In the *Indulgentiae urbis Romae*, the Veronica appears alongside images of pagan Roman history, portraits of the early popes, and contemporary papal *stemme* to project an image of Rome as a city of timeless power, both sacred and secular, with the popes as its imperial proprietors.

Chapters six and seven explore two further phenomena of Rome's emergence as a center of devotion and diplomacy in the late fifteenth century: first, the fates of "refugee relics," brought from Turkish-occupied lands to Rome and made to speak in favor of a new crusade and in praise of the pope (St. Andrew's head, the Holy Lance, and the Holy House of Loreto, all of whose cults were publicized in print), and second, the new tradition of the obedience oration, in which foreign diplomats made submission to a newly elected pope in elaborate rhetorical performances that were reproduced in print and circulated widely in Rome and abroad.

Chapter eight examines diplomatic and ecclesiastical printing under Julius II and Leo X. Two popes of very different character, each used the press to project a particular political *persona*, to prosecute rivals, and to suppress dissent. Erasmus famously pictured Julius hurling bulls against his enemies

like thunderbolts, and Julius's printed bulls of excommunication and interdict certainly amplified the bellicose persona this pope sought to project.
Innovatively, this was a visual as well as textual project: after the 1500 Jubilee, printers and papal secretaries collaborated on new ways to project papal
charisma in both woodcut and type; the design of Julius's early political bulls
laid the foundation for the uniform "branding" of the Fifth Lateran Council
and its dozens of printed decrees. When Leo succeeded Julius and presided
in turn over the conclusion of the council, he transformed both the form and
the content of its proceedings, laying new emphasis on conciliar, consultative, and irenic themes. The chapter concludes on the brink of the Reformation, with the 1520 publication of *Exsurge Domine*, Leo's bull condemning
the errors of Martin Luther. By then, fifty years on from the arrival of print
in Rome, the papacy had developed a sophisticated command of the medium
and its graphic and textual possibilities. Popes from Sixtus IV to Leo X had
used the press to engage in pamphlet wars, launch new devotional practices,
and rebrand the council as an exclusively papal initiative. As they entered the
confessional conflicts of the long sixteenth century, the Renaissance popes
had grasped the possibilities as well as the pitfalls of mass communication.

In all these cases, I start with the fact that a text was printed and ask why
someone—author, curial official, diplomat, printer, occasionally the pope
himself—would have chosen to send that document through the press. Although papal printing preserved and projected many long-standing practices of the manuscript chancery, nevertheless there was something different,
something momentous, about choosing to publish a text with movable type.
Ordering paper, setting up a text in type, planning its imposition in folio or
quarto, seeing it through the press, and arranging for its distribution were all
labor-intensive and expensive processes. The choice to print certain documents (and not others) tells us something about agency and intentionality in
the Renaissance chancery that the decision to copy a manuscript may not.[43]
It also tells us something about the increasing need—felt not just by the popes
but also by other actors around them—to project their words to a larger audience, a European public that was only just beginning to exercise its own
agency in public affairs.

This book does not attempt to provide a new history of Roman printing.
After the initial efforts of Sweynheym and Pannartz and their early contemporaries, printing in Rome became more of a group effort, and curial printing
was never limited to a particular, official printer or shop. At many points, the

chancery was consigning documents to two, three, or more different printers. Over time, certain names recur: Stephan Plannck, Johannes Besicken, Eucharius Silber and his son Marcellus, and Giacomo Mazzocchi all have many papal publications to their names. Few of these printers signed their products, and many modern attributions rest on the identifications of types alone.[44] In the later sixteenth century, individual printers would forge strong commercial relationships with the papal administration, but in this period, commissions were distributed widely across a network of presses, with printers even trading types and woodcuts of papal arms, meaning that individual attributions are often close to impossible. Although I untangle some bibliographical knots in the course of this study, for the most part my focus is on the texts themselves, the contexts in which they were composed, and the *fact* of their printing. The archive of printed material from Renaissance Rome reflects the unique political culture that brought it into being: an autocratic court, under near constant assault from external and internal forces, where particular political, theological, and historical claims were advanced in order to shore up the authority of the pope and help him weather a constant series of crises, large and small.

Nearly fifty years ago, Elizabeth Eisenstein identified the printing press as an agent of change and associated the technology with revolution, whether confessional, scientific, or political. A generation of scholarship has since modified that view, reminding us to look for continuity as well as change in European communications: scribal publication, epistolary networks, the handwritten newsletter, whispers at court, rumors in the piazza, the roar of the crowd—these "medieval media" still conveyed information as they ever did, long after the introduction of print.[45] Print in the papal city was a medium both revolutionary and traditional, one that broadcast papal authority to a wider audience than ever before, but one that also used a democratizing medium to elbow out rival voices and silence competing claims.[46] When Aeneas Sylvius Piccolomini heard news of the Gutenberg Bible in 1455, he quickly grasped the potential of the press to multiply texts with clarity and put them before larger audiences than ever before. Even so, neither then nor as pope did he ever abandon his sense of the papacy's peculiar claims to authority in matters of law and doctrine. Pius would not have been troubled by how Sixtus IV used the press to condemn and excommunicate challengers to his authority, he would have cheered Julius II as he published bulls condemning the unauthorized Council of Pisa, and he would have approved Leo's attempts to silence Luther through the medium of a printed papal bull.

The Renaissance popes, with their keen appreciation for ritual and tradition, their insistence on their ancient and absolute powers, and their concern to ground their authority in the spaces and sacred places of Rome, seized on the new technology of mass communication to project precisely those archaizing and sacralizing aspects of their monarchy that seem least congruous with our notions of revolution or modernity. The result was a paradoxical chapter in the history of the papacy and in the history of the book, a period when printed texts made studious reference to their manuscript originals, "publication" by posting on walls and doors competed with the circulation of printed texts, dense and legalistic Latin was mass-reproduced in cheap print and put up for sale on the open market, and brand-new graphic and visual vocabularies emerged to articulate the authority of a supposedly ancient and unchanging institution. In the chapters that follow, I trace these themes as they developed in the curia, in consistory, in the chancery, and in the printshop, but the first chapter starts—as the last chapter ends—with the posting of a document on a door.

Urbi et orbi

Just south of Piazza Navona stands Pasquino, a relic of Rome's classical past and the Renaissance city's most popular bulletin board. In the early sixteenth century, satirists and critics of the papal regime pasted anonymous libels on the statue's plinth alongside scraps of verse, political puns, and scabrous commentary. Standing in the heart of Renaissance Rome, Pasquino also stood at the intersection of the city's cultures of antiquarian taste, literary competition, and political satire. When he visited Rome in 1511, Martin Luther may well have walked past the raucous corner Pasquino occupied.[1] He certainly would have agreed with the critical jibes against papal luxury and corruption often pasted on the statue's sides. Before long, Luther would be known for his own anti-papal "postings." Though scholars dispute whether he really nailed his Ninety-Five Theses to the doors of the castle church in Wittenberg in October 1517,[2] Luther's radical theology would puncture papal pretensions and topple the entire economy of penitence and indulgence on which those claims were built.

Pasquino and Luther were both vectors for criticism of the papacy, albeit criticism of very different sorts. My concern is how each participated in the economy of early modern publication—the process of making a text public—an undertaking that had deep legal, social, and political significance. Publication in Renaissance Rome drew on long-standing medieval practices of communication and notification that were reshaped but in no way replaced by the invention of print. Critics of the Renaissance papacy published their attacks in a variety of media—in print and in manuscript, in books and on broadsides, on statues and on doors. But the popes were adroit publishers too, who used old and new media alike to promote their interests.

This chapter examines the systems of political communication and publi-

cation that operated in late medieval Rome and the political and legal assumptions that underpinned them. From the mid-thirteenth century, before
and during their time in Avignon, to the century after their return to Rome
in 1420, successive popes co-opted, changed, and deployed these older practices for their own political ends. Meanwhile, critics of the papacy and rival
European powers subverted these same conventions to attack the authority
of the popes. The communication practices they deployed—delivering manuscript documents, posting them in public, and reciting them aloud—persisted
well after the introduction of print. The new technology did not supplant
these older modes of political messaging but amplified them. Doors and statues continued to speak to one another, even as their conversations were also
recorded in movable type. Indeed, the very fact that these texts had been publicly posted was carefully noted in printed editions, lending a peculiar depth
and authority to the ephemeral pamphlets that reproduced them.

Even today, Pasquino is a popular site for the display of anonymous manifes-
tos, anarchic slogans, and political cartoons.[3] The ancient sculptural group,
thought to represent Menelaus with the body of Patroclus and dating to the
third century BCE, was unearthed from Roman soil sometime in the fifteenth
century and acquired by Cardinal Oliviero Carafa in 1501. He set it up outside
his palace near Piazza Navona, and on April 25, the feast of St. Mark, he
draped it in a toga and invited poets to festoon it with Latin epigrams, inaugurating an annual tradition that would see Pasquino decked with poems on
the same feast day (figure 1.1). Some verses were inconsequential, and some
trumpeted praise of the pope, but some were pointedly critical. Soon other
ancient statues in the city joined the conversation. In the sixteenth century,
Marforio near the Forum, Abbot Luigi at Sant'Andrea della Valle, and Madama Lucrezia at San Marco all became sites for posting the short satirical
verses that came to be known as *pasquinades*.

The phenomenon of the talking statues is well known to scholars of Renaissance literature. Less considered is the question of how Pasquino *worked*
as a site for publication in early modern Rome. What did it mean to post a text
on an ancient statue or some other vertical surface?[4] While it may seem clear
enough that Renaisance Romans wanted to celebrate newly discovered relics
of their city's ancient past and delighted in making them "speak" to one another, there were deeper cultural and legal resonances to the act of fixing a
text on or near an ancient object. To excavate these meanings, we have to go

Figure 1.1. After Nicholas Beatrizet, *Pasquino*, 1550. The Metropolitan Museum of Art, New York, 41.72(2.102).

down an unexpected lane of inquiry, namely, the legal codes of the late medieval commune.

Long before Pasquino was erected, and well before the papacy came back
to Rome in 1420, there was a communal government in Rome, similar in theory if not practice to the communal governments of Florence, Siena, Perugia,
and other Italian city-states. Although Rome was the see of the pope, in the
Middle Ages individual popes could hold court in other cities—Orvieto, Perugia, Anagni—for months or years at a time. Rome often fell to the control
of contentious baronial families. In the 1140s, during one of these periods of
papal absence, the Roman commune first emerged out of a violent popular
revolution.[5] Although the nobles quickly regained control over the city, certain municipal institutions like law courts and processes for resolving disputes persisted just as they did in other, more successful Italian communes.
In the 1340s, with the popes in distant Avignon, Cola di Rienzo led another
revolt against the barons and instituted a new republican regime. This, too,
was short-lived, but some of Cola's administrative reforms were codified in a
new set of communal statutes in 1363.[6] It is unclear whether these statutes
were ever fully in force, but as normative documents, they reflect a theory of
government that held increasing sway across central and northern Italy in
the later medieval period. Among other civic values, the statutes place a high
priority on communication based on the public delivery of documents and
the public display of information, whether on doors, on walls, on criminal
bodies, or near ancient statues.

Communal Communications

The 1363 statutes, based on a substrate of (now lost) legislation from the first
commune of the 1140s, underwent some revision in the fifteenth century
and were first printed in 1470; the last edition was published in 1519–23.[7]
Whether or not they provide reliable evidence for the lived experience of
late medieval Romans, they offer eloquent testimony for how the communal
government imagined itself and its dealings with constituents. One of the key
obligations the commune undertook was to keep the *popolo* informed about
what it was doing.

The official seat of the commune was the Senator's Palace on the Campidoglio, the flat, open space at the top of the ancient Capitoline Hill. The vast
brick palace, built on the foundations of the ancient Tabularium of the Roman
Republic, housed the chief magistrates, law courts, and various workers whose

job was to communicate information to citizens: bellringers (*pulsatores*) and trumpeters (*tubatores*), heralds or criers (*banditores*), and bailiffs and couriers of the court (*mandatarii, tabellarii,* or sometimes *fossores*). These officials conveyed a variety of messages through the city: edicts and proclamations, judicial and fiscal announcements, legal citations and summons. They made these public by reading them aloud, or by posting them in certain, set places, or by delivering them to the homes of private citizens.

Civil and criminal cases alike depended on the exchange of paper. Like a modern process server, the communal *mandatarius* took a summons (*citatio*) from the civil courts, usually having to do with unpaid debts, and served it at the home of the party concerned by putting it directly into his hands or into those of his landlord. If neither man could be found, the statutes say, the process server could affix the summons to the door of the person's house, in the presence of a witness, while reading the content of the document to his neighbors in a loud voice. Once the papers were attached to the door and their contents read aloud, a debtor, even if nowhere near the house at the time of delivery, could not pretend ignorance of a summons.[8] The courier was to bring his papers (or a copy of them) back to the court so they could be registered as having been served. Not everyone was happy to receive a court summons. The statutes instruct the agent on how to proceed if a debtor should rip the papers from his hand or otherwise destroy them: a *mandatarius* who returned to the Campidoglio empty-handed could swear an oath that he had delivered the papers, and they would be considered served.[9]

Criminal cases or cases involving confiscation of goods required further action, a process called the *fossura*, which seems to have involved nailing (as opposed to merely "affixing") a notice to the wall of the accused's house.[10] The statutes name a second official, sometimes called *cavator* or *supercavator* but more usually a *fossor*, distinct from the *mandatarius*, who performed this task. In the case of a criminal charge, a *fossor* should visit the home of the accused on at least three occasions and *fodere* him, summoning him to answer the charges against him. If, after three *fossure*, the accused did not appear, he would be considered in contempt of court. Whether *fossura* meant nailing some kind of notice to the exterior of the house or carving a mark of indictment into the wall, it was a more drastic way of notifying a malefactor than simply sticking a notice to his door with glue or wax.[11] Still, the statutes try to protect those accused of these graver crimes: *fossores* were told to make their mark where the least damage would be done to the wall, and one statute commands that if the *fossor* found a wall with paintings of military scenes or other images, he should not make his mark on them so long as there was anywhere else that he could make it.[12]

Even with such safeguards in place, performing the *fossura* was risky: penalties were imposed on anyone who attacked a *fossor* or prevented him from completing his task.[13] This is hardly surprising: for both *mandatarii* and *fossores*, the act of posting the documents in their charge was an act of social as well as legal significance. Posting a notice on a door or wall could be mortifying for a householder—why else would the statutes protect the *mandatarius* from theft of his papers or the *fossor* against attack?

In their analyses of trial reports from sixteenth-century Rome, Thomas and Elizabeth Cohen explore the shame and dishonor that attended an attack on the door of a private home. With the facade standing in for the owner's own face, any mark, whether painted or carved or scorched, would be read by the neighborhood as an assault on the honor of the *casa*, a term encompassing both the domestic structure and the family it housed.[14] A court summons slapped on the door could sting as much as a smack across the face. At the same time, though, public posting was considered necessary: it protected the citizen against secret denunciations or unexpected confiscations. Statutory insistence on the personal delivery of paperwork thus reflects the expanding apparatus of an urban regime that sought to regulate its citizenry, on the one hand, and the conviction that citizens had a right to clear, public interactions with that apparatus, on the other. Long before the introduction of printing, then, Rome had a sophisticated system of official publication. In many ways, print would replicate—but also amplify—this sense that publication was not only a duty of government but also an articulation of its power, an obligation but also a threat.

The statutes warn *mandatarii*, *fossores*, and *supercavatores* not to shirk their duties; they could not refuse to handle certain cases and could be punished for leaving papers undelivered. Notably, any agent who was overzealous in the pursuit of his office—who harassed a debtor by making a *fossura* on his house even after he had satisfied his debt, for example—could be charged with disobedience. The penalty was public ridicule: on market day, the delinquent officer would sit astride an ancient marble lion that sat at the top of the steps of the Senator's Palace, before the doors to the hall of justice known as the Capitoline court. The officer's face would be smeared with honey, he would wear a paper mitre or dunce's cap with his offense written on it ("I disobeyed my orders"), and thus arrayed he would "ride" the lion for as long as the market was open that day. In this way the social shame intended for the debtor rebounded onto the offending official: the errant publisher became, himself, a site for legal publication.[15]

This ritual punishment took place at the most potent site in the urban topography of communal Rome: the steps of the Senator's Palace at the top of the Capitoline Hill. The palace housed the Capitoline court and the communal prison; nearby stood the Palace of the Conservators, the communal gallows, and the city's marketplace. Presiding over the entire precinct was the ancient stone lion at the top of the courthouse steps.[16]

This scruffy beast (figures 1.2 and 1.3), part of a Hellenistic sculpture originally depicting a lion attacking a horse, was installed on the Campidoglio sometime before 1347.[17] Like the SPQR monogram, a potent political symbol of the ancient Roman government that the commune adopted in the twelfth century as its own,[18] or the title of "Senator" bestowed on its chief officials, the lion asserted a direct link between the communal regime and ancient Rome, a pedigree older than that of the pope. Later, the lion would be supplanted by the bronze statues of the "Capitoline" wolf and Marcus Aurelius, but in the fourteenth century those treasures were still papal property, installed at the Lateran. The marble lion was the medieval commune's most visible link with its ancient past, a totem of antiquity that dominated the political and economic center of communal Rome.[19]

The lion and its steps figured prominently in the commune's judicial work and communications. The steps were a liminal place, between the seat of communal authority inside the palace and the public marketplace outside. They were an ideal location from which the decisions of one body could be transmitted to the other. It was here that the ancient assembly of the people or *parliamentum publicum* convened when summoned by the palace bell.[20] The steps were also a common site for public punishments. Cola di Rienzo condemned two malefactors to death while standing at the top of the steps, the spot his biographer calls "the place of the lion."[21] Later, when the Roman mob turned on Cola, it was here that he made his final stand and his assassins struck him down.[22] The 1363 statutes specify less severe penalties to be imposed at the place of the lion: notaries convicted of forgery were presented here for ridicule by crowds summoned by the palace bell;[23] a man convicted of striking another with an open hand could choose to pay a fine or suffer the same blow, to the same part of the body, while standing on the steps.[24] And it was here that delinquent communal employees had to take their shameful ride on the lion for the length of a market day.

Within the image, handwritten caption: *Fragmentum e marmore mire arte celebratum in capitolio Romae.*

79

Figure 1.2. Giovanni Battista de' Cavalieri, *Lion Attacking a Horse*, from *Antiquarum statuarum urbis Romae* (Rome, 1585–94). Hesburgh Libraries, University of Notre Dame, Special Collections, Rare Books Large NB 115.C38.

Figure 1.3. *Lion Attacking a Horse*, Musei Capitolini, Rome. Photo courtesy of author.

The steps were also a strangely potent site for the delivery of proclamations and summons, with the lion serving as a sort of miraculous amplifier of the communal voice. The 1363 statutes usually insist that legal citations be delivered to individuals or onto their doors, but when a party could not be located, or if they dwelt outside the city, the statutes allow that they could be summoned by a *mandatarius* reading the citation from the palace steps. His voice was deemed to carry far and wide by virtue of emanating from this spot.[25] The legal fiction that this was so would seem to contradict the commune's usual insistence on the physical delivery of documents. But if delivery to a particular location was impossible, then the physical seat of the commune—the whole complex of senatorial palace, judicial court, public steps, and ancient lion—by virtue of the political authority that rested there, had the power to broadcast a message to wherever the commune held sway. In this curious ordinance the commune's insistence on public notification intersected with an almost fetishistic regard for the power of its ancient architectural setting.

Thus the commune handled court papers. Municipal announcements, called *banda* or *bandimenta* in Latin or *bandi* in Italian, were carried through the streets by a different set of communal employees, the *banditores*, whom the statutes also instruct.[26] The town criers of Rome, *banditores* were to spread their *bandi* on the same day on which they were given them. They were to *bandire* from the steps of the Capitoline court but also at various intersections (*in capo crucibus*; *in quatriviis*) that were established by ancient custom for the reading out of notices. Frustratingly, the statutes identify none of these places but refer to them only as the *loca consueta*, the "usual places." *Banditores* made fiscal and municipal announcements about tax deadlines, exchange rates, changes of office, new building regulations, and so on.[27] Notices went out when it was time to register flocks of goats and sheep with the fiscal office; these were announced every year "in a high voice" and with the help of a trumpeter in the fortnight before Christmas.[28] Regulations about riverbank fishing and the levying of tolls and customs duties were also proclaimed throughout the city.[29] Political news—the conclusion of a treaty with Ladislas of Hungary in 1412, or peace between Eugene IV and Emperor Sigismund in 1433, or peace among the Italian powers in 1455, or the striking of treaties between Rome and Venice in 1487—was also announced by a *banditor* with trumpeters.[30] In December 1501, six pipers were deployed to convey a fairly routine notice, namely, that Carnival festivities would begin, as usual, the following New Year's Day.[31]

Although the statutes do not say whether *banditores* were expected to post copies of their announcements in the *loca consueta*, they probably did post some notices on the doors of the Capitoline court. In cases where goods pledged or owed to the court were at risk of confiscation, notice was to be given by a *bandimentum generale* on the steps of the court, which gave the interested party eight days to appear and redeem the goods;[32] the imposing of a time limit suggests that the *banditor* must have posted some kind of paper notice either at the palace or in the *loca consueta* (or both). Likewise, when a magistrate investigated a senator at the conclusion of his term, he would "issue a public announcement from the steps of the Capitol and at the other usual places of the city," inviting anyone with a grievance against the senator to come forward; the judge was to issue four such notices altogether.[33] It seems that these notices must have been posted with an indication of the date of posting or of the deadline imposed.

Jobs on the communal communications staff were not high-status appointments. *Mandatarii, tubatores*, and *banditores* were paid with income from the city's annual Carnival games, along with the communal bell ringers, porters, ushers, and grooms; the communal barber surgeon; and the keeper of the city lion.[34] This same pool of income was used to pay for a suit of livery (*indumenta*) for each of the workers, monogrammed with "SPQR," and for matching caparisons for the horses they rode from the communal stable. The workers were to present themselves in this array every year before the senator and the Roman people at the Carnival games in Testaccio and in Piazza Navona, "for the honor of the Roman Republic."[35] In this way, Rome's communications personnel, the clothes they wore, the documents they carried, and the locations from which they broadcast their messages all drew on—and further advertised—the authority of the popular commune through the long fourteenth century.[36] Whether the pope was in Rome or Avignon, whether the city was ruled by barons or by Cola, by a papal vicar or by some other power, when there was schism in the Church and when it was resolved, communal messengers carried on with their bureaucratic business, knitting together the civic fabric with their papers and wax, their trumpets and their cries. They kept doing so even as the commune began to cede its political authority to the resurgent Renaissance papacy.

Papal Promulgations

Martin V returned the papal court to Rome in 1420. During its long absence in Avignon, the curia had grown in size and complexity. Alongside scores of secretaries, chancery officials, financial clerks, chaplains, guards, cooks, grooms, and domestic servants, Martin brought with him his own communication staff, courtly counterparts to the communal *mandatarii* based on the Campidoglio. In the papal curia these officers were called *cursores*, or couriers; they delivered documents both within the papal city and abroad. An ancient office of the Church (originally called *viator apostolicus*), the *cursor* was first recorded in his modern form in the thirteenth century. A combination of herald, messenger, usher, and ceremonial guard, the *cursor* accompanied the pope in procession; summoned cardinals to consistories; took part in liturgies, processions, the opening and closing of a Holy Year, and funerals for the pope and major cardinals; assisted during conclaves; and delivered private as well as public communications.[37] *Cursores* took an oath on installation, and, like the communal *mandatarii*, they wore livery; theirs was embroidered with the papal coat of arms,[38] and on ceremonial occasions they carried a silver mace or a rod of briarwood.[39] The office of *cursor* could be bought, but for considerably less than more highly skilled posts inside the chancery like secretary or abbreviator.[40]

There were thirty *cursores* in the *familia* of Clement V in 1305, and fifty in that of Gregory XII in 1406, but Nicholas V reduced their number from thirty-eight to nineteen. Twelve of them formed an honor guard for the pope in ceremonies and processions.[41] At least some were stationed outside the papal palace in the Cancelleria Vecchia (now Palazzo Sforza-Cesarini, in via dei Banchi Vecchi). Built in 1458 by the vice-chancellor Rodrigo Borgia, the building served as the papal chancery until the office was transferred to Raffaele Riario's neighboring palace (now Palazzo della Cancelleria) in 1521. The old chancery still remained in use, however: Tempesta's 1693 plan of Rome marks the building with the caption *Li Cursori*. An undated *motu proprio* secured by the papal master of ceremonies, Johann Burchard (in office 1483–1506), reminds *cursores* to deliver their documents discreetly, without making a lot of noise, and whenever they were instructed to do so. They should be ready to perform their duties at all times and should not steal away from the court without permission—not even during holidays, nor in the month of August.[42]

The work of papal *cursores* and that of communal *mandatarii, banditores*, and *fossores* were roughly comparable. All delivered certain documents by hand, whether civil citations or papal briefs and curial messages. They also handled more formal documents: informational *bandi* and legal *fossure* on the communal side, and on the curial side, papal bulls, documents that, like their civic counterparts, required both a *viva voce* reading and posting in public.

In formal terms, a bull was simply a letter sent by the pope to one or more people and marked by special formulas of address and subscription that established its authenticity (figure 1.4).[43] A papal bull was also distinguished by certain material signs. These included the exaggeratedly tall letters with which the first line of the document was written, yellow and red silk cords with which the parchment was sewn shut, and the lead seal that fixed the cords together, imprinted with the name of the current pope and the heads of SS. Peter and Paul, the *bulla* that gives the document its name (figure 1.5).

These elements had assembled over the course of the Middle Ages in order to assure the recipient of a bull that it was genuine.[44] They presuppose the

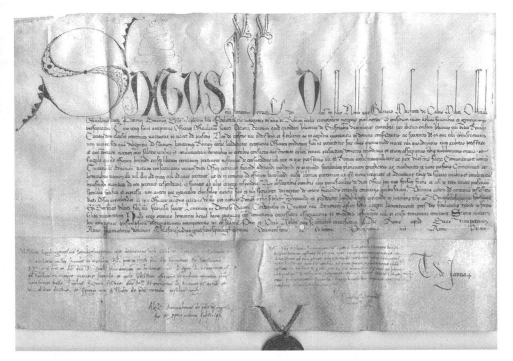

Figure 1.4. Bulla of Sixtus IV, 1471. Victoria, British Columbia, University of Victoria Libraries.

Figure 1.5. Bulla of Sixtus IV, 1471. Victoria, British Columbia, University of Victoria Libraries.

delivery of the document to a particular person, whether a local agent charged with implementing its instructions or the person receiving whatever appointment, dispensation, pardon, sentence, or other decree the bull conveyed. The idea of delivery was crucial to the development of a bull as a legal instrument. Rather like the court papers of the communal process server, a papal bull might not be considered binding until the physical document had been handed to its target. A pope could also have paper copies of a bull drawn up and certified when he wanted its contents to be widely known, and the texts of all bulls were supposed to be copied into registers in the papal archive; however, it was the original parchment, with its lead seal and silk cords, that carried legal weight. Over time, the process of drafting, copying out, correcting, registering, signing, sealing, and promulgating papal bulls was distributed among different papal offices, creating work for an entire class of educated literary men and a valuable income stream for the popes who sold them their offices.[45]

As medieval popes came into conflict with other political actors, especially after the Gregorian reforms of the eleventh and twelfth centuries, delivering a papal bull became harder, and popes began to use other methods to promulgate their decrees.[46] Medievalists well know that in the later Middle Ages papal bulls were often posted on church doors, but the origins and early chronology of the practice have rarely been explored. In 1254, Innocent IV issued a bull that specified at its conclusion that it should be posted on the door of his palace. Once posted, the bull should be considered legally published.[47] Neither Innocent nor his successors invoked this clause again for several years, but in 1273 Gregory X summoned the English knight Guy de Montfort to answer for the murder of his cousin; Gregory posted the summons on the doors of the cathedral in Orvieto, where he was holding court at the time, and directed that the public display of this *citatio* should be considered as effective as if it had been handed to the Englishman directly.[48] In 1281, Martin IV issued a bull of excommunication against the Byzantine Emperor Michael VIII Palaeologus. This was to help Martin's protector, Charles of Anjou, in his project to capture Constantinople; the pope declared there was no legitimate emperor in the East, the throne stood empty, and Charles was free to take it. The Greeks would hardly allow a papal *cursor* to deliver such a bull, nor were there Latin-rite bishops in Constantinople to pronounce the sentence on the pope's behalf. Martin ordered the sentence posted on the doors of the Duomo in Orvieto, where he was staying; with the bull "so openly published," he declared, there was no way the Greeks could pretend that the contents of the bull had not reached them. Its posting in Italy was enough to render the emperor excommunicate in Greece.[49]

Thereafter, the papal chancery increasingly held that a bull should be considered legally published if it was posted on the doors of a palace or important church in the city where the pope was residing. Physically posting the bull made it public, and therefore binding, whether a copy ever reached the hands of the bull's addressee or not. The principle was remarkably long-lived: not until 1909 did publication in a printed journal—"publication" in the modern sense of the word—become the legal requirement for the promulgation of pontifical laws.[50]

Not every bull required public posting. The procedure was used for major announcements (like the convening of a council); legal summons where the party concerned was unlikely to respond to a private citation (*bulla citationis*); bulls warning the recalcitrant to respond within a certain time or face penalties (*bulla monitoria*); and, after the passing of a deadline, bulls imposing actual penalties on lay or clerical figures (*bulla excommunicationis, bulla privationis, bulla suspensionis*, and so on). In the thirteenth century, the papal court was peripatetic; bulls were posted on church doors in Orvieto and Anagni,[51] as well as in Rome.[52] The practice migrated with the curia to Avignon. A bull of Clement V excommunicating the people of Venice was posted on the doors of the cathedral and the Dominican church of Avignon in 1310.[53] A bull reserving papal rights during an imperial interregnum appeared on the cathedral doors in 1317.[54]

In 1323, John XXII summoned the emperor, Ludwig of Bavaria, to appear before him within three months or else lose his crown and face excommunication; this, too, was posted on the cathedral doors in Avignon.[55] Pope and emperor remained at odds for a long time. In 1328, Ludwig entered Rome with armed troops and issued a decree deposing the pope. He appointed his own antipope, Nicholas V, and put a stuffed effigy of John XXII on trial for heresy and other crimes. The effigy was pronounced guilty and burned at the stake. To punish the Romans for tolerating this spectacle, Pope John issued a bull from Avignon putting the city under interdict and had a Colonna ally post it on the doors of the Roman church of San Marcello.[56] Ludwig and his antipope soon moved on to Pisa, where, with the support of local Franciscans, the emperor reissued his decree deposing John and had it posted on the doors of the cathedral there. The Franciscans then read out a long denunciation of the Avignon pope and posted it on the same doors, and then another dummy dressed in pontifical vestments was tried, stripped of office, and consigned to the flames.[57]

As these remarkable events suggest, posting notices on the doors of churches was fast becoming a feature of Italian political life. On Ash Wednesday 1347, Cola di Rienzo launched his revolt against the Roman barons by posting a placard on the door of San Giorgio in Velabro.[58] Within months, support for his regime faded as the barons chafed at his popular regime. In December, a battle of placards unfolded on the doors of Sant'Angelo in Pescheria, Cola's parish church and headquarters of his movement. On December 14, a Savelli baron posted a notice on the church door calling on his allies to rally to his side. The tribune had the notice torn down and posted another in its place, a citation summoning Savelli to appear before *him* within three days. The exchange of notices brought buried tensions into public view. The next day, Cola tried to muster his supporters on the Campidoglio but was forced into exile.[59]

During the Great Schism, with a pope ruling in Rome at least some of the time, papal officials began to post documents fairly regularly on the doors of Roman churches and palaces—not only major legislation like bulls of excommunication but also local ordinances and administrative decrees. In 1390, Boniface IX issued a bull threatening anyone who held goods belonging to the Apostolic Camera. The bull was to be posted on the doors of St. Peter's and other churches around the city for a month: anyone found with papal property after that would incur automatic excommunication.[60] In 1418, Innocent VII published bulls of warning and excommunication against Ladislas of Hungary on the doors of St. Peter's.[61] In 1425, Martin V issued a decree forbidding barons from shielding murder suspects in the countryside. He had it posted on the doors of the Capitoline court at the Senator's Palace—the spot to which malefactors were to be remanded.[62]

Eugene IV used doors to publish documents momentous and quotidian: his *cedula* announcing the transfer of the Council of Basel to Ferrara on January 9, 1438, was posted on the doors of the cathedral in Ferrara;[63] on February 10, a general invitation to prelates to participate in the new council was posted on those same cathedral doors, the doors of the papal palace in the city, and the doors of various urban convents.[64] Later, Eugene had a bull outlining new tax policies posted on the doors of the Capitoline court.[65] Pius II had his bull forbidding the looting of antiquities read out at various points in the city and posted at the Capitol as well.[66] Gradually, papal officials were staking claim to not just the functions but also the central political spaces of old communal Rome.

Along with this development in practice, new clauses entered the legal formulas stacked up at the end of every papal bull, stipulating how the bull should be published and what its display in public signified. The wording is first found in Martin IV's bull against the Byzantine Emperor of 1281, but even in the fifteenth century, the formula could vary from document to document—indicating how recent an innovation in chancery practice this was. The message was consistent: since it would be difficult to publish the bull abroad (for reasons sometimes stated, sometimes not), the pope was publishing it near himself, by posting it on a particular door or set of doors:

> Since this letter cannot be safely published in Venice on account of this same
> Venetian regime . . . we wish it to be known that the letter was posted to the
> doors of the basilica of the prince of the apostles and by its posting there, it
> was made public, and we decree that the publication of this letter, performed
> in this way, renders those whom it concerns notified and bound just as if it had
> been personally delivered and read to them on the day of its publication. For it
> is not plausible to argue that a document could remain unknown among them
> after it has been made so public in this way.[67]

There was a potency to a Roman door that amplified the message of a document posted on it, so that its contents were considered to have been made public everywhere. Papal bulls might still be published on doors in other cities.[68] In times of conflict, a bull could be posted in a friendly city that was close to a hostile regime.[69] Bulls published in rural districts could specify multiple churches as potential sites of publication; so long as it was posted on the doors of one or two of these, it could be considered published throughout the district.

When it came to matters of ecclesiastical justice, it was always preferable that a citation or summons be delivered in person or, failing that, that it be posted on the doors of the church nearest to the defendant.[70] But by the late fifteenth century, major pieces of pontifical legislation were routinely posted in Rome as an integral and necessary part of their promulgation. A bull issued by Julius II in 1505 even specified that posting in Rome was *all* that was required: "This letter should be posted on the doors of St. Peter's, the Chancery, and in Campo de' Fiori, nor is any other form of publication necessary, nor should it be expected; posting in this way will suffice to make it published and put it into effect."[71]

The English pilgrim William Brewyn, in Rome for Easter 1469, attended the by-now annual reading of the bull *In coena Domini* on Maundy Thursday at St. Peter's. This bull imposed excommunication on all manner of people who disregarded or interfered with papal authority in any way: heretics and schismatics but also pirates and highway robbers, especially those who harrassed pilgrims on their way to Rome or merchants supplying the papal city.[72] It had become customary to reissue it every year after the liturgy commemorating the Last Supper. William was impressed by the ceremony of the event:

> The denunciation was pronounced by two cardinals, who stood near the pope, one on his right hand, and the other on his left, and that denunciation lasted for a full hour by the clock of St. Peter. And many of them, I am not certain about all, held in their hands great lighted wax candles, which they cast down between the feet of the bystanders in token of their contempt and reprobation of those who were thus excommunicated. . . . And I, the aforesaid William, copied the excommunication which here followeth, from the bull of the lord Pope Paul as it was hanging over the door of the church of St. Peter the apostle, on Easter Eve.[73]

So the bull was read out in its entirety to the crowd, as well as posted; it was left on the basilica doors for several days (till at least Holy Saturday and probably through the entire Triduum); and it was posted in such a way that a pilgrim could approach and read it and even stand before it long enough to copy its contents by hand.

The claims made in papal bulls—that public posting of a document in Rome automatically created notification in Florence, Venice, Ferrara, or somewhere farther afield—seem to borrow from, and at the same time reconcile, two contradictory practices that had developed in the laws of the Roman commune: first, the requirement that documents be posted on doors or walls in order to make them public, and second, the assertion that the contents of a document could be broadcast to distant places by virtue of the location in which it was posted.

For the commune, this had always been "the place of the lion," at the top of the steps of the Senator's Palace, before the doors to the Capitoline court. It took longer for the papacy to establish its *loca consueta*. Papal bulls from the first half of the fifteenth century tend to specify different places of posting depending on their content. Bulls setting out bureaucratic procedures might be posted inside the papal palace; on the doors of the administrative courtroom called the *Audientia*, where judicial and legal pronouncements were read out to assembled listeners (hence its name); or on the doors of the chancery itself.[74] After the bull was read out, it or a paper copy would be left posted on the door for a set amount of time. Bulls having to do with municipal business might go up on the doors of the Capitoline court, as we have seen. Some bulls were posted on the doors of the building their contents concerned: the bull granting privileges to the Confraternity of Santo Spirito was to be posted, annually, on the doors of the hospital the confraternity served.[75] A bull indicating an indulgence for pilgrims to a particular church could be posted there,[76] and bulls announcing new taxes or fees on ecclesiastical services were posted on the doors of parish churches where their stipulations applied. The papal master of ceremonies, Johann Burchard, mentions the curious requirement that any priest wishing to celebrate mass in St. Peter's had to have a bull of authorization to do so: the priest was supposed to post the bull on or near the altar he was using to perform the rites. (We know this because Burchard complains so often about priests celebrating mass without displaying their bulls.)[77] Of course, most bulls did not require posting at all. Dealing with private matters that did not need publication, most functioned more like a license or privilege, to be produced by the holder when necessary, rather than a public notice or decree.

Bulls communicating important ecclesiastical business—the announcement of a Jubilee, a major new doctrine, or a sentence of interdict or excommunication—would be posted in multiple locations. By the middle of the fifteenth century, four sites around the city came to be used more or less exclusively: the doors of St. Peter's, the doors of the basilica of St. John Lateran, the marketplace at Campo de' Fiori, and the doors of the chancery, the building where the documents were produced. The doors of St. Peter's (always identified as *valvae*) were probably the central pair facing into the atrium of the old basilica.[78] The main entrance to the Lateran basilica in the fifteenth century was by the (liturgical) south transept, facing north into the center of the city. The original bronze doors that guarded that entrance are still preserved in the cloister. We know less about the space described as *in acie Campi Florum*, but it seems that either a portion of a building's facade or a special hoarding was set aside for the posting of notices on this central piazza. The last location, the chancery palace, had an entrance facing onto via de' Banchi, an important artery connecting Campo de' Fiori to the Tiber and the Vatican, and notices were probably posted on the doors facing that street.

The choice of these four sites echoes both age-old claims (apostolic succession at the tomb of Peter, the Donation of Constantine at the Lateran) and new civic realities, specifically, the steady extension of papal authority over the city. When Martin V brought the papal court back to Rome in 1420, he took up residence in his family palace, Palazzo Colonna, near the Forum. The Lateran Palace was in ruins, as was much of the city. The Venetian Eugene IV, for the few years he was in Rome, ruled from Palazzo San Marco. Nicholas V had no family palace and moved the papal court across the Tiber to the Vatican. His decision was practical, but it was also a tactical retreat: the city center was baronial territory, not always safe or entirely under papal control.

Richard Ingersoll has described the urban development of Renaissance Rome as a struggle between papal authority, now firmly lodged in the Vatican, and communal or baronial authority, based on the left bank of the river and above all on the Campidoglio.[79] The *possesso* route that ran through the heart of the city, for example, was rife with staged or ritual expressions of popular hostility to the pope.[80] Over the last decades of the fifteenth century, however, as the authority of the commune continued to decline, the popes began to assert their presence in the districts within the Tiber bend. Cardinals and curial officials built palaces around the Banchi and via Papalis. Some of these became the seats of administrative departments. Rodrigo Borgia's palace, as we have seen, housed the chancery in the fifteenth century, its operations expanding to Riario's palace in 1521. Papal building projects—Sixtus IV's works at Santa Maria del Popolo, Santa Maria della Pace, Ponte Sisto, and the Ospedale di Santo Spirito—also extended papal influence into new territory. Certain communal operations migrated away from the Campidoglio to join them. In the 1470s, the city market moved from the Campidoglio to Piazza Navona. The grain office and the customs house soon followed. Executions moved from the gallows on the Capitoline Hill to the bridge at Castel Sant'Angelo. In the early sixteenth century, Julius II floated a plan to build a new palace of justice, replacing the Capitoline court, opposite the chancery between his new via Giulia and the Tiber.[81] Papal documents followed their makers, now appearing at the Vatican, the Lateran, Borgia's chancery palace, and the marketplace at Campo de' Fiori. Thus, the papacy enveloped Rome in an ever-expanding web of documentation. The places where documents were posted lent the pronouncements legal weight, and the public display of the documents themselves reinforced the papacy's claims to control over the entire city.

How were papal bulls posted? *Cursores* almost certainly used glue or wax, not nails, to affix their documents to their appointed doors.[82] They were instructed to take the same bull from one door to the next, leaving it posted in each place for a set amount of time—if at one of the basilicas, for the duration of a mass—before moving it to the next location.[83] At Campo de' Fiori and at the chancery, *cursores* were to leave a bull posted "for a time" and sometimes post it repeatedly, on separate occasions—and they often read the contents of the bull aloud as well, just as their communal counterparts, the *banditores*, did with their announcements.[84] The *cursores* returned to the chancery and wrote down an account of just how they had published their document: On May 23, 1492, one reads, "This bull was posted and published on the doors of the basilica of the prince of the apostles while a solemn mass was celebrated there, and as a great crowd of people was coming in and going out, by me, Bartolomeo de Miranda, *Cursor*."[85] It was not just the exposure to eyes and ears, however, but also display at a potent site of apostolic sanctity, papal authority, or popular commerce that made the document truly public and therefore published. While communal *banditores* read their notices from the steps of the Capitoline court and the unnamed *loca consueta*, papal documents specified all the places of their posting, recording an expansive process of distribution and broadcast, reaching out from the chancery to the major basilicas east and west and important civic points in between. Furthermore, papal heralds did not post multiple copies of a bull in these different places at the same time; rather, it was important that *one* bull made the rounds of the different Roman sites, and its circuit was carefully recorded. The uniqueness of the posted copy was important, somehow, to its legitimacy or efficacy.

From the foregoing, we can imagine fifteenth-century Rome as a city crisscrossed by communications personnel: process servers delivering citations, *fossores* carving their marks on domestic doors and walls, *banditores* blowing trumpets and announcing their news, and papal heralds posting their bulls on church or palace doors and standing watch over them while they hung. But to local observers, the differences between communal and papal agents would have been obvious—differences that went beyond the competing sets of livery they wore. For one thing, while communal *mandatarii* and *fossores* were cautioned against abusing their power—issuing invalid citations, defacing murals, or harassing debtors—papal *cursors* operated with no such safeguards. Papal documents evince no concern for the rights of their targets. A bull of Innocent VIII imposes stiff penalties on those who would prevent curial officials, including *cursores*, from delivering documents or promulgating their instructions,[86] but there is no acknowledgment that papal *cursores* should exercise their office with similar care. They owed the pope's public nothing. The formulas within the bulls allowed for no doubt: once a bull was posted, no one could pretend ignorance of its contents, nor find recourse from its prescriptions.

The view from within the curia, incidentally, was rather different. Critics of the papacy like the republican Stefano Infessura, from whom we will hear more shortly, saw a growing curial apparatus and a centralizing policy that sought to control the processes of justice and communication. But to Burchard, the master of ceremonies, the apparatus seemed creaky and hard to control. A harried middle manager, Burchard complains frequently about his staff of heralds, who lose messages, mix up their instructions, interrupt important meetings, and tend to disappear around Ferragosto. When *cursores* did deliver their messages, not everyone was happy to receive them. Infessura relates how a cardinal sent a *cursor* to serve court papers to a Roman baron, but the baron beat the *cursor* so ferociously that he died three days later.[87] Burchard notes another extraordinary case in which a herald, having delivered a bull containing some unwelcome news, was forced by the recipient to eat the offending parchment. Papal officials issued yet another sentence, this time adding obstruction of justice to the original charge; one wonders how that summons was received.[88]

As papal *cursors* carried, delivered, and posted their documents across the city, they moved in parallel with ambitious architectural and urbanistic initiatives. One example is the practice of permanently affixing bulls in churches as stone inscriptions. This was not a Renaissance innovation: older inscriptions recording the texts of bulls granting privileges or endowments are found in many Roman churches. The church of SS. Giovanni and Paolo preserves an inscription recording a grant of land near Velletri by Gregory the Great (590–604), and St. Peter's has a similar plaque recording a gift by Gregory II (717–31). An inscription in Santa Sabina reproduces a bull granting indulgences to the church by Gregory IX in 1238.[89] The text of Boniface VIII's bull *Antiquorum habet* of February 22, 1300, announcing the first Jubilee year, was carved into a stone slab now high inside the porch of the new St. Peter's. In 1372, Gregory XI issued a bull declaring the Lateran the preeminent church in the world; a copy of this text was inscribed onto a stone plaque still preserved in the Lateran cloister. Exceptionally, the stone carver reproduced elements from the manuscript original on the stone tablet, including two circles representing the two sides of the original *bulla*, one with the heads of Peter and Paul and the other with the name of Pope Gregory (figure 1.6).

A bull of Pius II granting privileges to the basilica of Santa Maria Maggiore is carved in monumental square capitals on a stone plaque now in the nave of the basilica, partially obscured by a confessional.[90] A wordy bull of Sixtus IV granting privileges to the Lateran cathedral is inscribed on a massive stone slab originally displayed in the nave.[91] And at Santa Maria del Popolo, the church that received more della Rovere patronage than any other, Sixtus issued two bulls granting indulgences to the faithful who attended mass there on certain days. The text of each was carved in Roman capitals onto monumental marble plaques that were installed—and remain—on the facade of the building.[92]

What meanings did these epigraphic bulls convey? In part, they were imperial gestures. The three inscriptions of Sistine bulls certainly fit with Sixtus's program of renovating the city: they record restoration work or endowments or the issuing of indulgences that brought the churches pilgrim funds. Other famous Sistine inscriptions in the city commemorate the construction of Ponte Sisto, the clearing of roads and streets, building works at Santo Spirito, and the pope's gift of antiquities to the Roman people, the nucleus of the Capitoline Museums.

Sistine bulls, whether carved on stone or written out on parchment, added to the larger web of epigraphic text then spreading across the city. In a vivid passage describing the proliferation of inscriptions across Renaissance Rome, Armando Petrucci imagined how epigraphy, "clothed in the rediscovered and reelaborated antique style, was busily conquering outdoor urban space, boldly appearing on the facades of buildings and churches, weaving itself about the city, dividing and accenting its surfaces."[93] Petrucci refers only to stone inscriptions, but I would include paper and parchment bulls in this phenomenon as well. Across the city, papal texts were proliferating, broadcasting messages of munificence and authority. The pope could post bulls on doors and carve them into walls; his promulgations could give graces but also demand payments or impose penalties; pontifical inscriptions recalled acts

Figure 1.6. Stone Bull of Gregory XI, Cloister of San Giovanni Laterano, with detail of engraved *bulla* (*opposite*). Photo courtesy of author.

of civic generosity but also branded the city as the pope's, metropolis of his ecclesiastical empire.

Communication and Justice in Renaissance Rome

The Renaissance popes also took over the administration of justice from the commune, an activity that was itself richly communicative. In the later fifteenth century, Roman diarists recorded instances of judicial actions both everyday and extraordinary: proclamations issued, notices posted, sentences pronounced, and punishments inflicted. They document both standard practices and moments when procedures were not followed, norms were flouted,

or justice was not served. In the following analysis I draw on their accounts, above all on that of Stefano Infessura, a high-ranking official of the communal senate, a client of the baronial Colonna family, and a fierce critic of successive Renaissance popes.

Justice in late medieval Rome was a public affair, as violent and laden with symbolic meanings as in any premodern city.[94] Malefactors were exposed to public view for sentencing and punishment on the steps of the Capitoline court, on the steps of St. Peter's,[95] or on the steps connecting the church of Aracoeli to the Campidoglio, a place both ecclesial and communal.[96] Criminals might be confined in cages hung high over the market at Campo de' Fiori,[97] or dragged through the streets on carts. Those condemned to die were hanged on the gallows on Monte Caprino, the southern spur of the Capitoline Hill, or from a window or off the loggia of the Senator's Palace,[98] or from an ancient elm tree in front of the Lateran,[99] or on gallows thrown up for the occasion in Campo de' Fiori or Piazza Navona.[100] Later in the century, the city's judicial center of gravity shifted from the communal Campidoglio toward the Vatican, and hangings occurred on the battlements of Castel Sant'Angelo, on gallows erected across Ponte Sant'Angelo,[101] or, later still, at the new papal jail constructed upstream at Tor di Nona.[102] The bodies of the condemned were left to dangle for hours or days, sometimes upside down.[103] They might be subjected to further humiliations: burned on elevated pyres,[104] or drawn and quartered, their heads paraded on pikes over the crowds,[105] with other parts dragged through the streets and then nailed up at city gates and bridges.[106]

The hands of executed criminals, the most potent agents of criminal action, might be cut off and "affixed" (like a bull!) to the facade of a nearby church, especially if the crime involved that church in some way,[107] or they might be nailed along the twin parapets of a bridge,[108] or beside the bronze wolf in its niche in a brick wall at the Lateran,[109] or on various city gates. Noble offenders had the privilege of a private decapitation behind the walls of Castel Sant'Angelo, but their corpses, too, were made available for public inspection in the neighboring churches of San Celso or Santa Maria in Traspontina.[110] Particularly notorious criminals might also be memorialized in *pittura infamante*. A painting of the condemned (usually shown after death, sometimes hanging upside down) would be applied to an exterior wall near the site of execution: there are records of such paintings on the Senator's Palace, at Aracoeli, and at the Lateran.[111]

These rites of humiliation, violence, dismemberment, and display might seem communicative enough, but the execution of justice also relied on more explicit signs and indications, both oral and textual. Offenses were described and sentences were pronounced loudly and publicly, often accompanied by the ringing of bells.[112] Offenders were made to wear paper mitres, attire that was inherently shameful but that might also bear an inscription detailing the nature of their crimes.[113] Strips of paper describing offenses were wrapped around necks or pinned to clothing. A thief hanged in Campo de' Fiori had gold foil attached to his noose to indicate his greed.[114] Some curial employees convicted of forging bulls were paraded through the streets on a cart with examples of their handiwork—four phony bulls in all—dangling over their heads from a pole. They were hanged and their corpses burned together with their bulls in Campo de' Fiori.[115] Many of these punishments were meted out to workers whose job it was to communicate with the public. Notaries, heralds, couriers, and chancery scribes were all meant to convey messages and decrees clearly, not to obstruct or distort them. Their punishments included public ridicule and graphic penalties keyed to the crime they had committed, whether forgery or mere dereliction of duty or, sometimes, some graver crime.[116]

One of the most famous judicial spectacles of the fifteenth century was the "reverse canonization" of the lord of Rimini, Sigismondo Malatesta, in the spring of 1462. Pius II not only excommunicated the rebellious baron but also, finding this penalty inadequate to his crimes, decreed that Sigismondo's soul was already in Hell. On April 27, 1462, on the steps of St. Peter's, a dummy dressed as Sigismondo was charged and convicted of a host of mortal sins and then burned at the stake. Pius himself describes the event:

> A great pyre of dry tinder was piled before the steps of St. Peter's, on top of which was placed an effigy of Sigismondo that reproduced the wicked and accursed man's features and dress so exactly that it seemed a real person and not a dummy. But lest there be any doubt as to the depiction, an inscription issued from the figure's mouth that read, "Sigismondo Malatesta, son of Pandolfo, king of traitors, hated by God and man, consigned to the flames by vote of the sacred senate." A great crowd read this text. Then, while the people stood by, the pyre was set alight together with the effigy, which at once burst into flames. Such was the mark that Pius branded on the godless house of Malatesta.[117]

No fewer than three effigies of Sigismondo were burned in Rome that day. Identical dummies, each with the same inscription emerging from its mouth, were cast into fires at St. Peter's, Campo de' Fiori, and Campidoglio—all sites for the publication of decrees and the execution of justice, as we have seen.[118] Describing the nature of an offender's crimes—with text written on a mitre, or on a placard hung around the neck, or in a banderole emerging from a dummy's mouth—was a standard part of the ritual of judicial humiliation, but these communicative acts, like the *citationes* and *fossure* of the commune, also served to justify the actions taken; they came to be expected, demanded even, by observers. Infessura took special notice when—as he saw it—the popes began to disregard these norms. In 1453, the republican rebel Stefano Porcari and his co-conspirators were executed after "the sentence was read to them, that is, the reason why they were being hanged."[119] By contrast, under Innocent VIII, executions no longer took place at the communal gallows near Campidoglio; instead, Romans were hanged secretly in the night at the new jail of Tor di Nona. Their bodies would be discovered in the morning, dangling from gallows "without indication of their name or crime."[120] Noble prisoners were still decapitated, but now without due process: a papal order, briefly flourished, was enough to seal their fate, without any public announcement of their offense.[121]

Infessura never questions the judicial logic of these punishments; rather, it is the transparency of the process (or its lack) that concerns him. The pope had the authority to punish, he granted, but the people had a right to know why. Papal officers could use information to humiliate, but citizens could also take offense at its withholding—or worse, at its deliberate perversion. Infessura records how, in the summer of 1489, Falcone de' Sinibaldi, apostolic protonotary and papal treasurer, used the machinery of justice to arrest a rival. Marching on his house at dawn with a crew of *fossoribus domorum* armed with clubs and carrying lanterns, he dragged the man off to Tor di Nona and had him hanged without charge or verdict, while the man's brother looked on in horror. This action was outrageous enough, but Falcone then left the body to dangle from the battlements of Castel Sant'Angelo for two days despite the protests of the victim's mother. Worse still, "on the second day after his death one of Lord Falcone's servants placed a false and mendacious placard at the man's feet, in which the deeds [of the real criminal, who, according to Infessura, was one of Falcone's own men] were attributed instead to him."[122] The customary gesture of communication was not just withheld but perverted and falsified.

Pure coincidence, but suggestive nonetheless: Infessura recorded, just a few years later, the miraculous discovery of the Titulus Christi. In 1492, workers in Santa Croce in Gerusalemme discovered a box immured in a wall containing the very board on which the Romans had inscribed "INRI," or "Jesus of Nazareth, King of the Jews," and which they had affixed to his cross. The *titulus* was, in other words, a judicial *cedula*, issued in mockery to be sure but still a public statement indicating why the prisoner had been condemned to die. The object was installed with great pomp in the basilica alongside other relics of the Crucifixion held there—the True Cross, a nail, the sponge, and a piece of the good thief's cross. This was one of a host of miraculous discoveries and transfers of relics related to Christ's life and passion that proliferated across Italy in the late fifteenth century as access to the Holy Places was cut off by the Turks. Our main source for this particular discovery is Infessura, and the episode appears in the text at a moment when he is preoccupied with the *lack* of such judicial notifications from the civic authorities of his own day. Even Pontius Pilate had exhibited more transparency than they.[123]

Infessura's contempt for the papal style of justice reaches a climax in his account of the Roman summer of 1489.[124] It was, he says, a season of extraordinary disorder and lawlessness, with robbery, rape, and murder on the rise. In the midst of it, the pope began issuing bulls to his allies absolving them and their servants of any crimes they might commit, up to and including murder, granting them safe conduct wherever they went in the city and rendering them immune from judicial action. Protected by a bull, the son of Francesco Buffalo was able to murder his pregnant stepmother, for example, and face no penalty. Eight other Romans convicted of murder flourished their bulls and were pardoned and spared the gallows in the same stretch of time.[125] "These days," Infessura remarked, "no one who has the means to redeem himself is sentenced to die. But if someone does not have the means . . . he hangs."[126]

Infessura is often dismissed as an unreliable source for Roman history precisely because of his bitter animosity for the pope and his insistence that the restoration of papal authority over the city was a tragic blow to communal liberty. His purpose in relating these various anecdotes, none of which may be completely true, is hardly disinterested: he means to paint the pope as contemptuous of long-standing custom and the rule of communal law. But his reliability is beside the point. What matters is *how* Infessura paints his picture of papal contempt, and what is striking is that he focuses so frequently on how the Renaissance popes spurn the requirements of open communication.

Sixtus and Innocent both trade in false accusations, secret judgments, and summary executions, favoring their noble friends with bulls of immunity and putting the machinery of justice up for sale to the highest bidder. They show particular scorn for the offices of the old communal government, letting curial favoritism, or baronial privilege, or sheer greed ride roughshod over ancient law and custom. The usual machinery of communal communication was thoroughly perverted: the *cursor* with his *citatio* beaten and left for dead, the *fossor*'s doorbusting services appropriated for private vendettas, the publication of capital crimes, traditionally written out on a placard and placed at the foot of the condemned, either omitted or falsified to nefarious ends, or else the criminal freed on payment of some undisclosed sum without ever making public satisfaction.[127]

And yet, even as Infessura complained about the lack of papal transparency in judicial matters, he could also bridle at the ubiquity of papal postings on ecclesiastical and fiscal affairs. After Sixtus died in August 1471, the diarist gave free rein to his hostility. God had liberated his people from a wicked and impious ruler, he claimed; Sixtus had been tyrannical, vainglorious, a shameless nepotist, a sodomite, and a lover of boys; he had sold offices and imposed new taxes and fines, only to squander the treasure of the Church on pointless wars. There were no limits to his fiscal chicanery. Among other things, he would routinely slap outrageous notices on the doors of the city's churches: "He would levy contributions (which he called 'tithes') both from clergy and officials alike without leniency. For he would fix an order onto the door of a church requiring that a certain amount be paid within an extremely short period of time, under pain of excommunication and loss of office; if the deadline was not met, additional fines of fifty or a hundred ducats must be paid; and if the sum was not paid as demanded, the church itself would be placed under interdict and the canons deprived of their positions."[128] Thus, the pope, in the eyes of his critics at least, imposed his authority on the city of Rome, its institutions, its buildings, its people, and their bodies. Texts issued and texts withheld together built an ever more powerful documentary regime.

Placards, Libels, and Counterpublications

Once the papacy staked its claims to the doors of Roman churches, the doors themselves became vulnerable to attack by other parties. Libels and placards of protest, rebellion, or mere ridicule had long been a feature of late medieval urban life. Successive papal regimes attracted these sorts of critiques, some merely satirical, some posing a real legal threat to papal authority. Their authors adroitly used the doors of churches and palaces—elsewhere at first, and then later in Rome itself—as sites for publicizing their views. In 1329, as we have seen, Ludwig of Bavaria posted his decree deposing John XXII on the doors of Pisa Cathedral; the Franciscans of the city posted their denunciation of him on the same doors, condemning him as worldly and corrupt.[129] In Avignon in 1352, a letter from the devil was found posted on the door of a cardinal's *livrée*. The prince of darkness congratulated the pope and his court for doing so much diabolical work on his behalf: oppressing the poor, squandering the treasure of the Church, selling offices to servant boys, sowing discord, fomenting war, and weakening the faith of the common man.[130]

The controversies surrounding the Council of Basel prompted much posting and counterposting. Eugene IV used doors in Ferrara to announce his transfer of the council from Basel to that city, as we have seen. A few years earlier, in 1432, the antipope Felix V issued a bull of his own, which he had posted on the doors of the cathedral of Geneva and the Dominican church there. Felix noted that if his courier could safely travel to Rome, he might post copies on the doors of various churches near the city, but legal publication of the bull did not depend on it.[131] Posting was always more difficult for the council and its adherents: when the fathers who remained in Basel issued their decree deposing Eugene, they sent copies to be posted on the doors of the cathedrals of Strassburg, Speyer, Worms, and Mainz. The first three cities allowed the document to be displayed, but in Mainz the bishop forbade it or any others like it to be affixed to his cathedral's doors.[132]

In the second half of the century, by contrast, critics of the papacy had easy access to Rome and its walls and doors. Infessura notes several Latin couplets that were found posted on Roman walls after the death of Sixtus IV.[133] The *condottiere* Virginio Orsini was said to be the author of certain *cedulae* that circulated through the city in the winter of 1485 encouraging the Romans to rebel against Innocent VIII.[134] The letter of "Silvio Savelli," which circulated both in manuscript and in print during the reign of Alexander VI, did much the same.[135] Around the same time, a Venetian chronicler reported that eight hostile couplets had been found pasted on the doors of Alexander's library in the Vatican Palace, prompting the pope to increase the size of his bodyguard.[136] Other verses circulated after Alexander's death, murmuring against his greed and immorality.[137]

The popes themselves could authorize the circulation of such material. In Bologna in 1429, as the armies of Martin V besieged the city, "many *cedule* were found in various places about the city, tacked up to the walls, especially at the corners of the piazza and at the house of the [noble family] Canetoli." The scraps of paper had verses in Italian encouraging the people of Bologna to submit to Church authority and predicting the flight of the traitors occupying the city.[138] In 1506, after Julius II expelled Giovanni Bentivoglio and his family from Bologna, another pro-papal poem appeared, "publicly posted in various places around the city," this one crowing over the misfortunes of the exiles and poking fun at their name (once "Bentivola," now "Malivola," "Multivola," "Mortivola," and so on).[139] Later that year, when the matriarch of the clan, Ginevra, died from the shock of her son's disaster, verses were posted throughout the city mocking Ginevra as "impia, avara, tenax, horrida, terribilis," and so on.[140] Yet another set of anti-Bentivoglio verses was posted in public after the family's palace was destroyed and their tower pulled down.[141]

More often, the popes were on the receiving end. Burchard records that in the hot summer of 1501, Alexander VI was considering leaving Rome for a hill town to safeguard his health, but placards appeared around town predicting that if the pope left town he would surely die. The first of these, Burchard says, was attached to *Maestro Pasquino*—this is, in fact, the first reference to the statue as a site for anonymous posting. News of the prophecy spread rapidly, and that same morning similar placards were found posted around the city.[142]

The popes had borrowed the practice of affixation, or *publicatio in valvis*, from the more quotidian legal practices of the medieval Roman commune. Now, by insisting on publication on Roman doors and walls as a legal requirement for ecclesiastical promulgation, they turned those same surfaces into potent targets for anti-papal propaganda. Doors could amplify statements of papal authority, but also challenges to it, particularly challenges reviving the vexed question of the council, the church assembly that successive popes from Martin V on had promised to convene, as a condition of their election, only to disavow once they came to power.

In 1430, German critics of Eugene IV posted placards critical of the pope on the doors of the Vatican Palace and other buildings around the city, calling on him to convene the church council he had promised.[143] In the midst of a dispute with Sixtus IV in March 1476, Louis XI of France had a call to a general council affixed to the *valvae Sancti Petri*.[144] In 1482, in the old council city of Basel, Archbishop Andrija Jamometić condemned Sixtus as an agent of Satan—a nepotist, warmonger, embezzler, and tyrant—and issued a call to a new council. He had his appeal printed and posted copies on various doors and walls around Basel; a papal legate later complained that the watchmen of the city stood guard over the copy posted on the main bridge across the

Rhine, so that no pious defender of the pope could tear it down. The legate, who had been sent to prosecute the archbishop and secure his arrest, struggled for more than a year to get his own sentences of excommunication and interdict posted anywhere in the region.[145]

In 1483, Venice went to war with Sixtus; after the pope put the city under interdict, a document rejecting the sentence and calling the pope to account for himself before a new council appeared on the doors of San Celso in the Borgo.[146] Back home, Venetian chroniclers exaggerated the success of the exploit, recording incorrectly that the Venetian agent dispatched to post the notice had stuck copies on the doors of such key Roman monuments as the Pantheon and Castel Sant'Angelo, as well as on the parapets of Ponte Sant'Angelo. In their version, the pope awoke to news of the postings and, in a fury, had the night watchmen of the Borgo hanged, while the Venetians rewarded their agent with a pension.[147] Very little of this was true,[148] but the important thing is that the Venetians told themselves it was: they knew what it meant to post a hostile notice on the doors of a major Roman monument, under the very nose of the pope. Meanwhile, the same Venetian chroniclers tell us that *their* internal security forces took care not to allow any *scrittura* to be posted on any church door in the Venetian state.[149] Watchmen were set at the borders to intercept any *cursor* coming from Rome; the patriarch was told not to open any letters sent from the city.[150] Thus, the apparatus of papal publication could be subverted by a state with enough boldness. The mischief continued into the sixteenth century. In 1501 anonymous poems criticizing Alexander VI appeared on the doors of the Vatican, the Lateran, and the chancery.[151] In 1509, after Julius II threatened Venice with another interdict, Venetian spies infiltrated Rome and, in the night, posted another call to a general council on the doors of St. Peter's.[152] The French did the same in 1511, posting a call to a council on the doors of San Francesco in Rimini when Julius was staying there.[153]

The second decade of the sixteenth century brings us, finally, to the most famous poster of notices in early modern history, Martin Luther, who affixed his Ninety-Five Theses to the doors of the castle church in Wittenberg—or, at least, is said to have affixed them—in October 1517. Whether Luther really nailed or glued his theses to the church door remains unclear, but like the story the Venetians told themselves about their spy in the Borgo, the legend tells us much about how contemporaries perceived the practice of *publicatio in valvis*, its power to shock, and its potential for subversion.

Posting and Printing in Renaissance Rome

Outside Rome, the recipients of papal bulls often had them published in print, especially when it served their interests to publicize their contents: an appointment to a new office, for example, or the grant of a new privilege to a church or shrine.[154] But the papal chancery took longer to see the value of printing bulls it had already published in the usual way. Gradually, starting with Paul II's printed bull of 1470 announcing the Jubilee of 1475 and followed by Sixtus IV's Jubilee bulls and bulls for the Ospedale di Santo Spirito in the later 1470s, papal documents began to proliferate in print. The contents and chronology of these publications are reviewed in later chapters. Here, it is worth considering how these printed papal documents assured their readers that they had been properly published—in the earlier sense of the term.

Despite their embrace of the new technology, the popes continued to insist that public posting constituted formal promulgation of their decrees. And they took pains to record in print the fact that the older process had been followed. Many printed bulls record information on where the original, manuscript document was posted, when and how long it hung in the public view, and who had done the posting. The more solemn the bull, the more likely it was to include such information. Sixtus IV's bull excommunicating the Venetians, issued in May 1483 during the War of Ferrara, was printed with a statement by the master of the heralds, who had posted it at St. Peter's: "On Saturday the 24th of May, 1483, this was posted on the doors of the apostles Peter and Paul in the City, by Ranaldus Cavix and Iohannes Nicolai, Heralds: G. Donatus."[155]

Paradoxically, the longer print was in use in the city, the more emphasis printers (or those who commissioned them) came to place on recording these details. Nearly every one of the dozens of bulls printed under Julius II includes this same information, set apart from the text of the bull itself and sometimes flagged with the heading *Publicatio* (figure 1.7):

> In the year of our Lord 1509, in the twelfth indiction, on the 27th day of the
> month of April, in the sixth year of the pontificate of our most holy father in
> Christ and lord, Lord Julius II, by divine providence Pope, the present apos-
> tolic letter (written overleaf) of the aforesaid our most holy lord the pope was
> posted and published on the doors of the basilica of the prince of the apostles
> in Rome, and the doors of the apostolic chancery, and on the edge of Campo

de' Fiori by both the undersigned *cursores*, namely Eurardus du Vivier *cursor* of the aforesaid our most holy lord the pope and of the Roman Curia, and so confirms Michel le Boux, *cursor* of our most holy lord the pope and present master of these *cursores*.[156]

The complete process of publication was recorded: the manuscript bull, prepared by the chancery, was taken by the *cursores* on their rounds to publication sites in the city; they returned it to the chancery in via de' Banchi, where they noted down on the reverse the places and times they had posted it. The chief herald signed his own name below, confirming their testimony. It was only then that the bull was sent to one or more of the printers whose workshops stood near the chancery, in the streets around Campo de' Fiori. Once a printer received the bull, he set not only the text of the document in type but also the notes that the heralds had written on the reverse of their manuscript sheet, recording its process of urban publication. Some printed bulls include even more information: the name of the secretary who had prepared or registered the bull; the names of cardinals who had been present at its signing; the names of recipients to whom it may have been sent; or some other evidence or authenticating statement intended to reassure the readers of future copies, whether manuscript or printed, that a genuine original stood behind the reproduction. The bull of warning that Julius issued against a group of dissident cardinals in 1511 reproduces such a formula of authentication at the start of the text, by the cameral auditor Girolamo Ghinucci of Siena (figure 1.8):

> Let all and sundry about to read this public copy know that I, Girolamo
> Ghinucci, canon of Siena, doctor of both kinds of law and auditor of the apostolic camera, have held, seen, and carefully inspected this apostolic letter of
> warning, written by our most holy father in Christ and lord, Lord Julius II,
> by divine providence Pope, and I certify that there I saw the true leaden bull,
> with silken cords of red and yellow color, hanging from it in the manner of the
> Roman Curia, and therefore above all suspicion, and written on the back of it,
> the *publicatio* of the same letter, signed by three heralds of the aforesaid most
> holy lord our pope, noting that publication had been duly performed in the
> places designated in the letter itself. And the contents of the letter are as
> follows.[157]

Ghinucci's statement recapitulates the entire process of *publicatio*: the material preparation of the bull, its sealing with silk and lead, its circuit around

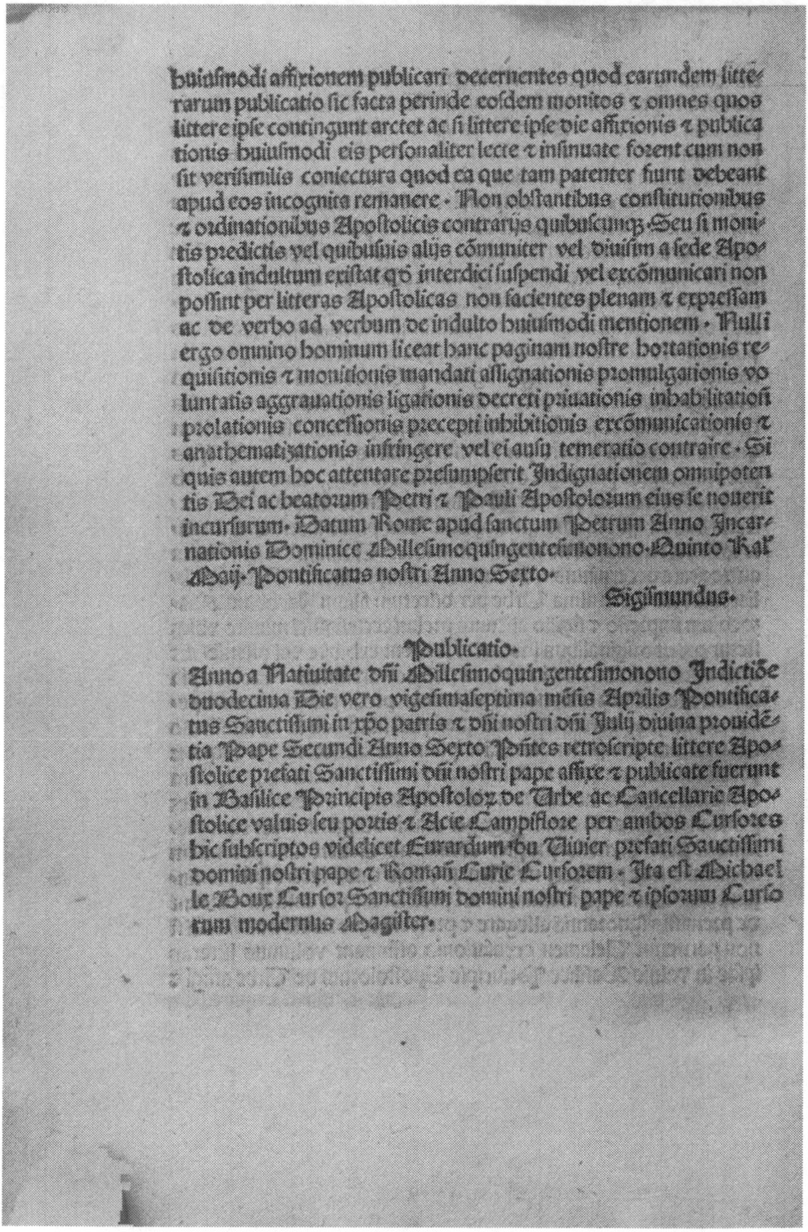

huiusmodi affixionem publicari decernentes quod earundem litterarum publicatio sic facta perinde eosdem monitos z omnes quos
littere ipse contingunt arctet ac si littere ipse die affixionis z publica
tionis huiusmodi eis personaliter lecte z insinuate forent cum non
sit verisimilis coniectura quod ea que tam patenter fiunt debeant
apud eos incognita remanere · Non obstantibus constitutionibus
z ordinationibus Apostolicis contrarijs quibuscunqz·Seu si moni-
tis predictis vel quibusuis alijs communiter vel diuisim a sede Apostolica indultum existat qd interdici suspendi vel excommunicari non
possint per litteras Apostolicas non facientes plenam z expressam
ac de verbo ad verbum de indulto huiusmodi mentionem · Nulli
ergo omnino hominum liceat hanc paginam nostre hortationis re-
quisitionis z monitionis mandati assignationis promulgationis vo-
luntatis aggrauationis ligationis decreti priuationis inhabilitation
prolationis concessionis precepti inhibitionis excommunicationis z
anathematizationis infringere vel ei ausu temerario contraire · Si
quis autem hoc attentare presumpserit Indignationem omnipoten
tis Dei ac beatorum Petri z Pauli Apostolorum eius se nouerit
incursurum· Datum Rome apud sanctum Petrum Anno Incar-
nationis Dominice Millesimoquingentesimonono·Quinto Kal
Maij·Pontificatus nostri Anno Sexto·

Sigismundus·

Publicatio·

Anno a Natiuitate dni Millesimoquingentesimonono Indictie
duodecima Die vero vigesimaseptima mensis Aprilis Pontificatus Sanctissimi in xpo patris z dni nostri dni Julij diuina prouidē-
tia Pape Secundi Anno Sexto Pntes retroscripte littere Apostolice prefati Sanctissimi dni nostri pape affixe z publicate fuerunt
in Basilice Principis Apostolorz de Urbe ac Cancellarie Apostolice valuis seu portis z Acie Campiflore per ambos Cursores
hic subscriptos videlicet Eurardum dn Uiuier prefati Sanctissimi
domini nostri pape z Romañ Curie Cursorem · Ita est Michael
le Bour Cursor Sanctissimi domini nostri pape z ipsorum Curso
rum modernus Magister·

Figure 1.7. Statement of *publicatio* in Julius II, *Monitorium contra Venetos* (Rome, 1509). Munich, BSB, Res / 4 Eur. 330.22, sig. a8v.

N NOMINE DOMINI
Amen• Nouerint uniuersi & singuli
hoc præsens publicum Transumptum
ispecturi qd nos HIERONYMVS
de Ghinucciis Canonicus Senen̄. V.
Iu: Doc .Curię causarum Camerę Apo.generalis au
ditor: habuimus uidimus & diligenter inspeximus
litteras apostolicas monitoriales Sanctissimi in xp̄o
patris & dn̄i nostri dn̄i Iulii diuina prouidentia Pa
pę Secundi infrascripti tenoris: eius uera Bulla plū
bea cū cordulis sericeis rubei croceięᴈ coloris/more
Ro. Cu. penden̄ bullatas: & omnimoda suspitione ca
rentes: unacum earundem litterarᴈ Apostolicarᴈ pub
licatione & affixione a tergo per tres prefati• Sancti
ssimi dn̄i nostri Papę Cursores in locis in dictis l it•
teris Apostolicis designatis: debitę executiōi deman
datas. Quarum quidem tenor sequitur & est talis•

 VLIVS EPISCOPVS SERVVS
seruorum Dei . Ad futurā rei memo
riam Celestis altitudo potentię/diui
na prouidentia cuncta disponens ut
ecclesiæ triūphanti militantē ecclesiā
simile faceret/uoluit summo Pontifici xp̄i Vicario/
dominici gregis pastori adesse in regenda & guber•
nanda uniuersalis ecclesiæ Monarchia Consiliarios
& ministros sanctæ Ro.ecc. Cardinales ut ipsi tanᵹ
eiusde Pontificis & ecclesiæ mēbra suo capiti defer•
uirent & submissis humeris ad regendū uniuersalis

Figure 1.8. Girolamo Ghinucci, authentication statement, Julius II, *Bulla monitorii contra tres reverendissimos cardinales* (Rome, 1511). Munich, BSB, 4 J.can.f. 161, fol. 2r.

the city, the signatures of the *cursores* who had taken it on its pilgrimage, its safe arrival back at the papal court, and the subsequent production of *transumpta* (manuscript copies), which Ghinucci further authenticated, one of which would have then made its way to the printer. It was then reproduced in dozens of uniform printed copies. For all its mechanical efficiency, however, the press still recorded the older, more immediate, and tactile form of *publicatio* the original document had experienced: the moment when a particular piece of sheepskin, inscribed by hand, pierced with silk, and sealed with molten lead, had hung for a day on a Roman door for the public eye to inspect.

Such formulas had long been attached to the manuscript *transumpta* (copies) of official documents that were released for public circulation. But the inclusion of these formulas in *printed* pamphlets bears examining. Why not just assert that the printed copy was officially approved, that it was an accurate transcript of the original? The presence of these older formulations may seem a sign of insecurity, as if those responsible for their printing thought that printed copies could not stand on their own merits without reference to their manuscript originals. Certainly, it is ironic to think of a printed text taking pains to establish its authenticity by invoking these archaic procedures of *publicatio* that were now hardly necessary thanks to the advent of the press. "Publication" in an entirely different sense was now possible. Uncertainty about the status of the new medium is one possible explanation for these typographical choices, but another is this: a conscious awareness of the power of urban *publicatio* as a form of political ritual. The older process of *publicatio* was inherently authoritative. It had both legal force and a social and political valence of its own. By invoking the material practices of the papal chancery, the publication routes of heralds as they traversed the papal city, and the particular, ancient, and holy doors on which the documents had hung, early printed bulls projected not just their contents but also the authority of the institution that issued them and its command of the city—and world—they addressed.

Finally, the development of these administrative practices illuminates the almost exactly contemporaneous phenomenon of Pasquino. Poems were first posted on the statue in 1501, as we have seen. Starting in 1509, Roman printers began to publish annual editions of the verses posted the previous year. Like the papacy's printed bulls, the Pasquino volumes take pains to emphasize that their texts were first posted in Rome and that this is why they deserve attention. The first edition of *pasquinades*, entitled *Carmina quae ad*

Pasquillum fuerunt apposita or *Poems Posted on Pasquino*, appeared just a few years before Ghinucci's carefully authenticated papal bull.[158] Giacomo Mazzocchi, whose printshop stood practically across the street from Pasquino, issued this and subsequent collections of the verses. Still other printers, including Johannes Besicken and Etienne Guillery, published their own editions; they were much employed printing papal bulls in these same years, as we will see. Many of their editions feature a woodcut image of a fancifully restored Pasquino on the title page, shown in the costumes of various classical gods. The editions gave the ruined statue a voice that carried far beyond the city center. Like the commune's marble lion and the *valvae Sancti Petri*, Pasquino had the power not only to display but also to broadcast, to amplify, to *publish* the texts he raised to the city's view.

Long before the coming of print, the medieval papacy had a sophisticated apparatus for the publication of documents, developed in tandem with and, sometimes, in conflict with the corresponding legal processes of the Italian city-republics. The idealistic legislation of the communes declared—and attempted to realize—a contractual relationship between government and people. In Rome, communal officials undertook to put documents in the hands of recipients when they could, and to broadcast their contents from a particular civic site, "the place of the lion," when they could not. Papal procedures for the posting of official decrees, dating to the thirteenth century but codified in the fifteenth after the papacy's return to Rome, paralleled those of the commune but were more one-sided, relying on the assumption that public posting of a text in the papal city, rather than physical delivery beyond it, was the more effective and indeed efficacious means of publication—that publication was more important than delivery. This conceptual adjustment primed the bureaucracy to adopt the press as a tool of publication, and thus of policy.

As the papacy took over the machinery of Rome's communal government, papal acts—bulls, proclamations, inscriptions, and performative acts of justice—multiplied and spread across the Renaissance city. But even as the popes used publication as a tool to reclaim their metropolis, they faced political challenges both at home and abroad. Critics of the papacy published libels and complaints at the same sites the popes used for official promulgations. Agents of rival states infiltrated the city and posted counternotices on the same doors and walls as a way of embarrassing the pope and challenging his claims. Their activities paralleled those of the local wags who, as the Roman Renaissance reached its climax, adopted Pasquino as a site for posting critiques of papal pretensions. As the lion of the Campidoglio fell mute, Pasquino picked up the chorus of communal voices and broadcast them far beyond the Tiber. The doors, walls, and statues of Rome became a noisy battleground where competing political actors launched their attacks. In far-off Wittenberg, the doors of the castle church stood waiting to join the fray.

Although printing reached Rome in the 1460s, it would take a long time—centuries in fact—before printing *counted* as publication in the legal sense. Roman printers began to publish the texts of papal decrees quickly enough, but in the technology's early years, printing seemed to occupy a completely separate category in the minds of Roman citizens and officials alike. The text of a bull was still copied by hand into the (manuscript) papal registers; it was still considered officially published only when it was copied out on parchment and posted on a circuit of important doors; printing, when it was deployed at all, was for distributing the contents of these notices as booklets or broadsides that would pass from hand to hand, but not be posted themselves.[159] (Not until the 1520s or 1530s do we see broadside *bandi* or *editti* printed for public posting in Rome.) In earlier decades, bulls were printed as pamphlets that made reference to their manuscript originals and to the fact that those originals had traveled to and spent time at specific Roman sites. Under Julius II, bulls regularly advertised this expansive *cursus* of publication, underscoring the primacy of the manuscript, posted notice in Roman legal and political thought.

The potency of this process may explain why the papacy was slow to adopt print at first, and why, when it did, it was so traditional in its use of the medium. With an expansive and effective communications apparatus already in hand, papal officials seem not to have detected any advantage to print, at first, beyond its ability to *multiply* a text that had already been published—because it had been posted—and to *advertise* the fact of its posting at a specific time and place. The question of who in Italy did seize on print as a new means of political communication in the technology's early years is the subject to which we now turn.

CHAPTER TWO

Humanists, Printers, and Others

Rome in the middle of the fifteenth century was a bustling city of pilgrims and diplomats, ambitious curial officials, indigent scholars, cardinals and their households, bankers, merchants, old aristocrats, and assorted hangers-on, ruled by a series of autocratic popes who left their marks on the city in various ways—in architecture and epigraphy, in ceremonial and in law. As the Renaissance popes took over the work and the authority of the old medieval commune, they sidelined Roman barons and communal administrators alike, while also asserting their right to legislate on ecclesiastical affairs across Christendom. And yet, despite the size and efficiency of the papal chancery—or, perhaps, because of it—the papacy was at first slow to adopt the new technology of printing. With a chancery that could churn out dozens or even hundreds of copies of a single bull, a network of couriers who could deliver them across Europe, and a legal framework that prioritized the posting of bulls in the city over any other means of publication, the papal bureaucracy saw little need for the new technology of mass reproduction in its earliest years.

Despite decades of detective work, historians of printing have found little evidence of a papal invitation summoning printers to Rome, or papal patronage of the earliest printers who worked in the city. Sweynheym and Pannartz set up their press at Subiaco in 1464, either at the close of Pius II's reign or at the very start of Paul II's. Scholars have found signs of Cardinal Nicholas of Cusa's patronage in Subiaco, and both Cardinal Juan de Torquemada and Archbishop Rodrigo Sánchez de Arévalo seem to have been involved in some way in the printers' earliest projects in Rome. Even better known is the assiduous work of Bishop Giovanni Andrea Bussi, their editor, protector, and advocate in the city. Bussi and other humanist editors and correctors in Rome guided German printers toward the production of editions that matched their intellectual interests—mostly classics and patristic texts, alongside a small number of works by contemporaries like Cardinals Bessarion and Torquemada. Before long, these printers turned to more lucrative projects, publishing legal handbooks, theological manuals, and the liturgical texts that were in demand by clerics everywhere.

The papal chancery did make some use of the press for its administrative needs. Two early printers in Rome, Ulrich Han and Sixtus Riessinger, published between them eight editions of chancery rules and regulations during the reign of Paul II. In 1470, Riessinger issued Paul II's bull announcing the Jubilee of 1475;[1] the bull would be reprinted six times in Rome in the next five years. Under Sixtus IV, yet more administrative publications, including books of chancery regulations, tables of offices, and schedules of fees, appeared in print (some twenty-seven editions over the course of his pontificate), as well as three bulls concerning the 1475 Jubilee, six concerning his refoundation of the Ospedale Santo Spirito in 1478, and some seven bulls and indulgences encouraging contributions to the crusade against the Turks. All told, these account for less than 10 percent of the editions printed in Rome during Sixtus's reign.[2] The vast majority of texts printed in Rome in the early years were not official publications but speculative literary and scholarly productions.

When Francesco della Rovere was elected pope on August 9, 1471, there were five printers at work in the city.[3] At first, these entrepreneurs were more concerned with what the pope could do for them than vice versa. In March 1472, Sweynheym and Pannartz wrote to Sixtus begging for aid after Bussi's humanistic publication program had brought them to the brink of bankruptcy.[4] Around the same time, the Sicilian printer Giovanni Filippo de Lignamine published two theological tracts written by the pope earlier in his career. He supplied each with a flattering letter of dedication in a clear attempt to win Sixtus's notice and patronage.[5] Bussi and de Lignamine were ambitious men of a certain social standing. Bussi, bishop of Aleria, would be appointed Vatican librarian in 1475; de Lignamine, a Sicilian nobleman, had been appointed a *scutifer* by Paul II and later won a lucrative appointment to sell indulgences in Sicily. Both used the press as a means of attracting attention to themselves, a tool for raising their profiles.

It would take some years for Sixtus and his advisors to adopt the press as a medium for advancing political or ecclesial policy, and the pope was not alone in this regard. As printing spread across the Italian peninsula, popes, princes, and republics alike showed little interest in the technology at first.

The Italian powers had long-established means of communicating on political matters, whether with one another or with their subjects: the posting of notices, public proclamations, formal entries and other civic ceremonies, and the circulation of open letters, pamphlets, and even poetic libels in manuscript afforded the powerful many ways to broadcast political messages to their late medieval publics.[6]

In the early years of print, it was literary men of more middling social status—clerics, jurists, scholars, courtiers, poets, and gentleman editors and printers like Bussi and de Lignamine—who seized on the new technology as a way to promote themselves and their works. Many of them chose to publish on controversial or pressing political themes, including the problem of Turkish expansion in the Mediterranean and the perceived threat of religious minorities from within the borders of Christendom. Presenting themselves as expert commentators on these issues, their goal was to capture the attention of their superiors—above all, of the pope. In the first decade or so of Italian print, the conversation was decidedly one-sided.

Two sensational episodes from the early to mid-1470s especially caught the attention of humanist poets, scholars, and political commentators, who composed letters, orations, poems, and tracts about these events and quickly put them into print. The first, the fall of the Venetian colony of Negroponte to the Ottoman Turks in 1470, prompted the publication of some of the earliest political pamphlets in all of Europe, over a dozen tracts printed in cities up and down the peninsula (and beyond), many of them the first books to be printed in their respective cities. The second was the gruesome blood libel case of Simon of Trent, a Christian child found dead in the Alpine city of Trent on Easter 1475, whose death was blamed on the local Jewish community. Here, too, ambitious clerics, jurists, and humanists seized on the event as an occasion to promote themselves, their literary talents, and their city as they celebrated the miracles of the supposed martyr and urged Sixtus IV to declare the boy a saint. The press would play a powerful role in spreading news of the boy's death and publicizing his cult, even as the pope urged caution and professed himself skeptical of the campaign and its media. In this chapter, I survey the printing history of these two episodes not only to illustrate the wide range of political literature, written in poetry and prose, in Latin and in various vernaculars, that was put into print in Italy in the first decade of printing but also to highlight the absence of official state publications from the mix. Political print in the first decade was directed at the pope, not generated by him.

The Fall of Negroponte

Negroponte fell to the Turks on July 12, 1470.[7] The city, commanding the island of Euboea, was one of Venice's most important possessions in the eastern Mediterranean. After the Turks captured Constantinople in 1453, Venice had fallen back on Negroponte; the city was a major commercial center and a vital military outpost, the site from where, for the past seven years, Venetian galleys had pursued an unsuccessful war against the Turks. In the late spring of 1470, the Ottomans laid siege to the city. Although a Venetian fleet under the command of Admiral Niccolò Canal sailed to its relief, the inhabitants were powerless to stop the Turkish attack. With the loss of this crucial forward base, the way lay open, or so European observers believed, for an Ottoman assault on Italy itself.

When news of the disaster reached Italy at the end of July, it set off a chain of political and popular reactions that would dominate the civic discourse of the peninsula for months to come. The loss of the colony precipitated diplomatic correspondence and negotiations among the Italian states, especially Venice and Rome, while the Italian public responded to the news with a mix of panic, self-recrimination, and prurient interest in the grisly details.

The fall of Constantinople in 1453 had provoked similar fear and fascination in Italy.[8] Both catastrophes gave rise to an enormous body of texts: sermons, poems, orations, letters, and apocalyptic tracts. Popular interest in the topic was intense, reflecting traditional dread of the infidel foe and a keen interest in the way the Italian states were managing that threat and, by extension, their relations with one another. The loss of the city provided new material for the ballad singers who entertained audiences with topical compositions in the marketplace; it also provided a new topic for learned commentators hoping to raise their literary profiles and gain the ear of a powerful patron. The fall of Negroponte differed from any previous event in Italian history, however, in that it coincided almost exactly with the spread of printing through the major cities of the peninsula. Late 1470 and 1471, the year and a half immediately after Negroponte's fall, saw printing spread from Venice and Rome to a further eleven Italian cities, including Bologna, Naples, Milan, Ferrara, and Florence.[9] More than a dozen Italian editions relating directly to the disaster at Negroponte survive from these years, along with two editions printed north of the Alps.[10]

Throughout 1470 Venice had tried to publicize the danger that threatened its colony and to agitate for aid from the Italian states. In July, the Senate, with the support of Paul II, persuaded Naples, Milan, and Florence to renew the mutual defense pact they had agreed on at Lodi in April 1454.[11] After the fall of the city, Doge Cristoforo Moro sent open letters (in manuscript) to the Italian states in which he described the horrors of the city's sack, demanded help against the Turks, and warned against inaction.[12] At the same time, Paul II issued a series of open letters (also in manuscript) exhorting the princes of Christendom to settle their differences and mount a new crusade against the Turks.[13] In Rome, Cardinal Bessarion privately circulated a collection of manuscript letters and orations on the topic among his prominent and influential friends.[14] For a brief moment, a spirit of cooperation took hold among the Italian states. In mid-September, Paul II invited ambassadors to Rome to discuss possible responses.[15] Negotiations continued through the autumn, as rumors of an impending Turkish attack swirled through city streets and squares.

The fall of Negroponte, the Turkish threat, and the Christian response all remained topics of interest in Italy and north of the Alps through 1471 and 1472. Discussions revived whenever changes in government suggested new political alignments or new support for the crusade. The death of Paul II and the election of the more militant Sixtus IV in the summer of 1471; the election in December that same year of Doge Niccolò Tron, whose son had died fighting at Negroponte; and Sixtus IV's dispatch of papal emissaries seeking crusade support in northern Europe in the spring of 1472 all kept the story alive in the public imagination; so, too, did the spread of the printing press.

Printed Responses

Two poetic *lamenti* for Negroponte were printed a total of five times in the decade after the city's fall.[16] The *Lamento di Negroponte* and the *Piante de Negroponte* represent a literary tradition far older than the invention of printing.[17] For decades such ballads had been recited or sung to live audiences in Italian piazzas, sometimes by the poet himself, sometimes by a professional crier or peddler, who might also sell paper copies of the text alongside devotional images, patent medicines, and charms.[18]

Both texts employ the dramatic inflections of this oral poetic tradition to recount Negroponte's fall. They are packed with dramatic descriptions of the sultan's wily consultations, his formidable fleet and bloodthirsty troops, the smoke and thunder of his guns, the bravery of the defenders (sometimes expressed in direct speech), and the bloody aftermath of the sack. The information comes with frequent authorial asides: appeals to the audience; apostrophes to God, the Virgin, or the poet's muse; rhetorical questions; and dark commentary on the Christian negligence that had allowed the catastrophe to happen. The poems participate in an established tradition of vernacular poetic commentary on current events that would persist well into the sixteenth century.[19] Worth noting here is the speed with which they were published. Between 1470 and 1472, at least three, and possibly as many as five, editions of these two texts were issued by newly established printers in Venice, Milan, Florence, and Naples, some of them the first ever to work in their respective cities.

Nearly a dozen other editions relating to the fall of Negroponte survive from the period 1470–72, editions that might seem to belong to an entirely different world from that of the anonymous, popular laments. Most of these works were written in Latin—some in quite elevated style—and with one exception they were all signed by their authors. All were dedicated to prominent political figures. The first was by Paolo Marsi, scholar, poet, and an important figure in the Roman Academy of Pomponio Leto, the leading classical scholar in Rome. In 1468, after the antiquarian group was charged by Paul II with heresy, paganism, and conspiring to overthrow the papacy, several academicians were detained and tortured in Castel Sant'Angelo. Marsi and Leto were both in Venice when the affair erupted; after Leto was extradited to Rome as the assumed ringleader of the plot, Marsi attached himself to the household of the Venetian diplomat Bernardo Bembo, then embarking on an embassy to Castile.[20] Returning from Spain the next year, Marsi transferred his services to a new patron, Niccolò Canal, then setting out with the Venetian fleet to the eastern Mediterranean. Canal would prove no great soldier; he was a man of letters, a patrician doctor of law and an amateur scholar of Greek who had a library of philosophical texts installed in his cabin before setting sail. He hired Marsi as a sort of onboard panegyrist to record his maritime exploits for posterity.[21] Marsi witnessed the fall of Negroponte from the decks of the Venetian flagship, and the long poem he composed immediately afterward did not celebrate the admiral's virtues but chronicled the disaster he helped to create.

Marsi finished the work on August 26, 1470, while still at sea.[22] When his ship reached Venice in September, he wasted no time finding a printer. The first edition of his poem had been identified as a rough "first attempt" at printing by Federico de' Conti, who printed classical and humanist works in

Venice until 1472, stopping abruptly as so many early printers did after their early attempts proved unprofitable.[23] As Conti's first dated work appeared on January 5, 1471, Marsi's poem must have been printed in late 1470. A second edition appeared in Rome in early 1471, issued by an anonymous printer whose entire remaining output consists of classical texts edited by Pomponio Leto. By then, Leto was out of trouble: after a stint in the papal prison of Castel Sant'Angelo, he had resumed teaching at the Roman university in 1469 and started editing texts for this otherwise-unknown Roman press.[24]

Meanwhile, in Naples, the Genoese humanist Giorgio Fieschi produced another Latin account of Negroponte's fall: 725 epic hexameters that cast the siege and sack of the city as a reenactment of the first two books of the *Aeneid*.[25] It was printed in Naples by Sixtus Riessinger, who had left Rome in late 1470 or early 1471 to bring the art of printing there, and who would soon issue an edition of the *Lamento di Negroponte* and another work in Latin, a *Lamentatio Nigropontis*. The state of the type marks it, like Conti's Venetian edition of Marsi's poem, as Riessinger's first attempt at printing in Naples. It was likely the first book printed in the southern city.[26]

Other humanist authors meditated on the story of Negroponte in more oblique ways, often setting the catastrophe in a broader historical, theological, or moral frame. In Rome Giovanni Alvise Toscani, a jurist at the papal court, and in Venice the poet Raffaele Zovenzoni each wrote long Latin poems exhorting the princes of Europe to action against the Turks; both recounted details of Negroponte's sack as a way of underscoring the need for a new crusade.[27] In a different vein, Rodrigo Sánchez de Arévalo, bishop of Zamora and a prominent prelate in Rome, addressed a long consolatory letter on the fall of Negroponte to Cardinal Bessarion, a personal friend. It was one of the last things the bishop wrote before his death on October 4, 1470.[28]

Sánchez was a renowned theologian. Drawing on passages from Isaiah and Jeremiah, his letter insists that defeat in this world is temporary and inconsequential. True victory remains to be won in the next life, where the infidel will soon know true death. The bishop was also alert to the uses of humanist learning. He starts with a précis of ancient geographical lore concerning Euboea, its cities and coastlines, drawing on favorite humanist authorities like Pliny's *Natural History* and the geographies of Strabo and Pomponius Mela. Sacred and secular concerns are likewise woven together in an anonymous *Lamentatio Nigropontis* first printed in Naples in 1470 and reprinted in Rome around 1472, and also addressed to Bessarion.[29] (The cardinal, as leader of the exiled Greek community in Italy and a forceful advocate for a new crusade, was clearly seen as the Roman prelate most interested in the Turkish problem.) The author of the *Lamentatio* consoles Bessarion on the loss of the city; compares the plight of the Greeks to the tribulations of Israel under Pharaoh; likens Mehmed II to Xerxes, Hannibal, Caesar, and Alexander the Great; and castigates the princes of Europe for leaving the cardinal to lead the struggle on his own. Intriguingly, Pope Paul II figures among the princes the author targets: he excoriates the pope for piling up riches at the papal court when he could have been outfitting a new fleet to fight the Turks.[30] The fall of Negroponte is here exploited as a tool of political agitation: the critical tone of the work probably explains why it was printed anonymously.

Many of these same rhetorical commonplaces were invoked by the vernacular *lamenti* as they recount the fall of the city. Here, too, the poets pause frequently to castigate Christendom for allowing the disaster to happen, to invoke God's mercy on the dead, to pray for vengeance upon the Turks, or to exhort their listeners to take up the crusader's cross.[31] Classical allusions and exempla appear in the Italian works as well, with the sultan declared a new Hannibal, a raging Pyrrhus, and/or Attila reincarnate; Negroponte falls like Troy or Carthage; the Turks are new barbarian hordes, threatening to overrun Italy as the Huns and Goths before them. Most striking is their frequent editorializing, especially on the topic of the European states' complicity in the disaster. Eighteen stanzas of the *Lamento* (in its Florentine edition) are written in the voice of Negroponte herself. As the Turks breach her walls, she calls for help from the princes of Europe—from Naples to Scotland, Portugal to Norway, she beseeches each king and duke in turn and then weeps at their collective neglect of her plight. She concludes with a catalogue of eastern Christian territories now lost to the Turks.[32] The speech reveals a level of political and even geographical sophistication on the part of the poet (and of his audience?) that rivals much humanist commentary of the day.

For authors writing in Latin and Italian alike, it seems, a calamitous event like the fall of Negroponte could prompt a rupture in generic categories— between historical commentary and deliberative rhetoric, epic artifice and pious homiletics. As generic categories blurred, so did distinctions between popular and high culture. There is more literary polish and political sensibility in the vernacular works than might first be expected, and a good deal of genuine passion in the humanist accounts as well.

Nowhere is the hybrid character of these texts more apparent than in the poetic collection of the humanist poet Antonio Cornazzano, published under the title *Vita di Cristo* by an anonymous Venetian printer in 1472.[33] Straddling the ground between the vernacular and Latinate cultures of Quattrocento Italy, the text starts with a long account in *terza rima* of Christ's birth, miracles, passion, resurrection, and ascension, followed by an account of the growth of the Church and poetic confutations of heretics, false Christians, and other enemies of the faith. It is under this last heading that Cornazzano adds an *exhortatione* to Italy to come to the defense of the Christian religion under assault by the Turks: Venice has always striven to defend the faith, but the time has come for the other Italian powers to rally to her aid. The poem concludes with an account *Dell'expugnatione et ruina di Negroponte presa da Turchi*, a vernacular verse lament for Negroponte and its fall. The poet links each of these events to his larger theme of the trials of the faith and the faithful, while also comparing Mehmed II to ancient barbarian conquerors and urging the princes of Christendom to respond to the sultan's attacks by launching a new crusade.[34]

Cornazzano then switches to Latin, embarking on a *carmen heroicum* praising Venice as the perennial bulwark of Christendom against the infidel threat. There can be no doubt that the two texts were printed as a single unit; one stops and the other starts in the middle of the same quire, and the Latin poem begins with Cornazzano in high Horatian mode, boasting that he has been the first to sing the life of Christ in Aonian (or Italian) meters. But now he takes on an even greater subject, the perennial struggle of Venice against the Turks, and to do so he will employ the more exalted language of his classical forebears.[35]

Getting into Print

All the Negroponte texts—whether Latin, Italian, or some combination of the two—were printed relatively quickly, suggesting a single, spontaneous outpouring of texts that early printers were eager to print and Italian readers were eager to buy. One way the humanist Latin accounts differ from the vernacular ballads is their strong sense of authorial identity, articulated in the texts and in their material presentation. In each of the humanist editions (with the exception of the Latin *Lamentatio*, left anonymous for reasons just suggested), the author is identified on the first page, and every text, including the *Lamentatio*, is dedicated to an important political leader—the pope, the king of Naples, the doge of Venice, or Cardinal Bessarion, the champion of the crusade.[36] Dedication to a public figure, long a way for men of letters to attract attention to their work and enhance their reputations, raises the question of readership: did the Latin texts really have one? They may have been written to advertise the importance and erudition of their authors just as much as they were meant to inform the "reading public" of important news or to involve them in political debates of the day. In a few cases we even know why their authors hoped to attract attention. Cornazzano, who dedicated his poem to the Venetian Senate, was looking for work in the Republic in 1472; he had recently left the employ of the *condottiere* Bartolomeo Colleone, whom he had served as court poet between 1466 and 1470.[37] In 1471 Paolo Marsi, too, was looking for a job. After escaping the papal crackdown on humanists of 1468 by accompanying Niccolò Canal to the East, Marsi now needed to distance himself from the stain of the admiral's disgrace. The Venetian edition of Marsi's text includes lavish praise for Canal's successor in the post of captain general, Pietro Mocenigo;[38] the Roman edition, dating to early 1471, carries a dedication to Paul II.

Several humanist authors had connections to the early printing industry in Italy and likely helped see their works through the press. Zovenzoni had worked as an editor for the first Venetian printers, the de Spira brothers, between 1469 and 1472;[39] Cornazzano corrected text for Nicholas Jenson in 1470.[40] Both poets ended up printing their works on Negroponte with other presses—much newer presses, as it happens. Did the poets offer newly established printers their services as editors or proofreaders in exchange for getting their own works into print?[41] Fieschi's text was the first that Sixtus Riessinger printed in Naples; Cornazzano's was also his printer's first. Marsi's poem was the first printed by Conti at Venice, in an edition laced with congratulatory distichs by several of Marsi's Venetian friends, including Ermolao Barbaro and Raffaele Zovenzoni. Did Zovenzoni, the first Venetian humanist to work with and for the press, help his friend Marsi find a printer in Venice on his arrival back from Negroponte? It seems almost certain that Pomponio Leto helped Marsi get his text printed at Rome, as the press used there was closely associated with the Roman professor and Marsi was still in Venice when it appeared.

Most of the other Latin authors had connections to printing as well: Toscani—an accomplished poet, orator, doctor of law, and a good friend of Pomponio Leto[42]—edited and corrected legal texts for Roman printing firms throughout the 1470s, eventually becoming a legal publisher in his own right.[43] Rodrigo Sánchez, an even higher-ranking official, had already seen another of his works in print: his *Speculum humanae vitae* was printed by Sweynheym and Pannartz in 1469. In 1470–71, however, it was Ulrich Han who printed both Sánchez's consolatory letter on Negroponte and another of his works, a history of Spain. It is possible that all these printers simply happened upon copies of the works in question and published them without their authors' knowledge. But given the precarious economic position of many early printers in Italy, their perennial search for new and marketable texts to print, and the tendency of humanist scholars to attach themselves to their operations, the decision to print likely originated with the authors themselves or their friends, who hoped to bring themselves to the attention of the great and the good.

A brief episode in the career of Francesco Filelfo—perhaps the fifteenth century's most prolix commentator on Turkish affairs and an early enthusiast of printing—lends support to this hypothesis.[44] After the fall of Negroponte, Filelfo, a tireless self-promoter, wrote letters to prominent political figures across Italy expressing dismay at the news and hopes for a European response. In 1470, Filelfo also kept up a private correspondence with the Greek scholar Theodore Gaza. Through Gaza, Filelfo had come to know Giovanni Andrea Bussi, editor for Sweynheym and Pannartz in Rome.[45] The three were planning to publish an edition of Filelfo's translation of Xenophon's *Cyropaedia*. In February 1471, while negotiating over the text and some six months after news of the fall of Negroponte broke in Italy, Filelfo sent Bussi a short letter expressing regret that, despite Gaza's request, he could not supply a new crusade exhortation to the princes of Europe. Given the intransigence of the Christian powers, he wrote, the exercise would be about as useful as washing a brick.[46]

At this time, Bussi was looking for a short text to publish while his printers awaited delivery of a stock of royal paper for a long-planned edition of the Bible.[47] In the end, Bussi found a brief text of Cyprian to produce; it seems he and Gaza first asked Filelfo for copy to fill the gap in their production schedule—perhaps in the hope that a topical piece on the Turkish problem would sell better than Bussi's classics, which were slow to move off the shelf.

This apparent collaboration between a humanist author, editor, and corrector, working together to arrange the publication of an anti-Turkish text at "their" press (and then failing to do so), underscores an important point: the unpredictable and contingent nature of much incunable publication. This needs to be stressed if only to explode a persistent assumption regarding the printing of political information in Italy in its earliest phases: namely, that it was government sponsored—propaganda strictly defined. How unlikely this is can be determined by a glance at how the Italian states and the men who led them publicized their thoughts on the fall of Negroponte in 1470: they did not use print.

Cardinal Bessarion, for example, was a leading crusade advocate, an early link in the chain of reports about Negroponte, the dedicatee of two Negroponte texts printed three times in 1471–72, and the author of a series of letters and orations urging a European response to the disaster.[48] When he wrote these short texts, Bessarion was well acquainted with the printing industry in Rome. In 1468–69, Bussi assisted Sweynheym and Pannartz in publishing several Neoplatonic texts that Bessarion had championed at the papal court, as well as the first edition of the cardinal's philosophical tract *In calumniatorem Platonis*. But Bessarion never called on Sweynheym and Pannartz, or any other printers in Rome, to print his orations against the Turks. The most respected living authority on the Turkish problem, with the services of a great printing house at his disposal, took no steps to put his Negroponte text into print. His authority was such that he did not need the press to secure an audience for his text. Rather, in August 1470 Bessarion sent a manuscript copy of his Negroponte works to Doge Cristoforo Moro of Venice. Later that autumn the cardinal shared other copies with friends and former students.

It was Bessarion's humanist friends, occupying a lower rung on the social scale, who took the initiative to see his text into print. They did so without the cardinal's knowledge.[49] The humanist Ludovico Carbone, a sometime client of Bessarion, received a manuscript copy of the cardinal's orations in Venice in late 1470. Carbone was then working as a corrector for Christoph Valdarfer's newly established Venetian press. Carbone translated Bessarion's text into Italian; added a preface of his own recommending Bessarion and himself to Borso d'Este, duke of Ferrara (with whom Carbone hoped to find a job); and had the whole collection printed by Valdarfer before the end of 1471. Bessarion seems to have had nothing to do with its publication.[50]

That winter, the cardinal sent another copy of his orations to another humanist associate, the Parisian theologian Guillaume Fichet. Bessarion asked Fichet to circulate the manuscript at the court of Louis XI. The cardinal was quite unaware that Fichet had helped to establish and was now editing and correcting texts for the first press to operate in Paris. In the spring of 1471, Fichet had Bessarion's text printed, prefaced by his own effusive introduction. He then sent copies with further letters of introduction in his own name to lay and ecclesiastical leaders across northern Europe. The Holy Roman Emperor, the kings of France and England, the dukes of Bavaria and Burgundy, and various French, German, and Spanish cardinals, bishops, and abbots all received printed copies of Bessarion's orations on Negroponte from Fichet, though Bessarion had no idea that Fichet had even published his text. The cardinal was delighted when he finally heard about Fichet's efforts on his behalf.[51]

The actions of both Carbone and Fichet support the model of humanist publication seen at work in the other Latin editions of Negroponte texts. In each case, a humanist scholar with connections to the press took the initiative for publication of his own or another author's text. Publication was not state sponsored, but it did not necessarily cater to existing public demand either. The purpose of most of these publications was to promote the interests and reputation, and perhaps to advertise the social and political connections, of a relatively little-known author, translator, or editor. It is unlikely that a cardinal or any other senior political figure would have employed such middling types to broadcast a topical message when he could have done so just as easily himself. (Indeed, Moro and Bessarion did just that when they exchanged open letters to one another and to the Christian princes during the late summer and fall of 1470. The cardinal and the doge issued their communiqués in manuscript.) Despite the close ties linking many of these humanist authors to political leaders in Rome and Venice, there was no organized or official push to publish their Negroponte texts. Men of letters, not governments, were the first to recognize the advantages that print could bring, and for them it meant the possibility of patronage and fame as much as the ability to broadcast the latest news or sway popular opinion.

In 1470, printing was not yet an instrument for official propaganda. The Italian states, and above all the popes, had a well-tuned apparatus for communicating with one another and with their publics; in these earliest years of Italian print, neither Paul II nor Sixtus IV, nor the Venetian doge for that matter, saw the need to exploit the new invention in their struggles against the Turks. In 1475, as we will see, the same relationship still obtained: humanist poets, scholars, and jurists were still using the press to publicize their opinions and expertise—and themselves—while the same governments in Rome and Venice kept their distance, at times even legislating against the publication of unauthorized or inconvenient views.

The Prince Bishop's Press Campaign

Two-year-old Simon Unverdorben went missing on the evening of Thursday, March 23, 1475, in the Alpine city of Trent.[52] On Easter Sunday, three days later, his body was found in a sewer that ran under the house of a local Jewish family. The *podestà* arrested a group of Jewish men and had them tortured until they began to agree to the version of events their inquisitors proposed: they had abducted the boy and subjected him to terrible abuse, strangling and bleeding him until he died; they had collected his blood and used it to make *matzoh* for their Passover meal; afterward, they had uttered terrible curses over his corpse and thrown it into the sewer. In extracting this confession, the officials of Trent were guided by the late medieval conviction that Jews regularly sought Christian blood to use in rites that mocked the Last Supper, perverted the Eucharist, and drew on dark magic to create powerful curses.[53] The people of Trent were equally inclined to believe that their Jewish neighbors had secretly plotted to murder a young child and desecrate his corpse. It seemed little Simon was a modern Christian martyr. The boy's body, laid out on the altar of the church of SS. Peter and Paul, soon began to perform miracles; rumors spread that the boy might be a new saint, and pilgrims began appearing in the city from nearby valleys and towns.

Like the fall of Negroponte five years before, news of this sensational event prompted a flurry of printed publications across Italy and Germany: woodcut images of the new martyr circulated in and around Trent, followed by dozens of printed broadsides and pamphlets of poetry and prose in Latin, German, and Italian. Because Trent did not have a press of its own in the spring of 1475, these editions first appeared south of the city in towns of the Veneto— Treviso, Sant'Orso, Verona, and Vicenza. Within months, however, a German printer, Albrecht Kunne, had set up shop in Trent, attracted by the thriving market for texts about Simon; he would print at least four different texts about the city's new martyr in his first year of business. Pamphlets about the boy's death were also published north of the Alps in Nuremberg, Augsburg, Ulm, Vienna, and Cologne, as well as farther south, in Naples and Rome. Altogether some thirty-five editions of sixteen different texts appeared in print between 1475 and 1498.[54] No other news event was documented in more editions in the fifteenth century.

The authors of these tracts were humanist poets and jurists who approached their topic from different angles: some wrote verses lamenting Jewish perfidy, celebrating Simon's martyrdom, and promoting his sanctity; others wrote legal treatises defending the bishop's prosecution of the Jews and demonstrating their guilt. Most popular was a letter by a Brescian physician, Giovanni Mattia Tiberino, who examined the boy's body the day after its discovery and claimed firsthand knowledge of the Jewish conspiracy; his letter would be printed and reprinted more than fifteen times over the next seven years.

Like Negroponte, the affair of Simon of Trent offers another case where professional men—legal scholars, a physician, and various humanist poets—used the press to circulate their literary pronouncements on a sensational recent event. These authors deployed the same range of *genres*—Latin letters and histories, epic and elegiac poetry, technical *consilia*, and vernacular laments. Some of the same figures even played a role in both affairs: Raffaele Zovenzoni, author of a Latin poem on Negroponte, used his connections with Venetian printers to publish letters and poems by himself and others celebrating Simon and condemning the Jews. Bartolomeo Platina, who wrote a tract against the papal legate sent to investigate the case, was a member of the Roman Academy and a friend of Pomponio Leto, who also wrote a letter celebrating the Jews' trial and execution.[55] As in the case of Negroponte, the short texts these authors composed were especially attractive to itinerant printers setting up shop in various cities. Simon tracts were bestsellers, reliable picks for printers testing out a new market whether with the guidance of a humanist advisor or without.

Unlike the publications that appeared after the fall of Negroponte, however, in 1475 there *was* an authority directing much of this publishing activity. The prince bishop of Trent, Johannes Hinderbach, from the start encouraged the local faithful to think of Simon as a martyr. Anticipating a canonization case, he had meticulous records kept of the murder trial and the miracles attributed to the child.[56] He was also the driving force behind a voluminous publicity campaign intended to promote devotion to Simon beyond Trent by means of sermons, images, and printed texts.

Hinderbach's activities were born out of relative weakness, not strength, for his martyr's case was not as clear as he supposed. Within weeks of the Jews' arrest in the spring of 1475, authorities both north and south of the Alps had intervened in the case, voicing skepticism about the reported miracles and concern for the Jews of the town and the fairness of their trial. Archduke Sigismund of the Tyrol and Sixtus IV each weighed in on the trial at different points, suspending inquiries and banning discussion of the case. In response to these threats to the new cult, Hinderbach sent letters to neighboring princes and prelates celebrating his martyr, excoriating the Jews, and defending his conduct of their trial.[57] He commissioned paintings depicting Simon's passion and martyrdom and circulated woodcut images of the victim.[58] He encouraged his literary friends to compose poems and tracts of their own in praise of the little martyr. He hired jurists to compose *consilia* justifiying his legal actions. And he turned to the printing press. He underwrote the publication of poetry in Trent and paid for at least one of his legal *consilia* to be printed in Rome. His actions provoked outrage in Venice and Rome alike: Doge Giovanni Mocenigo and Sixtus IV each responded with orders forbidding preaching about Simon and banning the creation or circulation of images and texts that celebrated him as a martyr.

In their efforts to counter Hinderbach's press campaign, officials in Venice and Rome did not themselves turn to the press. Instead, they used traditional means of communication, like manuscript briefs to local authorities and edicts read aloud and posted in public places. Unlike the prince bishop, they did not see the matter as a contest for public opinion. Rather, it was a case of clerical insurbordination and judicial malpractice, something to be handled by traditional bureaucratic processes.

But Hinderbach's use of the press to promote Simon's cause may have influenced the papacy's own publication practices. In 1475, Sixtus tried to put a check on Hinderbach's activities by banning the circulation of printed images and texts about the little martyr of Trent. In June 1478, by contrast, just a few weeks before he issued his final ruling on Simon's case, Sixtus released a series of bulls condemning Lorenzo de' Medici in the wake of the deadly Pazzi conspiracy. After publishing these in Rome in the traditional way, he immediately had them printed.

First Reports and Responses

The physician Giovanni Mattia Tiberino was the first to report the story of Simon's death at the hands of the Jews. As Hinderbach's personal physician, Tiberino was one of two doctors summoned to investigate the body of the boy after its discovery in the sewer running through their cellar. In his letter he drew on details from the interrogations that had not yet been made public; Hinderbach or his subordinates likely provided him the information. On April 17, he sent his letter describing the Jewish "plot" to his compatriots in Brescia, where it arrived two days later.[59] Tiberino's letter would serve as a key source for most later ones.[60]

According to Tiberino, in the days before Passover, the men of the Jewish community had decided to kidnap a Christian child and use his blood to make *matzoh* for their holiday feast. The leader of the community, named Samuel, lured Simon into the house and muffled his cries as the rest of the

Jews fell upon him. Stretching his limbs out as though affixing him to a cross, they tortured the boy with needles and pincers, collected his blood in a bowl, and used it to make their unholy bread. Once he was dead, they laid him on a bench in their synagogue and recited curses over his corpse. Then they threw his body into the open sewer that ran under the house so they could pretend they had discovered it there. As soon as the body was found, however, suspicion fell on the Jewish householders, and city officials arrested six of them then and there. The next day, Easter Monday, Tiberino examined the body and concluded that the boy had not drowned but been murdered. The remaining Jewish men and one woman were then arrested and interrogated, and with this development Tiberino's letter concludes.

Tiberino's letter was not the only source of news about Trent's new martyr. Pilgrims traveling to or from Rome in the Jubilee year carried news of Simon and his miracles along the main roads over the Alps, and Franciscan preachers in neighboring towns began to propound on the perfidy of the Jews and the possibility that Simon was a saint. The tale was a sensational one, provoking devotional enthusiasm and hostility against the Jews in equal measure. By the middle of April, Jews in the cities of the Veneto were suffering harassment and violent attacks.[61] For their part, the Jewish captives and their families sent messages beyond the borders of Trent seeking help from contacts in Austria and Italy. Their actions probably explain why, on April 21, a message arrived from Archduke Sigismund of the Tyrol, Hinderbach's temporal overlord, ordering that the trial be suspended.[62] The following week, Doge Mocenigo issued instructions to the governors of Venetian cities in the Terraferma to protect the Jews in their jurisdictions and prevent preachers from making inflammatory statements against them.[63]

Hinderbach was in a difficult position. On the one hand, it seemed that Providence had sent a new saint to his city, with all the prestige and ecclesiastical revenue such a miraculous development could bring. On the other hand, authorities north and south of the Alps were pressing him to give the Jews a fair trial. This could mean backing off the more outrageous claims being made about Simon and shutting down his nascent cult. Hinderbach's letters and the dates given in some of the printed editions reveal how quickly he moved to defend himself and the story of his martyr. On April 30, he sent Raffaele Zovenzoni a copy of Tiberino's letter of April 17 and asked him to mine it for material for a poem in honor of Simon, which he hoped would spread his story far and wide. Hinderbach may also have written about the case to another literary friend, the poet and scribe Felice Feliciano, for on

May 22, an Italian text about Simon, attributed to Felice, is recorded as having been printed in Verona.[64] On June 19, Tiberino's letter was published in Rome; the next day, an edition appeared in Treviso. On June 24, an Italian lament for Simon was published in the town of Sant'Orso, just down the road from Trent. On July 24, a second edition of Tiberino's letter appeared in Rome. In September, a German prose account of Simon's death and the trial of the Jews was printed in Trent. In October, Hinderbach urged Zovenzoni to circulate woodcut prints of Simon in Venice, and Zovenzoni in turn thanked Hinderbach for sending him one such *sanctissima imago*. Sometime after December 6, Silvestro da Bagnoregio's legal *consilia* about the trial were printed by Albrecht Kunne in Trent; the same printer brought out a revised edition of Tiberino's letter with additional poems in February 1476.

Printed Editions of Tiberino's Letter

Whether Hinderbach encouraged his physican to write his letter or whether he merely appropriated it after the fact, the bishop seems to have taken the initiative to distribute it. A curiosity that helps trace his efforts is the fact that in some redactions the date of the letter has been changed from April 17 to April 4—perhaps to make the text seem closer to the events it described and therefore more authentic or trustworthy.[65] The letter would be printed in at least three different redactions, with different dates, or with none, reflecting different stages in its diffusion.

On April 30, Hinderbach sent a copy of Tiberino's letter, with the date changed to April 4, to Zovenzoni in Venice. Hinderbach had known Zovenzoni when he was a student in Ferrara and had corresponded with him in the past.[66] The bishop now asked his friend to draw on Tiberino's letter to compose a new poem in honor of Simon, which would help to spread the martyr's fame.[67] Hinderbach needed Zovenzoni's help, he said, because Archduke Sigismund, acting on bad advice from greedy Christians who were themselves in the pay of perfidious Jews, had suspended the inquest into Simon's murder. Even more alarming, Hinderbach feared that the emperor himself might be persuaded to intervene in the case on the Jews' behalf. Recently, however, an agent sent by Sigismund with a safe-conduct for the Jews had suffered a horrific accident: spending the night in a house in nearby Rovereto, he had tripped and fallen into a kitchen fire and been badly burned. This was divine justice in the bishop's opinion, proof that God did not want to see the Jews rescued, but condemned. Hinderbach sent Zovenzoni a brief poem celebrating this gruesome event, with the title *Miraculum*.

Hinderbach did not ask Zovenzoni to print any of these texts—not Tiberino's letter, or the *Miraculum,* or the new poem he hoped Zovenzoni would compose. He only asked the poet to write verses in honor of Simon and to send them back to him in Trent. Zovenzoni complied with this request, producing seventy-four hexameters celebrating Simon and condemning the Jews of Trent for their bloodlust and cruelty. Zovenzoni shared Hinderbach's concern about the political opposition to Simon's cause: his verses call on the emperor and princes of Christendom to avenge Simon's death by slaughtering the Jews—not just the Jews of Trent but all of them. The entire race should be wiped from the face of the earth. The Jews had killed Christ; they had long hated Christians; and most recently, they had lured the innocent Simon to his miserable death. But the boy's body was recovered and even now was performing miracles from its resting place in St. Peter's church; the Jews had been arrested and would be kept under guard till they made satisfaction for their crimes. The Christian princes, especially Archduke Sigismund, must allow justice to take its course: the murderers should be burned.

Zovenzoni took the various texts he had in hand—Hinderbach's letter of April 30, Tiberino's letter, the *Miraculum,* and his own poem against the Jews—and assembled them into a single collection to which he made some further modifications. He changed the salutation of Tiberino's letter, which Hinderbach had clearly identified as a letter from Tiberino to his fellow citizens in Brescia, so that it now read as a letter from Tiberino to *him.*[68] (He also added a flattering sentence at the end, making Tiberino bid farewell to Zovenzoni, *Musarum decus,* or "pride of the Muses.") And he entrusted the compilation to the press—or rather, to two presses: the Venetian printers Gabriele di Pietro and Nicholas Jenson each published an edition of Zovenzoni's compilation. Zovenzoni had worked for Jenson as a corrector in the past, and he seems to have been involved in preparing the text for each printer. Each starts with a pair of distichs in which Zovenzoni exhorts the printer ("Print this, Gabriele!" "Print this, Nicholas!").[69] Both editions are undated; since they ask Sigismund to condemn the Jews to death, it seems likely they were published before their trial resumed in Trent in the middle of June or, at least, before the first executions, which took place on June 22.

It is worth emphasizing that Hinderbach did not ask Zovenzoni to publish his text. It was Zovenzoni who took the initiative. Acting rather like Fichet and Carbone had when they each published Bessarion's orations on Negroponte, Zovenzoni took a text sent to him by a social superior, added words of his own that highlighted his connection to the author and his own thoughts on the issue, and sent the whole through the press. His edition served Hinderbach's interests while advertising his own status as the bishop's favored poet and trusted correspondent (and *Musarum decus* to boot).

Apart from the two editions Zovenzoni was responsible for in Venice, a further thirteen editions of Tiberino's letter appeared in other cities north and south of the Alps. In most of these, Tiberino's letter is still addressed to the Brescians, the letter bears the fictive date of April 4, and the *Miraculum* poem is included at the end. It was Hinderbach who backdated Tiberino's letter and appended the *Miraculum* when he wrote to Zovenzoni, but Zovenzoni further changed the letter so that it was addressed to himself and not the Brescians. Those editions printed with the false date and the *Miraculum* but still addressed to the Brescians likely stem from other manuscript copies of the doctored letter and poem that Hinderbach sent abroad to other recipients than Zovenzoni.

One of these was printed on June 20, in Treviso, by the Flemish printer Gerardus de Lisa; this same printer produced another, undated edition of the letter, either before or after June 20.[70] Another edition appeared from the press of Johannes de Reno in nearby Sant'Orso.[71] Both printers would publish other texts about Simon in the coming months and years. Another edition of the letter, bearing no internal date at all, nor the *Miraculum*, appeared in Mantua; a local editor or the printer himself may have stripped these elements from Hinderbach's compilation, or they may indicate that yet another redaction was in circulation.[72]

On the other side of the Alps, editions of the Latin text of the letter, with the false date and followed by the *Miraculum*, were printed in Augsburg and Nuremberg.[73] The same Nuremberg printer also published a German translation of the text,[74] while a different German translation was printed at Augsburg by Gunther Zainer and supplied with fourteen woodcuts illustrating Simon's abduction and torture and events after his death.[75] In this edition, Tiberino's letter does not end with the Jews in prison; two more pages of text provide a continuation of the story up to the execution of the Jews in June, and the final woodcuts also illustrate these later events: the arrest and trial of the Jews, and their burning at the stake in June. (Oddly, and illogically, the letter still concludes with the date April 4.)[76] All these editions have been assigned by bibliographers to the year 1475. In the following years, this redaction of the letter would be reprinted in Rome, Naples, and Cologne.[77]

A third version of the letter was printed in Rome. Bartholomaeus Guldinbeck, a printer just beginning operations in Rome that summer, issued two editions of the text, one on June 19, 1475, and another on July 24.[78] The June 19 edition is Guldinbeck's first dated book.[79] In both editions, Tiberino's name and all reference to his correspondents in Brescia are removed. In Guldinbeck's hands, the letter becomes an anonymous *Historia* and addresses an indeterminate *vos*. Curiously, the date of the letter appears here as April 17 (the authentic date); neither Roman edition contains the *Miraculum*. Hinderbach may have sent an earlier redaction of Tiberino's letter to a correspondent in Rome, seeking to win curial support for his case.[80]

The publication of the letter in Rome had little effect on papal thinking. Over the summer, the Jews of Rome had petitioned the pope to review the case of the surviving Jews still in custody in Trent.[81] Guldinbeck brought out his second edition of Tiberino's letter on July 24, but already, a day earlier, Sixtus had issued an order to suspend the trial again. In a letter to Hinderbach, the pope declared that the conduct of the trial was so questionable that "many people, quite important people, even, have begun to whisper about it"; he ordered all judicial action to stop until his own representative arrived in Trent.[82]

Tiberino's letter on Simon traveled great distances and was printed multiple times in different locations. It is unlikely that Hinderbach requested or commissioned its printing in every case. Many printers will have judged that the text would sell well in their local markets and invested in printing it on their own initiative. In Venice around this time, an Italian lament for Simon appeared from another press, unconnected to Zovenzoni.[83] Other anti-Semitic tracts had already proved sucessful for printers in Venice, the Veneto, and southern Germany. Johannes de Reno, who printed Tiberino's letter and two more poems about Simon in Sant'Orso in 1475, brought out an edition of the *Epistola contra errores Judaeorum* attributed to one "Rabbi Samuel," supposedly an apostate Jew, that same year. De Reno had an intermittent partnership with the printer Leonardus Achates, with whom he had worked in Padua in earlier years. In 1474, Achates was also in Sant'Orso and had printed his own edition of the letter of Rabbi Samuel. In 1477, he would publish Pietro Bruto's *Epistola contra Judaeos* in Vicenza. In 1475, another edition of the letter of Rabbi Samuel appeared in Mantua from the same printer, Johannes Schallus, who printed Tiberino's letter there. An Italian translation of Rabbi Samuel's letter was printed in Bologna on June 17, 1475, and a German translation appeared in Augsburg at some point the same year. The calligrapher

Felice Feliciano was credited with a text about Simon, supposedly printed in Verona on May 22, 1475; the year before, he had copied out for Hinderbach an anti-Semitic text by Johannes de Lübeck from a printed edition produced in Padua.[84] In Venice, Nicholas Jenson had already printed a series of texts by Carthusian authors that blamed the Jews for Christ's death and a host of other ills.[85] Meanwhile Alessandro da Nevo, a legal scholar and editor of legal texts for Jenson,[86] composed a collection of *Consilia contra Judaeos foenerantes* ("arguments against the money-lending Jews") first printed in Venice in 1476 and reprinted at least nine times over the next decade. North of the Alps, Petrus Niger's tract *Contra perfidos Judaeos* was published twice in Esslingen, near Stuttgart, once in June 1475 and again in December 1477. In sum, the broad circulation of Tiberino's letter was an overdetermined event: Hinderbach or his local friends or agents may have sponsored publication, but Tiberino's short and prurient narrative was just the sort of text that printers wanted to publish and readers wanted to read.

The Papal Legation and Printing in Trent

While Tiberino's letter made its way through the presses of various European cities, other texts on Simon began to appear in print closer to Trent. Matthias Künig's poem about the murder was printed by Johannes de Reno in Sant'Orso, most likely before June 24.[87] On that day, the same printer issued an Italian poem ("The Torments of Simon") possibly by Giovanni da Conegliano but attributed by one modern scholar to Felice Feliciano.[88] This text would also be published in Treviso by Gerardus de Lisa, who, toward the end of the summer, also published a Latin poem by Tommaso Prato, *De immanitate Judaeorum in Simonem infantem*.[89] These works were published in Treviso and Sant'Orso; there was still no press in Trent.[90] By September, however, Albrecht Kunne of Duderstadt had set up shop in the city. Four of the nine editions he published over the next year would be texts having to do with Simon.[91] His first dated edition appeared on September 6: an anonymous German *Historie von Simon zu Trient* with extensive illustrations.[92]

At this moment, the start of September 1475, judicial matters were still unsettled. On July 23 the pope had ordered the trial suspended, and on August 3 he appointed a legate, Battista dei Giudici, bishop of Ventimiglia, to conduct an investigation into the case. Dei Giudici was to acquire the trial records and send them under seal to Rome, investigate the claims of ritual murder and the miracles attributed to Simon, secure the release of the Jewish women and children who remained in custody, and, if necessary, transfer the trial to a neutral site.[93] Dei Giudici reached Trent on September 2, four days before the *Historie* was published.

The book provides a vivid picture of the situation dei Giudici was approaching. The anonymous author of the *Historie* follows Tiberino's narrative of events up to the arrest of the Jews and adds details on their trial, the first round of executions, and several miracles performed by the little *beato* in the summer after his death. Intended to encourage belief in Simon's sanctity, its simple prose is illustrated by twelve dramatic woodcuts, depicting the initial meeting where the Jews plotted their deeds to Simon's abduction, his torments and death, the perverse rites of the Jews (shown in outlandish costumes both in the synagogue and at their grisly passover feast), the disposal and discovery of Simon's body, its examination by physicians, and the torture of the Jewish suspects and their execution at the stake or by the sword.

The ninth image in the series (figure 2.1) shows the scene inside the parish church of SS. Peter and Paul, where Simon's body had been kept since its discovery and where the faithful were flocking to venerate it. The boy's naked corpse lies on a bier between two lit candles. A crowd of pilgrims approaches—a woman on crutches, another clutching an infant, and a third, a crippled man with twisted legs, kneeling with outstretched arms before the little martyr. Above, a makeshift scaffold is festooned with *ex-votos* from grateful pilgrims. Three prelates surround the boy. These are Hinderbach, shown kneeling in prayer at the foot of the bier, with a shield containing his emblem on the ground before him,[94] and two bishop saints—St. Vigilius, the first bishop of Trent, and St. Peter, one of the patron saints of the church— who, the text explains, had made a miraculous appearance at either end of the bier after the boy's body was laid there.[95] Vigilius appears in his episcopal robes, holding a book in one hand and blessing his successor, Hinderbach, with the other, while Peter stands at the head of the bier, holding one of his keys and wearing the papal tiara.

The representation of Peter as pope is worth remarking on. The church of SS. Peter and Paul had undergone an extensive renovation, paid for by Hinderbach, just a few years before. In a contemporaneous panel painting that almost certainly served as the church's altarpiece, and which was likewise commissioned by Hinderbach, the two apostle saints appear together in rich red and green robes, but without a scrap of papal regalia for Peter. In another painting Hinderbach commissioned, Peter appears with a key, but no other signs of papal rank. Representing Peter as pope here, the woodcut suggests that Simon's cult enjoyed not just saintly but papal endorsement. The first

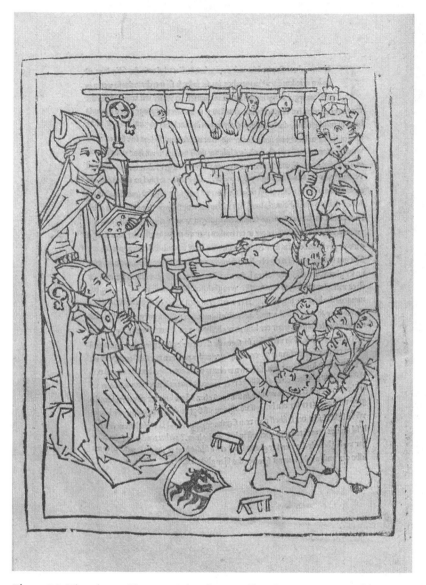

Figure 2.1. *Historie von Simon zu Trient* (Trent: Albrecht Kunne, September 6, 1475). Munich, BSB, 2 Inc.s.a. 62, fol. 9v.

and latest bishops of Trent—Vigilius and Hinderbach—were shown venerating the little martyr; if Peter, the first pope, stood in awe of the saint, shouldn't his current successor fall in line and do so too?[96]

The papal legate, dei Giudici, who by now had arrived in Trent, saw matters very differently. On September 7, the day after the *Historie* appeared from Albrecht Kunne's press, Hinderbach escorted the legate to the church to inspect the supposedly incorrupt corpse of the holy martyr. Dei Giudici was horrified to find a badly decomposed body laid out on the altar in the summer heat, unprotected by casket or coffin. When Hinderbach moved one of the legs, the stench from the corpse overwhelmed the legate and he struggled not to retch in the close air of the sanctuary. The scene was pathetic and disturbing, he wrote to the pope: there was something unhealthy about the entire situation. He immediately pressed Hinderbach to let him speak to the Jews and review the records of their interrogations.

Hinderbach replied that the trial records must remain sealed for security and that dei Giudici had no right to see them anyway since they dealt with murder, a civil crime. He also suggested that the papal legate could not be impartial since he had taken Jewish bribes.[97] After three weeks in the hostile city, housed in damp and uncomfortable lodgings, with the bishop uncooperative and the townspeople beginning to threaten violence, dei Giudici withdrew to Rovereto, in Venetian territory, to continue his investigation from a distance.

"Et imagines afficere"

Hinderbach continued to promote his martyr's case. On September 29, he sent an open letter to the princes of Germany defending his actions against the Jews.[98] He commissioned a German Dominican, Heinrich Institoris, to conduct research into previous trials of Jews in other German cities, presumably to turn up evidence that would support his own investigations.[99] Hinderbach also wrote again to Zovenzoni. On October 1, he asked the poet to argue his case with Venetian officials; he hoped they would take his side in the dispute with the pope, back Simon's cause, and communicate support for it to their subject cities. The bishop especially wanted Zovenzoni to lobby Venice about the ban it had issued in April forbidding preaching about Simon and violence against Jews. In August, the Senate had seemed to back down from this blanket prohibition, issuing a counterorder that now allowed preaching about the event. License was also granted to anyone who wanted to post images of Simon on walls or in other places, so long as there was no violence done to local Jews. In October, however, Hinderbach heard that certain parties in Venice wanted to rescind this order and reinstate the ban.[100] He asked Zovenzoni to pressure the Venetians to allow both preaching and the posting of images in public places ("et imagines afficere").[101] Hinderbach enclosed one such printed image as a gift for Zovenzoni. On October 7, Zovenzoni thanked him profusely for the picture.[102]

In Rome, the pope took steps to clamp down even more firmly on this kind of publicity. On October 10, Sixtus issued an instruction to the Italian states reminding them that questions of martyrdom and miracles were reserved to the pope, and he had not yet issued a ruling on Simon's case. Certain people were claiming that the boy had been martyred and was now performing miracles, and some were even painting pictures of him and composing tracts recounting his legend; others were setting these out for sale and putting them on display. The pope had already made a public announcement in Rome banning all such activities. Now he ordered the Italian powers to enforce a similar ban in their own jurisdictions.[103] In Venice, this gave Hinderbach's opponents the ammunition they needed. On November 5, Doge Mocenigo cited the papal brief in orders he sent to the governors of Venetian cities in the Terrafirma and added a line that also forbade the *printing* of such texts: "The pope has sent a brief to every prince and state in which he declares and orders, under pain of excommunication, that the person recently said to have been killed by the Jews in Trent should not be painted, nor should any texts about him be printed, nor should preachers say anything to slander or rouse up the mob against the Jews."[104] On receiving his copy of these instructions, the *podestà* of Brescia issued an order of his own, in Italian, to be posted both in the palace and in the "usual places" around the city: no one should paint, or pay someone else to paint, any image of the boy as a martyr or *beato*, whether on walls or on paper; no one should sell such an image; and no one should preach or recite verses or write letters about the child or sell any such texts.[105]

Since all these orders mention the creation and circulation of images, it is worth considering what early printed images of Simon may have looked like and how they would have functioned.[106] Surviving single sheet prints of Simon depict him in various ways. Some show the boy at the moment of his supposed martyrdom: either standing, with his arms stretched out, Christ-like, as wicked-looking Jews torment him from all sides,[107] or seated on the lap of one of them, in a grotesque parody of Christ's circumcision, as various figures lean in to wound his penis.[108] Others show the boy after death, stretched out on a ceremonial bench in the synagogue where the Jews said they had laid him while they prepared their ritual curses. In these woodcuts, Simon's body assumes the pose of the dead Christ, and he is surrounded by the instruments of his passion—nails, a knife, pincers, the bowl in which his blood was collected, and so on. But these images also blend time, for while Simon's body is represented at the site of his murder, he is also shown as an object of devotion. Votive objects hang above him, and in one image a group of pilgrims approaches his body as though it already lies on the altar of St. Peter's church.[109] A third type shows Simon in triumph after his death, either standing and grasping a fluttering banner, like the resurrected Christ, or seated in majesty on a throne, grasping a martyr's palm as angels hover overhead with a crown.[110]

These last two images appear on printed broadsides illustrating poetic texts. The others, entirely woodcut, contain just a few lines of xylographic text. All identify Simon as *beatus, martyr*, or both. Scholars have supposed they cannot have been made until after Sixtus issued his judgment approving Hinderbach's conduct of the case on June 20, 1478, which opened the way to calling Simon *beato*, and were probably produced even later, after the formal application for his canonization was made on May 20, 1479. But given the order *against* prints identifying him as a martyr that the pope issued in October 1475 (and which officials in Venice and Brescia further endorsed), this argument doesn't seem to hold. Prints of Simon certainly circulated well before 1479. Hinderbach sent one to Zovenzoni in October 1475 and expressed the hope that Venetian officials would allow many more to be posted in the Terrafirma. In April 1476, a man in Padua reported that he was miraculously cured of an ailment after praying to an image of Simon he had bought in Venice.[111] In March 1478, one of Hinderbach's procurators in Rome, Approvino degli Approvini, wrote to the prince bishop requesting a very particular image of Simon, one showing the boy at the moment of his martyrdom, surrounded by Jews holding his arms in a crosswise position. Approvino told Hinderbach that he had had such an image of Simon made in Rome but was unsatisifed with the result. He would rather have one from Trent, presumably because local artists knew how to get the scene right.[112]

Printed images of Simon circulated in elite circles and among the poor and middling. A "handbook for charlatans" produced in the mid-1480s lists various kinds of buskers and beggars, including *aconi*, or icon-men, who pinned sacred images to their shirts, or laid them out on the pavement in front of churches, and then begged for alms: "Like those who read a few stanzas of Simon of Trent and then ask for coins."[113] On the other hand, at least two learned humanist collectors also included images of Simon in their collections. The lawyer and print collector Jacobo Rubieri included a woodcut image of Simon in one of his legal miscellanies, now in Ravenna;[114] in Nuremberg, the humanist physician Hartmann Schedel owned a copy of the Augsburg edition of Tiberino's letter, with its wooduct images; tipped into the back of this copy is a single-sheet woodcut image of Simon's murder, which seems to have served as the model for the woodcut of Simon's death that Schedel later included in his *Nuremberg Chronicle* of 1493.

Hinderbach's press campaign was clearly inciting violence against Jews; equally worrying to Rome, the bishop was usurping the pope's authority to determine who was a martyr and whether miracles were genuine. The legate dei Giudici was also concerned about the danger to people's spiritual health; he reported that locals in Trent had started to say that Simon was a greater martyr than any other saint, surpassed only by Christ in his sanctity. This was close to blasphemy. Even if everything about the supposed plot was true, Simon had made no moral choice to die for his faith; promoting him as a martyr could cheapen the word.[115] Thus, the various authorities arrayed against Hinderbach had many reasons to want to restrict the circulation of printed images and texts about Simon. But popular demand was strong, and printed images and texts could go on to have second lives as devotional objects, effecting miraculous cures or securing alms, or as humanist curiosities, pasted into albums or books. Despite papal and Venetian prohibitions, texts and images of Simon continued to pour off the presses in Trent and farther afield.

Printing in Trent and Lobbying in Rome

Dei Giudici left Trent in mid-September 1475, intending to continue his investigations from Rovereto. Twice in late October he sent orders to Hinderbach and the civil officials of Trent asking them to release the Jews and transfer the trial to Rome, threatening excommunication if they did not, and in early November he sent Hinderbach a copy of Sixtus's brief forbidding sermons and publications on Simon, but the bishop continued to ignore his orders. Frustrated, the legate departed for Rome, arriving there on December 1.[116]

Sixtus then appointed a committee of six cardinals to investigate Hinderbach's conduct of the trial and the supposed miracles. They would take two and a half years to complete their work, leaving Hinderbach plenty of time to continue his lobbying and publicity campaigns. Sometime after December 6, the Trent printer Albrecht Kunne published a legal tract by the Augustinian theologian Silvestro da Bagnoregio exploring whether or not Simon ought to be canonized by the pope. (After weighing both sides of the question, the scholar concluded, not surprisingly, that he should.) The opening lines of the tract indicate that Silvestro had prepared the text for Hinderbach, and the last page includes his dedicatory letter to the bishop, dated December 6, 1475. The jurist, writing from Padua, informed Hinderbach that he had been working strenuously on his behalf, preaching and holding disputes on Simon's case even as local Jews and their supporters assailed him and subjected him to terrible "tribulations." A rival academic had even reported him under the Venetian statute that prohibited preaching about Simon; he had been summoned, on pain of excommunication, to appear in Rome to answer for his offenses. Before he went, he hoped to consult with Hinderbach about arguments he might advance at the papal court, and he took the opportunity to add that his legal difficulties had caused him considerable financial hardship. He asked the bishop to accept the little treatise that he had prepared and that he was dedicating to his honor, and to remember his services on behalf of Simon, the *beato* and "glorious martyr."[117] A year or so later, Hinderbach would authorize a payment to Silvestro for his trouble.[118] In the meantime, his text appeared in print in Trent, presumably at Hinderbach's expense.

More Simon texts were to come from Kunne's press in the coming months. On February 9, 1476, he printed a new edition of Tiberino's letter with additional text, now cast as the *Historia completa* of the case. In a new letter to Hinderbach, the physician recalled that the bishop had often encouraged him to revise his letter from April 1475 and to continue his story down to the conclusion of the trial and the executions of the Jews. A polished and expanded redaction of the original letter follows, along with a new collection of texts, also by Tiberino, all supporting the case for Simon's sanctity. These include a list of miracles performed by the boy both in Trent and farther afield and a new poem celebrating how a prayer to Simon saved Tiberino's own infant son from choking on a chestnut.[119] (This is followed by the *Miraculum* poem about Duke Sigismund's agent who fell into the fire on his way to help the Jews.) The edition also includes another new poem, entitled *Lamentationes beati Simonis innocentis et martyris*, written in the first-person voice of the child himself ("Sum puer ille Simon . . . "). Here, after describing his own torments and death, the child appeals directly to the pope ("Sixte, precor!") to award him the martyr's crown.

The colophon indicates that the book was printed "in order to make known the fury of the Jews," during the happy reign of Bishop Hinderbach, and at the instigation of one Hermann Schindeleyp (*auctore*). Early bibliographies identified Schindelyp as the book's printer, but in fact the types are Kunne's. Schindeleyp was Hinderbach's chamberlain. He appears frequently in the bishop's registers issuing payments for all kinds of expenses related to Simon's cult, for example, purchasing wax for the church and issuing a stipend to the boy's parents. It seems clear that he paid for the publication of the *Historia completa*, almost certainly on Hinderbach's orders. Later that year, Kunne reprinted the final poem from this new collection, Tiberino's *Sum puer ille Simon*, as a broadside, illustrated with a woodcut image of Simon enthroned in glory. Hinderbach likely sponsored the printing of this edition too.

By this time, Sixtus's commission of cardinals was performing its own investigation in Rome. Hinderbach now attempted to influence its members: he hired two procurators to represent his interests in the papal city, Approvino degli Approvini and Thomas Rottaler, who lobbied the cardinals on Hinderbach's behalf, relayed messages from Trent, and sought to discredit dei Giudici and other supporters of the Jews in Rome. Hinderbach also sent letters directly to his allies, including the papal nephew Cristoforo della Rovere, the Vatican librarian Platina, and the jurist Giovanni Francesco Pavini.[120]

Dei Giudici by now had been transferred to the episcopal see of Benevento, something of a backwater. Prompted by Hinderbach, Platina wrote an ugly invective gloating over the former legate's fall from grace; in this text he also expressed his loathing for the Jews and his hope that Simon would soon receive the justice he deserved. In reply, dei Giudici composed a brief letter to Platina and, later, a more extensive *Apologia Iudaeorum*. The entire exchange of texts, born of palace intrigue and intended above all for the cardinals and their retinues in Rome, not the general public, remained in manuscript.

Public discourse over the affair resumed in earnest when the commission delivered its findings in the late spring of 1478. On June 20, Sixtus issued a bull clearing Hinderbach of misconduct but also rejecting as unproven the claims that Simon was a martyr or performing miracles.[121] Partially vindicated and undeterred by the caveats the commission delivered, Hinderbach moved to open a canonization case for Simon; he filed a formal application in Rome in May 1479. These moves were accompanied by a further round of printed publications that appeared in Rome, in Trent, and in the Veneto, from the middle of 1478 until around 1482.

In 1478, the jurist Pavini composed a *Responsum super controversia de puero Tridentino a Judaeis interfecto* at Hinderbach's request; the bishop also paid to have three hundred copies printed by Vitas Puecher in Rome that same year.[122] In 1479 and 1480, the Brescian poet Ubertino Puscolo received payments from Hinderbach for an epic poem he was composing to honor the martyrdom of Simon. In 1481 Puscolo sent the work to Hinderbach with a letter asking the bishop to read and approve the text. Only then should it be printed.[123] Also in 1481, another poet, Giovanni Calfurnio, brought out a collection of poems on the "apotheosis" of Simon, with another set of poems by Zovenzoni, published by a new printer just recently arrived in Trent, Giovanni Leonardo Longo.[124]

The case of Simon of Trent was a publishing sensation. Authors composed and published their accounts out of genuine fascination with the event and an apparently sincere belief that something miraculous had occurred in Trent. In addition, those authors who signed their name to their works (Tiberino, Zovenzoni, Silvestro da Bagnoregio, Pavini, Calfurnio, and others) clearly wanted to be recognized for their efforts and hoped to win favor or patronage from Hinderbach. And Hinderbach himself was actively involved at every stage: he forwarded Tiberino's text to various correspondents, circulated images, commissioned Zovenzoni and Puscolo to write poems, paid for legal *consilia*, and very likely supported the two printers who worked in Trent between 1475 and 1482 and published Simon texts in great numbers.

But Hinderbach was a relatively minor figure in the larger political landscape of his day. The two major powers that sought to check his activities—the Venetians and the papacy—did not turn to the press to articulate their positions but sought instead to restrict publication about the case altogether. The Venetians posted orders forbidding abuse of Jews and the circulation of Simon images and texts. Sixtus sent letters forbidding preaching and the production of images. Venetian and Roman orders alike were conveyed via manuscript briefs and oral proclamation. When dei Giudici returned to Rome, he composed two tracts defending the Jews and his own investigation, but these, too, circulated only in manuscript. Neither dei Giudici nor his superiors in Rome thought that the argument *against* Hinderbach needed to be conducted in print.

The press was an instrument Hinderbach used to stir up popular opinion and put pressure on Venice and Rome to support his cause; Venetian and Roman authorities used their usual channels of communication and modes of publication to try to get Hinderbach to back down. As with Negroponte in 1470–72, still in 1475–78 neither the pope nor the chancery yet recognized that print was a medium that authorities themselves could exploit for the prosecution of political disputes. Perhaps the right occasion had yet to present itself. Whatever the case, these two episodes revealed the powerful role the press could play in Italian political discourse. The fall of Negroponte was a sudden flame that drew a swarm of Italian commentators into the printshop in search of literary fame. The affair of Simon of Trent showed that the pronouncements of these self-appointed experts could be a real nuisance, or worse, for authorities seeking to keep the peace in Italian cities. The two affairs between them certainly awakened Rome to the possibilities and perils of print. By the time Simon's case was settled, in 1478, Sixtus and his secretaries were ready to do more than just condemn the contentious publications of others. They were prepared to enter the fray with publications of their own.

CHAPTER THREE

Sixtus IV and His Pamphlet Wars

> La bolla nostra, quale è stata posta in stampa a tutto il mondo, dimostrerà nostra justificatione.
>
> Sixtus IV to Federico da Montefeltro, July 28, 1478

For the first decade after the arrival of print in Italy, the Italian powers seemed content to leave the technology in the hands of individual authors—poets, commentators, even agitators for particular causes. In the late 1470s, however, princes and republics alike began to see the potential of the press as a political tool and to exploit it for their own political ends. Sixtus IV played a central role in this process. In 1478, Sixtus was drawn into the first of a series of military conflicts with other Italian states that would endure for the remainder of his pontificate. Here, for the first time, his chancery used print to publish documents that threatened penalties, or actually leveled them, against the pope's political enemies. In these publications, the pope also tried to justify his actions to the Italian reading public.

Sixtus's aggressive policies, often aimed at advancing the dynastic ambitions of his extended clan, provoked bitter conflicts with other Italian states. In the Pazzi War of 1478–79 and the War of Ferrara of 1482–84, Sixtus attacked his secular rivals with the tools of his sacred office: bulls of excommunication and interdict, delivered in person to their targets in Florence and Venice but also put into print both in Rome and abroad. In a third contest (ostensibly, an ecclesiastical dispute, but one with political overtones), Sixtus was himself assailed in print by the contentious archbishop Andrija Jamometić, who in the spring of 1482 launched a press campaign against the pope, including a call to reconvene the Council of Basel. Sixtus and his agents took to the press to reply, issuing bulls of excommunication and interdict and letters of condemnation exhorting the European public to reject the arch-bishop's heretical claims.

These three episodes are treated over the next two chapters. In the Pazzi War, which broke out after a failed attempt to assassinate Lorenzo de' Medici in April 1478, Florence and Rome issued between them fifteen or so printed pamphlets, each side defending itself and attacking the other. Sixtus's dispute with Venice over the lordship of Ferrara in the early 1480s produced a similar run of disputatious publications. Here, it was not just official decrees but also private communications that were printed—sometimes without their authors' knowledge—and framed in a way that underscored the propagandistic claims each side hoped to advance. The Florentines and Venetians each printed at least one document that was fake. Exploiting and subverting one another's communiqués, diplomats and printers alike ushered in important conceptual changes in the way official (or seemingly official) texts were presented to the reading public. The Basel episode, finally, was the papacy's longest and most complicated print controversy yet. Andrija Jamometić was a prolific and provocative critic of the papacy, who used the press to assail Sixtus in cities across upper Germany in the summer of 1482. The papal legate sent to check his efforts found himself in a bewildering political landscape where printed broadsides proliferated across the doors of urban basilicas and country churches. His struggles to get his own message out to the reading public reveal some of the limitations the papacy faced in printing its decrees.

In all three cases, Sixtus and his agents used print to advertise ecclesiastical sentences the pope had pronounced, to justify his actions, and to publicize the objectionable behavior that had provoked him. In none of these cases was print necessary. The pope's legal decrees of excommunication, interdict, or privation were still delivered via manuscript bulls, either put into the hands of their targets or posted on doors in Rome and abroad. But issuing papal documents in print did allow the pope to broadcast his case far more widely.

Once published, papal texts provoked a variety of responses from the states and individuals they targeted. These were also put into print, not just in Rome, Venice, and Florence but in Naples, Padua, Basel, Strassburg, Nuremberg, Mainz, and even Westminster. The exchanges represent three of the earliest printed pamphlet wars in European history, certainly the first ones to unfold, wholly or in part, on the Italian peninsula. Yet even as Sixtus innovated in his use of the press to pursue political goals, the papacy's entry into the world of political printing was often constrained by its own administrative practices and traditions. When rival states and critics began to publish their responses to printed papal bulls, their replies tended to be more flexible and creative in both form and content. Arguably, they were more effective instruments for courting popular opinion.

The Sistine administration's early use of print was confined to promoting the Jubilee and other devotional events. In 1473, Sixtus issued a bull inviting pilgrims to Rome for the Jubilee of 1475 and advertising the spiritual benefits they would gain by making the trip.[1] Sixtus was following the example of Paul II, who had issued the first bull announcing the Jubilee in 1470. Three different editions of Sixtus's 1473 bull were printed in Rome.

The Jubilee of 1475 was a key element in Sixtus's plan to renovate Rome and impress the stamp of papal authority on the city: anticipating pilgrim crowds, the pope had streets cleared and derelict structures pulled down; he constructed a new bridge to help visitors make their way across the city, the Ponte Sisto, the first to be constructed across the Tiber since antiquity. A few years later, Sixtus undertook another urban renewal project, the refoundation of the Ospedale di Santo Spirito in the Borgo, an ancient hostel for pilgrims, which stood near the Tiber close to St. Peter's. In March 1478, Sixtus issued two bulls granting the foundation new privileges and advertising the spiritual benefits of donating to it. The Santo Spirito bulls were printed in Rome at least six times during Sixtus's reign by five different Roman printers.[2] Like the Jubilee bulls, they supported the pope's efforts to renovate the urban fabric of Rome, attract pilgrims to its precincts, and solidify its position as the metropolis of Christendom. One of the printers who issued Santo Spirito bulls, Johannes Bulle, published several more Sistine bulls in 1478, suggesting something like a regular commercial relationship between the chancery and his shop.[3]

The Pazzi War

In April 1478, just a month after the publication of the Santo Spirito bulls, the Florentines Jacopo and Francesco de' Pazzi launched their infamous conspiracy against the Medici. At Easter Mass in the Duomo of Florence, Francesco and his fellow conspirators fell upon Lorenzo, wounding him and killing his brother, Giuliano.[4] The Pazzi quarrel had its origins in a dispute over the city of Imola: in 1473, Girolamo Riario, nephew of Sixtus IV, hoped to buy the city from Milan and set himself up as a territorial lord in Romagna.

Lorenzo de' Medici opposed this move (Imola was close to Florentine territory), and the Medici bank refused to lend Riario the money. But the Pazzi, proprietors of a rival Florentine firm, extended a loan, thereby winning the pope's favor. Not long after, Sixtus shifted important financial contracts from the Medici bank to the Pazzi and ended the Medici monopoly on the lucrative alum trade.[5]

The financial wrangling in Rome masked older, civic resentments in Florence, where the aristocratic Pazzi opposed the Medici as populist upstarts. After the Imola dispute, Lorenzo found ways to deprive them of inheritances and exclude them from elective offices.[6] Francesco de' Pazzi conferred with Girolamo Riario in Rome: if they could overthrow the Medici, each might rise to greater heights in his respective state. Francesco drew in other conspirators: Francesco Salviati, archbishop of Pisa, from a family allied to the Pazzi, whose appointment to the see of Pisa Lorenzo had opposed; Jacopo Bracciolini, the humanist Poggio's disaffected son; the *condottiere* Giovanni da Montesecco, who was later captured and confessed many details of the plot; and Francesco's uncle and head of the family, Jacopo de' Pazzi. They drew in an unwitting papal nephew, seventeen-year-old Raffaele Sansoni Riario, a law student at Pisa recently appointed to the College of Cardinals. His visit to Florence in April 1478 offered the pretext for bringing Lorenzo and Giuliano together with the conspirators in the cathedral.[7] Before the party left for Florence, the pope was informed of the plot; Giovanni da Montesecco later claimed that Sixtus gave his blessing to the idea of a coup but not to murder. His words, as Montesecco remembered them: "I want no man dead . . . but a change of government? Yes."[8]

On April 26, amidst the Easter crowd in Santa Maria del Fiore, the assassins struck, killing Giuliano immediately and wounding Lorenzo. Archbishop Salviati, dressed in layman's clothes, led a group of conspirators to the Palazzo della Signoria, intending to overwhelm the priors who were lodged there. But nothing went according to plan. Francesco de' Pazzi was wounded and retreated to the family palace; Salviati faltered in his attack on the priors and was arrested in the Palazzo; Jacopo de' Pazzi rode through the streets calling on the Florentines to reclaim their liberties but was met with silence. Medici partisans hunted down the conspirators and brought them to justice. Salviati, Francesco de' Pazzi, and Jacopo Bracciolini were hanged from the windows of the Palazzo della Signoria. Their followers were thrown from the palace windows and ripped apart by crowds in the piazza below. Two days later, Jacopo de' Pazzi was captured and hanged; on May 4, Giovanni da Montesecco was arrested, interrogated, and put to death. The young cardinal was placed under house arrest. Meanwhile, the city rallied to Lorenzo, the Signoria pledged to protect him, and princes from across Europe sent letters of condolence and support. In July, Botticelli painted portraits of the executed conspirators on the walls of the Palazzo della Signoria and Bargello, with Latin verses commemorating their treachery painted beneath.[9] The remaining members of the Pazzi family were banished from the city.

The failure of the Pazzi conspiracy spurred Sixtus into a bitter feud with Lorenzo that ultimately led to war. The pope's enmity was fueled by equal parts frustration and embarrassment. As Lorenzo put it in a letter to King René of Anjou in June, Sixtus seemed most incensed by the fact that Lorenzo had survived: "I have committed no crime against the pope," he wrote, "save that I am alive . . . and have not allowed myself to be murdered."[10] The *stated* cause of Sixtus's anger was that the Florentines had hanged Archbishop Salviati and other priests in his retinue, violating their ecclesiastical immunities, and were keeping his young nephew prisoner. In May, the pope demanded satisfaction for these acts of sacrilege: the Florentines must banish Lorenzo, release Cardinal Raffaele, and beg forgiveness from the pope. When the Florentines refused, Sixtus issued the bull *Ineffabilis* on the first of June, which excommunicated Lorenzo along with the city's *gonfaloniere*, priors, and the eight members of its police commission (the Balía). Calling Lorenzo a son of iniquity and child of perdition, the pope declared that unless his fellow citizens surrendered him and others colluding in the cardinal's detention, Florence and neighboring territories would be placed under interdict.[11]

Ineffabilis was not short. A lengthy opening section outlined eleven charges against Lorenzo and his "accomplices," detailing offenses committed against the Church both before and after April 1478. Lorenzo and his associates had long aided rebels in the papal states,[12] they had harrassed "pilgrims" on their way to Rome,[13] and they had stubbornly resisted the appointment of Salviati as archbishop of Pisa. Finally, they had laid hands on that same archbishop and (*inauditum scelus!*) murdered him by hanging him from a window, then dropping his body into the piazza to suffer the outrage of the mob. They had hanged or beheaded other clerics in the archbishop's train and imprisoned Cardinal Raffaele without cause. Lorenzo, as the leading citizen of the state and a member of the Balía, bore chief responsibility for these crimes.

Ineffabilis was the first salvo in that summer's war of words; the papal chancery made sure its contents were widely known. A copy of the bull was posted on the doors of St. Peter's on June 4, and another was sent to Florence, where it arrived on June 10.[14] By then, the text had also been printed in Rome: in a letter of June 9, the Milanese ambassadors in the city described the bull as "published and even printed" (*hora che le sonno publicate et a stampa*), and on June 12 the Milanese ambassador in Venice reported that a copy of the printed bull (*processo delle . . . censure impresso a stampa a Roma*) had arrived there as well.[15] This was a remarkably quick turnaround, just four days from *publicatio in valvis* to publication in print. But the ambassadors were remarking on more than just the speed with which the bull had been published. The very fact that it had been printed must have impressed them, for this was a novelty in chancery practice.

The bull was entrusted to two printing houses for publication. As with the Santo Spirito bulls, the chancery used multiple presses to broadcast copies of a single document. Johannes Bulle printed the text as a quarto pamphlet, some ten pages long.[16] A second house, run by two printers, Johannes de Monteferrato and Rolandus de Burgundia, known only for their publication of this bull, issued an edition in folio, its dense text filling three large pages of a single folded sheet.[17] Neither edition does much to announce its author or contents, although the folio edition does print the words "Sixtus iii Pont. Max." as a subscription at the end. But as the length of the bull, its sensational rhetoric, and its patterns of distribution all suggest, the document was intended for a broader audience than just those Florentines whom it condemned.

Ineffabilis not only delivered a sentence of excommunication to Lorenzo and his colleagues but also underscored to his Florentine compatriots the danger of supporting him and argued to a wider reading public that the Medici had long been violating the rights of the Church. Sixtus shared the text with other heads of state both in Italy and abroad, sending a series of manuscript letters across the continent along with copies of the bull (some in manuscript, some in print).[18] A copy of the folio edition is preserved in Bologna, where it was sent by the papal chancery to ensure the loyalty of the communal government should war break out with Florence.[19]

Copying the "Christian princes" on major ecclesiastical acts was standard practice for the Quattrocento chancery,[20] but *printing* the bull was something new, an amplification of the pope's campaign for public opinion. The pope himself acknowledged as much. In a letter to Federico da Montefeltro, duke of Urbino, at the end of July, he explained, "We have sent many ambassadors to the king of France with our arguments, and likewise we have sent others to other princes, such as the emperor, the king of Hungary, the king of Spain, and all the rest. Moreover our bull, *which has been put into print before the whole world,* will demonstrate the rightness of our cause."[21] In August, the Florentine Gentile Becchi complained about this course of action, noting that Sixtus, "not content with the pen," had taken the extraordinary step of having his bulls printed as well. And "not content with the doors of churches," he had set them out for sale in the bookstalls of Campo de' Fiori in Rome, so eager was he to blacken the name of Florence to the rest of the world.[22]

After *Ineffabilis* was published on June 1, the Florentines released Cardinal Raffaele and sent him back to Rome, but they refused the pope's other demands. Accordingly, on June 20, Sixtus imposed an interdict on the dioceses of Florence, Fiesole, and Pistoia.[23] Two days later, he issued two more bulls, which were printed together by Johannes Bulle in another quarto pamphlet, this one only three leaves long.[24] The first bull restated the charges against Florence leveled in *Ineffabilis* (the confinement of the cardinal and the murder of Salviati and other clerics) and took the Florentines to task for ignoring its strictures and persisting in their disobedience. Not only had they refused to expel Lorenzo, but they had begun hiring mercenaries and preparing to go to war! The pope invoked his right to invite the "secular arm," in the form of his vassal, the king of Naples, to enforce his censures against Florence. He also extended a broader invitation to Christians everywhere to take up arms against the city, promising a plenary indulgence to anyone who fought against Florence on behalf of the Church. The second bull also addressed the military situation. Where the first encouraged Christians to take up arms against Florence, this one forbade lending the excommunicates any sort of help or entering their military service; no one, especially no subject of the papal state, should venture to assist them.

There was a clear rationale for printing these two bulls, directed as they were not against Florence but to an enormous audience, the fighting population of the entire peninsula. Even as they conveyed particular instructions to potential soldiers in the coming war, the printed bulls also provided the pope another chance to restate his case against Lorenzo. They had little effect, however: the Florentines soon hired Roberto Malatesta of Rimini and several other *condottieri* from the papal states to command their forces. In response, papal and Neapolitan troops under Federico da Montefeltro began advancing on Florentine territory in early July, and Ferrante sent his son, Alfonso, duke of Calabria, north with troops, which he billeted in neighboring Siena.[25] While the pope had Naples and Urbino on his side, Florence could count on

Venice and Milan, as well as Ferrara and, farther away, Louis XI of France, who had his own scores to settle with Sixtus. His ambassador, Phillipe de Commynes, would soon arrive in Florence.

The Pazzi War involved few military engagements. In August, the Florentine diarist Luca Landucci observed, "The rule for our Italian soldiers seems to be this: 'You pillage there, and we will pillage here; there is no need for us to approach too close to one another.'"[26] In this way hostilities dragged on another eighteen months till the end of 1479, when Lorenzo made a daring, secret journey to Naples to conclude a separate peace with Ferrante. The Florentines finally reconciled with Sixtus in December 1480, and the interdict was raised in the spring of 1481.

The printing of Sixtus's bulls in June 1478 marked the start of a propaganda war between Florence and Rome to which authors on both sides contributed documents that were quickly put into print. Nine editions of six different texts survive from the papal side, alongside six editions of four texts written in support of Lorenzo and the Florentines. (A few other texts known only in manuscript may well have been printed, but no printed copies survive.) For the first time, a pope exploited the press not as a promotional tool—to raise money for the crusade, to publicize the Jubilee, or to announce some other new initiative—but as a weapon against a political rival.[27]

The bulls printed in June conveyed a stark message: Lorenzo and his allies were cast out of the Christian community, beyond the pale. But Sixtus was also capable of a milder diplomatic tone, aimed at persuasion rather than provocation. On July 7, his secretary, Leonardo Grifo, drafted an open letter to the government and people of Florence inviting them to eject the Medici and restore their city to papal favor.[28] "If anyone thinks," the letter begins, "that we no longer wish to work for the good of Christendom, or that we are preparing to declare war on your city, they are sorely mistaken."[29] Sixtus assured the Florentines he meant them no harm; he had leveled his censures quite unwillingly. It was Lorenzo, not they, who was at fault. Medici machinations against the papal state had disturbed the peace of Italy and turned Christian princes from their common project of a crusade against the Turks. It was to safeguard that holy enterprise that the pope had taken action against Lorenzo. Sixtus bore the people of Florence no ill will; he only wanted to free them from the clutches of a tyrant who was doing them and all Christendom harm, to preserve their liberties, and to promote the crusade. In this holy intention he was supported by his ally, the powerful King Ferrante, and he hoped the Florentines would come to see the matter as he did.

With this last statement, the letter switched from persuasion to thinly veiled threat, and the menacing note was echoed by the manner of its delivery: on July 13, a herald employed not by the pope but by the duke of Calabria, Ferrante's son and commander of Neapolitan troops in Siena, arrived in Florence and, with a blast of trumpets, presented a manuscript copy of the letter to the Signoria. The priors received it with alarm.[30]

The papal chancery may also have had the text of this brief printed in Rome, though no copy of such an edition is known. What does survive, in a single copy, is an edition printed in Naples: *Copy of a Letter addressed by the most blessed and holy Pope SIXTUS IV to the government and people of Florence*.[31] Sixtus likely sent a manuscript copy of the letter to Ferrante in Naples, perhaps at the same time he sent the copy to Alfonso that the duke's herald delivered to Florence.[32] Whether the pope or the king commissioned its printing in Naples is unclear. The unsigned edition has been attributed to the Neapolitan press of Sixtus Riessinger and Francesco del Tuppo. Riessinger had briefly run a press in Rome in the late 1460s before moving to Naples, where he introduced the art of printing (with Fieschi's epic on the fall of Negroponte) and quickly became one of the city's most prolific producers of books. At some point in 1478, he moved back to Rome, leaving his types in the hands of his corrector and sometime partner, del Tuppo.[33] Riessinger's connections to Rome may explain the publication of the pope's brief at his press in Naples, but his and del Tuppo's shop was one that Ferrante's court often employed, and the commission may have come from the king.[34]

The publication of Sixtus's letter in Naples is significant for several reasons: formally, it was a brief, not a bull. It required no "publication" at all, so the decision to have it printed was purely propagandistic. As such, it took a softer tone than the earlier bulls, reflecting the pope's interest in driving a wedge between Lorenzo and his people. But the intended audience was larger than just the Florentines, as its publication in Naples underscores. The pope was trying to broadcast his "justifications" to readers not directly involved in the dispute.

In Florence, meanwhile, a small circle of government officials, jurists, and humanists of the Medici *famiglia* spent the summer drafting texts that protested the pope's censures, justified Florentine actions, questioned the legitimacy of the pope's behavior, and even called for a new council to resolve the dispute. Many of these texts, too, were put into print. The three principal authors of the project were Bartolomeo Scala, the Florentine chancellor; Gentile Becchi, bishop of Arezzo and Lorenzo's former tutor; and Angelo Poliziano, scholar, poet, and tutor of the Medici children. All three were closely connected to Lorenzo and stayed in contact with him as they wrote; all three had their works printed by the same Florentine printer, and the publication of Becchi's, at least, was paid for by Lorenzo.

The first texts the group produced were not printed (or rather, no printed copies survive) but circulated in manuscript: two letters written in response to Sixtus's brief of July 7. The herald who delivered the pope's letter with a trumpet blast on July 13 had caused an immediate stir, and the Signoria convened an emergency meeting the same day to debate how to respond.[35] Scala seems to have drafted something by July 15; on that day, Lorenzo made mention "d'uno brieve del Papa et della risposta" in a letter to Florence's ambassador in Milan. But, for whatever reason, the text as we have it is dated July 21.[36] Addressing Sixtus directly, Scala reviews and then refutes the claims of the earlier brief. Sixtus had urged the Florentines to banish Lorenzo as a tyrant and enemy of the Christian faith. By driving him out, the Florentines would regain their liberty. But how could they be free if they had to submit to the demands of the pope? Following his orders would merely bring the Florentine people a new master, one they were not sure they would like. As to the welfare of Christendom, there were many reasons the Christian princes were turning on each other instead of making common cause against the Turks. Lorenzo, like Cosimo before him, was ready to supply arms, ships, and funds to whatever Christian state would advance on the infidel. It was Sixtus who fomented war in Italy and kept the Christian powers from uniting for a crusade. His policies were so aggressive and warlike that it was hard to believe he was truly a pope.

This last clause was no throwaway line. For the past two years, Louis XI of France had been making noises about convening a new council to adjudicate his own disputes with the pope. Might Florence now call for a council as well, to impose some limits on the pope's power or even force his deposition? Here, Scala did no more than hint at the possibility that Sixtus was not a legitimate pope. But he was open about his state's alliance with France. Responding to the pope's invocation of the "secular arm" of Ferrante of Naples, Scala declared that the Florentines, with the protection of Christ, their allies, and *their* great patron, the king of France, would not hesitate to fight to preserve their liberty.

Eight copies of Scala's letter survive in manuscript, suggesting that some effort was made to spread its contents about.[37] A second letter, headed *Litterae Florentinorum* and probably composed by Gentile Becchi, also carries the date July 21.[38] Where the chancellor maintained an even tone, this letter is inflammatory and polemical. It mocks Sixtus's own letter by starting with the same words but twisting them to a different end: "If anyone thinks . . . that we have not always hated tyranny, then *they* are utterly deceived."[39] Like Scala, the author addresses the pope directly, but he uses the insulting title "Frater Franciscus" for the Franciscan pope. Sixtus might well be a friar (brother), but he was no father to the Florentine people. Though claiming to be the vicar of Christ, the pope had sacked cities, attacked Christians, and squandered the revenues of the Church on his nephews and their mistresses, always putting off his long-promised crusade. (Under the heading of mistresses, Becchi listed one in particular, "Teresa," the notorious mistress of the late Cardinal Pietro Riario; Andrija Jamometić would repeat these same charges, including a reference to Teresa, in his 1482 council appeal.)[40] Finally, while Sixtus could invoke Ferrante's support, Florence had an even greater ally, Louis XI, "who does not want to see a Judas sitting on the throne of Peter."[41] Frater Franciscus should take care lest he lose his tiara.

Both letters hint that a challenge to Sixtus's very pontificate might be in the works, both carry the same date, and both end by invoking the protection of France. The two authors likely conferred on their compositions. And yet it seems neither was printed. The following month, however, three longer and more carefully drafted Florentine responses did appear in print. On August 11, Scala issued a communiqué "to all and sundry whom this letter may reach."[42] Scala made a plea for sympathy and support from Christians everywhere, including Frederick III and Louis XI, whom he addressed in his conclusion. The chancellor pushed back against the papal claim that Lorenzo stood in the way of a new crusade. Nothing could be further from the truth. It was Florence, not Rome, that cared for the welfare of Christendom. The Turks stood on the very doorstep of Italy, and what did the pope do but get himself mixed up in criminal conspiracies, threaten the liberties of free people, level unjust sentences against them, and declare war on them instead of infidels? The pope's behavior was more barbarous than human.[43]

All this was delivered as a rapid prelude to the heart of Scala's tract: a carefully redacted transcript of the conspirator Giovanni da Montesecco's confession, presented in the original Italian.[44] The confession had been extracted from Montesecco under torture after his arrest by Florentine authorities in May. The *condottiere* implicated Sixtus himself as a partner in the plot; the pope had endorsed removing the Medici from power while adding a scarcely credible rider that the coup should happen without bloodshed ("I want no man dead").[45]

To authenticate these sensational claims, Scala included the statements of seven witnesses who heard Montesecco's confession on May 4, followed by the statement of a notary, dated August 11, which declared the transcript a true and accurate copy.[46] These authentications fill nearly four pages of the printed edition. Even so, they were not true: Scala carefully edited the confession to remove any references to the secular powers (Urbino and Naples) then threatening the city. In Scala's version of the confession, Montesecco blamed the Pazzi, Girolamo Riario, and the pope, and no one else. Scala's selective reporting reflected the larger Florentine strategy of turning public opinion against the pope without antagonizing his powerful allies.[47]

In the closing pages of the tract, Scala offered his own account of the events of April 1478 and Lorenzo's miraculous escape, describing how the people had rallied to Lorenzo's defense, Salviati and his accomplices had been arrested and killed, and Montesecco confessed the details. Out of respect for the young cardinal's rank and for his own protection, the government had placed him under guard and maintained him at their own expense until the *furor publicus* died down. Then they sent him back to Rome safe and sound. Even so, the pope had leveled the interdict on the Florentines, when in truth he ought to have thanked them! Scala could only conclude that Sixtus was using ecclesiastical censures to accomplish what he had failed to do by violent conspiracy. He had started a war against the Florentines "for no other reason than that we refused to allow ourselves to be killed."[48] Florence had responded correctly, as various *publica scripta* had already shown.[49]

Scala intended his tract to become another of these "public writings." At least one manuscript copy of the text was drawn up on parchment, signed by Scala and sealed with the communal seal, presumably to be sent to some foreign court.[50] In addition, Scala gave his text to Nicolaus Laurentii Alamannus, alias Niccolò della Magna or Niccolò Tedesco, one of the earliest printers to work in Florence and, in 1478, proprietor of one of just two printing houses in the city.[51] Tedesco printed his first book in Florence in 1474 or early 1475; by the summer of 1478, he had produced about a dozen books, mostly

Italian devotional texts. Between July and September 1478, however, his press was completely occupied with printing tracts in defense of Lorenzo: Scala, Becchi, and Poliziano all published texts with him in quick succession, most likely at Lorenzo's instigation and expense.

It is worth underscoring the difference between Scala's printed letter and Sixtus's publications. The Sistine bulls published in Rome were exact transcripts of their manuscript originals: anyone who read them would witness the *fact* of the sentences pronounced against Lorenzo and the Florentines, but the reader himself was not addressed. Even Sixtus's open letter to the Florentines, printed in Naples in July, offered the reader a view of a correspondence taking place between others—between the pope and the people of Florence. By contrast, Scala addressed the general public directly.[52] He used the rhetorical voice of a storyteller: "To all and sundry whom this letter may reach . . . I am about to tell a story so new, so unheard-of, so alien to human nature and every way of life, that I have no doubt that everyone who hears it will be simply amazed by the immensity of the atrocity and evil."[53] In publishing Montesecco's confession, he likewise addressed a general audience: the world needed to know that the pope was implicated in the plot, and Scala had the proof. "So let's listen," he says, "to what Montesecco has to say. He'll reveal how the whole thing happened."[54]

Scala conjured up a general readership for his tract, addressed those readers directly, and offered them a doctored version of Montesecco's confession. Gentile Becchi went even further down the road of invention. His tract, *Florentina synodus*, which also appeared from the press of Niccolò Tedesco, purported to communicate the deliberations and decrees of a synod or regional council of the Tuscan clergy, supposedly held in the cathedral of Florence in late July, which almost certainly never met.[55] Inventing a church assembly in order to distribute its fictive decrees was a bold step; it played into a larger strategy on the part of Lorenzo and the French alike to mobilize local clergy against the pope.

The Florentine synod, Becchi said, had convened in the light of the Holy Spirit in order to dispel the "Sistine fog" that had settled over the Church.[56] Becchi denounced the pope for poisoning the vineyard, infecting the sheep, casting pearls (the treasure of the Church) before swine (his nephews), and pimping out Holy Mother Church for personal gain.[57] A vicar of the Devil, not Christ, he had thrown the disciples overboard and filled the ship of the Church instead with "Girolamos and Teresas."[58] Most recently, he had fallen upon the Florentine sheepfold and slaughtered "the lamb," the innocent

Giuliano. "This latest shame," Becchi writes, "God wants us to reveal to all His faithful."[59]

Becchi, still speaking as "the synod," then provides another transcription of Montesecco's confession, this one even more tightly edited than the one Scala published. But Becchi finds it hard to maintain his own documentary conceit: he interrupts the transcript in several places to supply commentary and apostrophes ("See how blind, how lost is this old man!") and retells the bloody tale of the conspiracy itself: murder in the cathedral, confusion in the streets, the eventual failure of the plot. Next, he turns to the text of *Ineffabilis* and examines its eleven charges against Lorenzo. Here he sprinkles further extracts from Montesecco's confession and a copy of an earlier brief that Sixtus had sent Lorenzo, full of kind words and praise for his filial piety. The raving pope of today was much changed from this once-kindly pastor. Finally, he tackles the sentence of excommunication. It is, in the eyes of "the synod," an invalid pronouncement. Yes, the Florentines had executed the archbishop and arrested the cardinal. But why were they penalized for this? A pope should excommunicate one who betrays his guests, to be sure, but to excommunicate someone for defending himself and for fighting for his country? Someone who had, in fact, rooted a corrupt archbishop out of the Church? The pope should have thanked God for Lorenzo's services on behalf of his faltering institution. Likewise, the arrest of the cardinal was no sacrilege, but a protective move, as the young man himself said in a letter he sent the pope after his release. Becchi quotes it in full. The cardinal assured his uncle Sixtus that Lorenzo had treated him kindly, and he asked the pope to lift the sanctions against Florence. Rather than censures, these "best of men who deserve the best from us" ought to be showered with tokens of gratitude.[60]

Worst of all, Becchi continues, the pope spread his lies everywhere through the new technology of print: "Not content with manuscript, he had the bull printed; not content with the doors of churches, he had had it sold in Campo de' Fiori."[61] The remark provides important evidence for how the new printed papal bulls circulated, or were intended to circulate, or, at least, were *thought* to be circulating—not just sent as enclosures to other heads of state but sold on the open market. Becchi's remark also shows that in this war of words, one party's participation in the appeal for popular opinion could itself be a source of grievance for the other. At almost the same time that Becchi complained about the pope circulating his bulls in print, Sixtus was drafting a brief to the Christian princes that complained about the "calumnies" Lorenzo and his friends were disseminating about him.

After these several digressions, Becchi returns to the conceit that his text conveys an official, synodal decree. Since Sixtus was acting on bad information, in bad faith, and with bad intentions, it was clear that the Florentine clergy could not trust him to deal with Lorenzo fairly. Instead, they must seek another "light," another source of justice; they now appealed to the emperor, to the king of France, and to "all Christian princes and peoples" to examine this simoniacal and murderous pope and come to the council that they, the clergy, hereby invoked. The document closes with a prayer ("Lord, deliver us from false shepherds, from wolves in sheep's clothing!") and a subscription indicating that it was issued by the synod in the cathedral on July 20, 1478.[62]

Becchi's text was by far the oddest produced in the conflict between Florence and Rome. Becchi boldly claimed to convey the deliberations of an official synod, despite the fact that no such assembly had convened. He also failed, in places, to maintain the proper official tone, veering off into overwrought apostrophes. At times, the entire conceit seems to have been launched as a way to frame the presentation of other texts—Montesecco's confession, for example—that were too slight to be presented independently.

Peculiar as it was, the publication of this text at Tedesco's press was a key element in Lorenzo's publicity campaign against the pope. Although the edition has the text dated July 20, Tobias Daniels has shown that it was nowhere near complete at that point; the *Florentina synodus* underwent at least another month of editing and revision and cannot have been printed till the end of August or early September. A number of people had a hand in the process, including Lorenzo himself, his secretary Niccolò Michelozzi, several jurists in his employ, the Florentine ambassadors to Milan and France, and even the French ambassador to Florence, Philippe de Commynes.[63] Becchi completed a first draft of the text in mid-July. He sent it to Michelozzi, who added text and excised passages that seemed too inflammatory or irrelevant.[64] It was Michelozzi who then sent copies to others in Florence and Milan to read and review; Becchi's letters show him bristling at the number of people who were reviewing his work. In August, he asked Michelozzi to return the copies or at least stop circulating them. However, when Lorenzo raised questions about certain passages in the text, Becchi agreed to seek legal advice on the language he used to frame his appeal.[65]

Becchi may have tried to assert control over the contents of his text, but he certainly did not object to having it printed. In fact, he asked Michelozzi not to share the text with anyone else *except* the printer.[66] The decision to print rested with Lorenzo. On August 19, Becchi wrote to Lorenzo suggesting that if and when he decided to print his tract, he might append certain other documents that would further support his claims.[67] Lorenzo also had a team of jurists examining the legal grounds on which Sixtus had leveled the interdict and his other censures.[68] Poliziano, who had retreated to Pistoia with the Medici children earlier that summer because of an outbreak of plague in Florence, had been put in charge of coordinating their efforts. Poliziano started sending their *consilia* to Lorenzo in mid-August, forwarding the last of them on August 24.[69] Certain passages from these texts were then inserted verbatim into the *Florentina synodus*. This means that the text cannot have reached Niccolò Tedesco till the end of the month, or perhaps early September.

It is worth considering how Becchi's work fit into Lorenzo's larger strategy of seeking French backing in his conflict with the pope. The French crown had long opposed the increasing centralization and autocracy of Quattrocento papal policy. With the Pragmatic Sanction of Bourges, a document promulgated in 1438 during the Council of Basel, Charles VII had claimed substantial control over the "Gallican" Church, including rights over appointments and revenues. Louis XI had abrogated the Pragmatic Sanction in 1461 as part of a larger diplomatic settlement with Pius II, but he continued to seek ways to preserve and extend Gallican liberties. In 1476, Sixtus dismissed the papal legate in Avignon, Charles de Bourbon, a cousin of the king, and appointed his nephew Giuliano della Rovere in his place. Outraged by this and by Sixtus's ongoing support for Charles the Bold of Burgundy, Louis threatened to call a council of the Gallican Church, an assembly that had the potential to reinstate the Pragmatic Sanction or even call for the pope's deposition.[70] (It was for this assembly that French agents posted a call for a new council on the doors of St. Peter's in 1476.) Now, two years later, Lorenzo adopted the same attitude.[71] Repeatedly in June and July he declared he would "levar l'ubbidientia" if the pope did not withdraw his censures.[72] He stayed in regular contact with officials in Venice, Milan, and France whom he hoped would join him in convening a council to restrain or even depose the pope. Louis's ambassador Philippe de Commynes arrived in Florence on July 2 and let it be known that the king wanted a council; Commynes and his secretary both had a hand in revising the *Florentina synodus*.[73] After Venice and Milan offered only lukewarm support for the plan, Commynes and Lorenzo seem to have decided it would be easier to call a series of regional councils than to call for a general one. On August 1, French representatives in Rome announced that a Gallican council would be held that fall; on August 10, Louis informed Sixtus that it would meet in Orléans.[74]

It was during these same summer weeks that Becchi drafted the *Florentina synodus*. The idea, it seems, was to pretend that the Florentine clergy had already held their own meeting and voted for a council. When the French synod met, they would be adding their voices to a growing chorus.[75] The Florentines were using print to document an event that had never occurred and to publish decrees that had never been voted on. A few years later, Andrija Jamometić would publish the fictive decrees of another phantom council, supposedly convened in Basel.

Just as Becchi passed his drafts to Lorenzo, Michelozzi, and others for review, he and Scala probably conferred on their twin publication projects. The two pamphlets, composed just weeks apart and published by the same printer, invoke the same arguments and even repeat some of the same scriptural passages;[76] each presents excerpts from Montesecco's confession. The one set that text into the deliberations of a fictive church synod, while the other embedded it in a diplomatic communiqué from the chancery. They were pendants to each other: the one a sacred and the other a secular "publication" of the conspirator's incriminating text. Lorenzo probably paid for the printing of both.

Finally, Poliziano weighed in with his own, rather more literary contribution to the propaganda campaign, his *Coniurationis commentarium*.[77] Poliziano had served Lorenzo as a secretary, had tutored his children, and was a close friend of Giuliano (his famous *Stanze per la giostra* celebrating Giuliano were only half-finished when the conspiracy broke out). Poliziano had been present in the cathedral when Giuliano died, and he saw firsthand the confusion and uproar that followed. In the summer of 1478, secluded with the Medici children in Pistoia, he wrote up his recollections of the conspiracy.

Poliziano viewed the entire episode through a humanist lens. Taking Sallust's *Bellum Catilinae* as his model, he paints a portrait of the Pazzi brothers and Archbishop Salviati as deranged and decadent aristocrats, in contrast to the sober republican virtue of the Medici. From the start, the plot was the project of these arrogant, avaricious, dissipated noblemen; like Catiline, they were driven by a mix of patrician conceit and desperate penury, their hatred further inflamed by the innate generosity and goodness of the young Medici brothers. The text ends with a long encomium of Giuliano de' Medici, the "idol" of the Florentines, whose beauty, intelligence, grace, and goodwill had enchanted all who knew him. Poliziano had several aims in composing his tract: to record his own impressions of the events, to eulogize Giuliano, to assert that the Florentine people had defended Lorenzo in the moment and remained loyal to him still, and to demolish the reputations of the conspirators who had been cowards in life and rash at the moment of their death. All this being so, it is surprising that, unlike Scala or Becchi, Poliziano does not blame—or even mention—the pope; he refers to Girolamo Riario a few times as a distant connection of some of the lesser conspirators, but Sixtus's name never appears in the text. Even Giovanni da Montesecco barely appears; there is no discussion of his confession or of the conversations between Pazzi and the pope that he revealed.[78] His reasons for writing the story this way may have been purely literary: his model, Sallust, portrayed Catiline as a man driven by uncontrollable vice, and nothing else; Poliziano followed his lead.

Poliziano, like Scala and Becchi, saw his text published by Niccolò Tedesco,[79] but its publication seems less closely related to the coordinated efforts of Scala and Becchi and the advisors around Lorenzo. Poliziano's focus on the Pazzi as the only villains of the episode, taken together with his silence on the pope, may explain why the text was reprinted twice in Rome itself a few years after the conspiracy.

Meanwhile, the Florentine press housed in the convent of San Jacopo a Ripoli published a vernacular verse lament for Giuliano de' Medici, *Tradimento della morte di Giuliano*. The anonymous poet sings of the bloody events of April with breathless horror, cursing the conspirators and lamenting the death of a blameless young man, "a rose without thorns."[80] Like Poliziano, the poet places blame for the tragedy squarely on the shoulders of Salviati (whom he calls the *capo* of the conspiracy) and the extended Pazzi clan, a family plagued by gambling debts, envy of the Medici, and shameless ambition. The young cardinal is described as an innocent bystander, drawn into the plot by the Pazzi; there is no mention at all of Sixtus or even Montesecco. The Ripoli account book shows that this text was printed at least twice in late 1478, suggesting that vernacular readers and listeners were fascinated by the murder in the cathedral.[81] The poet's total silence with respect to the pope suggests that, like Poliziano, he was writing to cultivate local sympathy for the Medici, not to win international support against Rome as Becchi and Scala hoped their texts would do.

Other authors then weighed in with tracts in support of Rome, some of which were printed in order to counter the Florentine propaganda issuing from Tedesco's press. An anonymous tract, *Dissentio inter Papam et Florentinos*, was composed as a direct response to the *Florentina synodus* and was framed, like

Scala's *Excusatio,* as an open letter to the general public.[82] It was printed in Basel. Whereas Scala's letter *universis et singulis* was written as a statement from the Florentine government, the *Dissentio* claims no such corporate author, making for some awkward rhetorical moves.

The author is open in his hostility toward Lorenzo and the Florentines.[83] Speaking to "all Christians," he rehearses arguments by now familiar: Sixtus had been warmly disposed to Lorenzo and showered him with kindness, but Lorenzo had repaid this generosity with a thousand slights, political, fiscal, and military. These reached a head after the events of April 1478: an innocent young cardinal was detained; Giovanni da Montesecco, captain of the pope's private guard, was executed "with no respect for our lord pope or his dignity";[84] Archbishop Salviati was imprisoned for some eleven hours without benefit of trial and then hanged from the palace window; and the Florentines defiled his corpse. All this had been done outside the law. If Salviati was guilty, where was his confession? Where was the judge who had investigated and convicted him? The charges were trumped up to serve a nefarious agenda of which Lorenzo was the architect. Ambitious, disloyal, and scheming, he would stop at nothing to destroy his enemies and damage the Church. And now, he and the rest of the Florentines dared to challenge the pope with their "heretical and sodomitical synod."[85]

Here the author includes an indictment not just of Florentine actions in April 1478 but also of later Florentine spin. For too long the Florentines, "schismatic, heretical, excommunicated, anathematized persecutors of holy mother Church, public enemies, and false witnesses," had obscured the truth, which the author would now set out for all Christians to understand.[86] At the conclusion of the text, the author returns to this charge: Lorenzo and his colleagues had tried to destroy the pope's good name and reputation, daring even to call him a pimp and "the Devil's vicar." The author does not mention their use of the press, explicitly, but he certainly implies as much when he decries Florentine insults against the Church, "conveyed not just privately, but before the entire Christian world."[87]

With these arguments, the author addresses the claims of both Scala and Becchi head-on. Not shy about introducing himself into the narrative with first-person constructions, the author addresses Lorenzo directly and apostrophizes the horrors of Florentine policy. He sometimes directs his speech to God or to Sixtus, whose legitimacy he takes pains to underscore. He alternates between berating Lorenzo and castigating the Florentines in general. Again and again, he rebuts and rejects the charges of the Florentine "synod," drawing on arguments and documentation that can only have come from inside the papal curia. So the author presents himself as a spokesman for the papacy and his text as something like an official, papal response to the *Florentina synodus*. Nevertheless, it is a text composed not by the papal "we" but by a particular author, working beyond the formal constraints of a bull or brief. The first-person voice *seems* to present the opinion of an outside observer, albeit one who views the entire affair through a lens of ferocious partisanship. Combining these perspectives into a single document was no easy task, and the text does not quite succeed. It is too vehement and accusatory to sound like a formal condemnation of the *Florentina synodus*. But since the author does not identify himself, it would be hard for a reader to know how much credit to give his narrative of events or his interpretation of the "synodal" text. Still, the *Dissentio* represents an innovation in papal communications: here is a text that appeals directly to the broader reading public. Its publication is further proof of its reach: it was printed not in Rome, or even Naples, but in Basel, by the newly established printer Johannes Amerbach.[88]

Amerbach had only just started his career as a printer in Basel in 1478. The *Dissentio* was among the very first books he printed. Late in his life he recalled that he had been to Rome once, but we do not know when. In September 1477, just before he set up shop in Basel, he testified in a court case between two printers in Perugia.[89] It is not clear how Amerbach got hold of this pro-papal text and printed it in Basel, but these Italian connections may provide some clue.

At some point in August 1478, Sixtus himself composed another open letter to the Christian princes, justifying his military operations against Florence and again blaming Lorenzo for the entire dispute.[90] Sigismondo dei Conti describes how the pope was troubled by the "calumnies" that were being broadcast by his enemies and decided to write *ad omnes* to rebut them. In this letter, Sixtus acknowledged that he was the one escalating the conflict to open war, and he knew he would attract criticism for doing so. But he had his reasons: "We shall review the history of this affair and we shall explain why waging this war is necessary if we want Italy to enjoy peace, and why Italy cannot enjoy peace unless we wage this war."[91]

Sixtus then rehearsed the long story of Lorenzo's interference in the papal states and excoriated the Florentines once again for their harsh justice against Salviati and the young cardinal. Lorenzo might have paused after the attack on his life, to wonder whether he had brought it on himself by offending God in some way, but instead of showing penitence or humility, he had reacted with anger and violence; he had even disseminated *foedissima scripta* that had insulted the pope's honor and his good name.[92] Lorenzo's rage threatened the peace of Italy and might even open the peninsula to an attack by the Turks. It was for this reason, Sixtus told the Christian princes, that he had had to go to war. The pope closed by appealing to the princes for aid and support. With their help, Lorenzo could be defeated and the *impurissima pestis* of the Turks could finally be destroyed.[93] So, a repetition of by-now-familiar charges: Lorenzo had offended the pope, compounded the offense with his mendacious propaganda, and left Italy vulnerable to Ottoman attack.[94]

The Pazzi War dragged on for more than a year, with costs mounting on both sides. In the spring of 1479, Neapolitan troops captured some important Florentine forts, and anxiety mounted around the question of Easter and whether the clergy would offer the faithful their annual communion or not. All along, the Florentine clergy had continued to provide the sacraments in defiance of the interdict,[95] but to do so on Easter would be a high-stakes protest, and there were doubts that the clergy would continue to ignore the ban. Public opinion in Florence began to turn against the Medici. In March, libels critical of Lorenzo were posted on street corners in the city, and later in the year some hostile notes were found on the door of the Venetian ambassador.[96] In Rome, meanwhile, Sixtus was facing pressure from critics of his own. Since the start of the war, ambassadors from the northern powers (France, the Holy Roman Empire, Burgundy, even England) had come to Rome to urge the pope to make peace with Lorenzo and proceed with the crusade. Responding to their demands, and seeking once again to bring the Florentines to heel, the pope issued two more bulls in August 1479, both of which circulated in print.

The first took aim at Florence's military support; the pope excommunicated their *condottiere* Roberto Malatesta and three other captains in the pay of Florence.[97] Sixtus reviewed the honors, titles, lands, and benefits that he had bestowed on the men, all of them holders of papal fiefs or bound by papal *condotte*. Malatesta had sworn in church never to accept a foreign commission without the pope's permission, and never to fight against the Church, and yet here he was bearing arms against his rightful lord on behalf of the excommunicate, Lorenzo. The bull rehearsed Sixtus's complaints against the Florentines once more and reviewed the terms of the bull of June 22, 1478, that forbade lending the Florentines military assistance or entering their service. The *condottieri* had violated these strictures; now the pope denounced them as rebels, perjurors, and violators of oaths and condemned them for *lese majesté.* Their titles were void, their fiefs reverted to the Church, and they could have access to neither priest nor sacrament.

The second bull, dated the next day, was addressed to the Florentines themselves, though it spoke to a wider readership. Over the previous year, the pope explained, the Christian princes had urged him to make peace with Florence and lead a new crusade. The pope, eager to reconcile with Lorenzo and face the Turkish threat,[98] had sent conciliatory messages to the Florentines, hoping they would repent of their crimes and make the reparations he required. Still they refused, and accordingly, with this bull he renewed the interdict and all its sanctions against the citizens of Florence and any who gave them aid.

Both bulls concluded with telling instructions for their publication. In the first bull, against the *condottieri*, the pope acknowledged that "it would be difficult" to publish the sentence of excommunication in those cities and territories that the *condottieri* had once ruled, *propter illorum potentiam*, but it was important that their subjects should know that they were excommunicate and deposed. So that none could pretend ignorance, the bull would be posted on the doors of St. Peter's and the chancery in Rome, and every church and convent in the territories surrounding their lands should likewise post a copy; clerics in the vicinity were also instructed to preach on the contents of the bull.[99] In the second bull, against Florence, Sixtus again noted that "it would be difficult" to publish the decree in Tuscany, which was now a war zone; therefore, he ordered it to be posted on the doors of St. Peter's and the principal churches of Perugia, Bologna, and Siena: posting them there would provide sufficient notice to the Florentines, as clear as if the bulls had been delivered to them in person.[100]

Both 1479 bulls were printed as quarto pamphlets by a Roman printer, Georg Lauer, who had been at work in the city since the early 1470s.[101] In light of the obstacles to their delivery that the bulls themselves acknowledged, printing them could well be seen as another step in the process of making the sentence public, as a way of ensuring its validity. But the rhetoric of the documents, with their outraged review of the pope's past generosity, Lorenzo's offenses, the captains' rebellious ingratitude, the Florentines' obstinance, and the universal delay in launching the crusade, underscores that the bulls were propagandistic as well as legal in nature.

In December 1479, Lorenzo made his secret trip to Naples, where he made peace with Ferrante and effectively brought the war to an end. Sixtus agreed to a general peace in March 1480, although the interdict remained in place. The Turkish invasion of Italy and the capture of Otranto in August 1480 finally forced a resolution. Sixtus had been insisting that Lorenzo come to Rome and beg forgiveness in person, but now he agreed that an embassy of less prominent Florentines could perform the task instead.[102] On December 3, 1480, twelve Florentine ambassadors knelt before Sixtus on a stage set up before St. Peter's and confessed their guilt; the pope tapped each man lightly on the shoulder with a ceremonial staff, and the Pazzi War came to a close.

The variety of material put into print in the papacy's first printed pamphlet war is remarkable. On the papal side, the discourse expanded gradually, from the publication of three papal bulls of condemnation in June 1478, to Grifo's more persuasive briefs in July and August, to the wild vituperations of the anonymous *Dissentio* in September. All drew on the legal arguments that underpinned the original bull of excommunication: the Florentines had committed sacrilege against the clergy and deserved the highest censures. Along the way, papal writers also turned to the archives, drawing on earlier documents to highlight Lorenzo's interference in the politics of the papal states in the 1470s, and used rhetorical appeals to try to isolate Lorenzo from his fellow Florentines and allies in Italy who might lend him support. Print amplified these messages and ensured their broad circulation, as the Florentines themselves complained: papal charges against them were circulating not just on church doors but in the marketplace as well.

From the start, the Florentine responses to these printed salvos were more diverse. They ranged from the canon law reasoning of Becchi's fictive synod, to the diplomatic rhetoric of Scala's tract, to Poliziano's high-culture appropriation of Sallust, to the dramatic intonations of the vernacular lament. Varied as they were, they were still tightly coordinated. Lorenzo's jurists had worked out the basic line of attack: the Florentines had the right to defend themselves, and the pope had refused them due process. Both Scala and Becchi rang the changes on this set of claims while quoting from older papal briefs, Cardinal Raffaele's letter to the pope, and the Italian transcript of Montesecco's confession. Clearly there was a plan, guided by Lorenzo and Michelozzo in consultation with jurists and friendly diplomats, and a common stock of documents from which Scala and Becchi drew. By contrast, the compositions by Poliziano and the anonymous author of the *Tradimento* show little sign of being directed or coordinated by Lorenzo or his staff. They suggest that the Pazzi attacks had caused genuine outrage in Florence—a sentiment those authors shared and which they hoped to share with their readers.

Both sides took note of the others' pronouncements; as the conflict wore on, their publications swelled with complaints about both the original offenses of April 1478 and the calumnies, distortions, and lies that were spread in the meantime. Both parties also tried to reach out to a readership beyond their borders. Sixtus saw his letters printed in Naples and Basel, while the Florentines reached into Rome itself; in 1478, Poliziano's tract was printed by Johannes Bulle, one of the chancery's printers of choice for bulls and other publications. This seems an extraordinary coup, although by the time it was published, Rome and Florence were allied again, and the fact that Poliziano's account laid no blame at the feet of the pope may have made it an easier text to market in Rome.

The War of Ferrara

The Pazzi War might have continued but for the Turkish capture of Otranto in Puglia in July 1480, which threw all the Italian powers into a temporary truce. By January 1481, however, Girolamo Riario was again agitating for war, and Sixtus's recently reconciled allies—Milan, Florence, and Naples—observed with growing alarm that the pope supported his nephew's ambitions. By 1482, with Mehmed II dead and Turkish troops long gone from southern Italy, hostilities were brewing again. A new conflict, the War of Ferrara, brought Sixtus into a dispute involving Venice, Ferrara, Milan, Florence, and Naples; controversially, the pope changed sides halfway through.

Like the Pazzi War, the War of Ferrara featured a furious exchange of pamphlets issued by the pope and his opponents, printed in Rome, Venice, and the Veneto and even farther afield. The conflict also saw further developments in the rhetoric and presentation of political print. Whereas the Florentines in 1478–79 published legal documents in support of their cause (e.g., Montesecco's confession), in 1482–83 the Venetians went a step further and published the pope's own words against him, packaging papal briefs together with their own ripostes, some of them composed just for the press and never sent to the pope in person. They used print to create entirely new documentary collections in which they gave themselves the last word.

The War of Ferrara was the project of Girolamo Riario, who tried to use papal armies to secure his hold on the city of Forlì, which he had seized from the Ordelaffi family at the height of the Otranto crisis. If the war went his way, he meant to use that base in Romagna to strike at Ferrante of Naples.[103] Sixtus, irritated with Ferrante over the private truce he had negotiated with Lorenzo, was receptive to Girolamo's idea of consolidating papal power in the north. The first step would be to secure Romagna, for which Venetian cooperation would be necessary. Ferrara, as Pastor puts it, was "held out as bait."[104]

Riario traveled to Venice in September 1481 to persuade the republic to join a league against Ferrara and Naples. The following spring, the Venetians launched an attack on Ferrara, advancing infantry and a river fleet up the Brenta. In response, Ferrante's son, Alfonso, whose sister was married to Duke Ercole d'Este, marched north to her rescue. Three years earlier, Alfonso had led troops for the pope against Florence. Now, serving his own family, he advanced directly on Rome, pitching camp beneath the city walls while his troops rampaged through the countryside.[105] In July, Venetian troops under Roberto Malatesta (the *condottiere* whom Sixtus had excommunicated three years before, for leading Florentine troops against Rome) moved south to relieve the city. Venetian and papal troops together routed the Neapolitan army at Campo Morto, near Anzio, on August 21, 1482. But Malatesta soon died of fever, and the Venetian forces withdrew north again to concentrate on Ferrara. As the autumn wore on, Sixtus began to worry that Alfonso would return to ravage the Roman countryside again and that the Venetians, now closing in on Ferrara, would neglect their alliance with Rome. Now it was the pope's turn to make a separate peace. He negotiated a secret truce with Ferrante and his allies. On November 28, 1482, the terms of a pact were agreed: Rome, Naples, Milan, and Florence would together keep the peace with Ferrara; Ercole d'Este was confirmed in possession of his state; and the territory he had lost to Venice during the war would be restored to him. Somewhat disingenuously, Sixtus included a line in the treaty to the effect that the Venetians were very welcome to join this league and indeed would be wise to do so. The parties had fifteen days to consider the terms. When no objections came, the pact was ratified on December 12, 1482, and Cardinal Francesco Gonzaga was sent to Ferrara with a brief for Ercole outlining the new state of affairs.

Sixtus, having left his Venetian allies in the lurch, chose to promote his actions boldly. On November 28, the terms of the peace were publicly announced in Rome by heralds with trumpets.[106] On December 12, the pope himself led a solemn procession from St. Peter's to the church of Santa Maria della Virtù, near piazza Navona, which he rededicated as Santa Maria della

Pace in honor of the pact. The news was publicly announced in Ferrara on December 13. Back in Rome, on Christmas Eve, the entire text of the treaty was read aloud at the conclusion of the vigil mass at St. Peter's.[107]

The pope also turned to the press. Eucharius Silber, a German printer who had only recently set up shop in Rome but would have a long career printing papal documents, published a copy of the peace treaty of December 12, listing Sixtus, Ferrante, Gian Galeazzo Sforza of Milan, the Florentine Republic, and Ercole d'Este as cosignatories to the pact and including the invitation to the Venetians to join if they wanted to.[108] Printing the document served no legal purpose, for the pact was already signed; its terms had been circulated in manuscript to European courts and read aloud in Rome and Ferrara. Rather, printing publicized the new diplomatic alignment of the Italian powers more widely, putting pressure on the Venetians to accept the reality of the pope's *volte face* and relinquish their claims on Ferrara.

Sixtus may well have authorized publication of the treaty in print, but he did not abandon quieter and softer means of communication with his former ally. The Venetians had been surprised and distressed by his decision.[109] In late December, the pope sent a team of diplomats to Venice to justify his actions to the republic and persuade them to join the peace. His overtures and the Venetian response to them both found their way into print, but not by Sixtus's choice. This time, it was the Venetians who seized the initiative, using the press to frame the issue to their own advantage.

The story of the papal-Venetian correspondence of 1482–83 and its later *fortuna* in print unfolds, in part, as a story of traditional diplomatic communication in the late fifteenth century, with its reliance on rhetorical performance, face-to-face negotiations, and the exchange of handwritten communiqués, some secret and some intended for wider circulation. In addition, both sides used the press as another tool for applying diplomatic pressure. As in the Pazzi War, Rome seems to have used print as a relatively transparent medium, reproducing the texts of documents already drawn up and circulated in manuscript, but the Venetians, like the Florentines before them, took a more creative approach, turning Sixtus's words against him and inventing new texts in an attempt to further control the debate.

In mid-December 1482, Sixtus had two briefs drawn up, both drafted by his secretary Leonardo Grifo, who had also drafted letters for him during the Pazzi War. The first, dated a day before the treaty (i.e., December 11), was

written in Sixtus's name and addressed to the Venetian doge Giovanni Mocenigo; the second, dated December 16, was written in the name of the entire
College of Cardinals and addressed to the Venetian Senate. In each, the pope
urged the Venetians to give up their war, join the peace, or face ecclesiastical
penalties for disobedience and warmongering.[110] Sixtus gave the briefs to another secretary, Sigismondo dei Conti, who traveled to Venice to deliver the
letters in person and underscore their message. The Venetians did not receive him warmly. In his memoirs, Sigismondo described the difficulty of his
mission:

> The pope had embraced peace but had not informed the Venetians before
> doing so. He decided he ought to communicate his decision to them, and he
> also wanted them to excuse him for it. . . . I set out for Venice, but I did not
> find a single soul, neither friend nor stranger, who dared to speak to me. Over-
> night the pope had gone from ally and friend to enemy, and he was held in the
> greatest contempt by all. Still, the Senate received me on the second day after I
> applied for an audience, and I was granted the opportunity to deliver the briefs
> and say what I would.[111]

Sigismondo had a daunting task: to excuse the pope's behavior to an audience
convinced it was inexcusable. In his account of this awkward enterprise,
Sigismondo reveals how each side relied on both written texts and oral performance. At his audience, he delivered his written credentials, read out the
text of his two briefs, presented them to the doge, and finally delivered an
oration of his own urging the Venetians to join the peace and leave Ferrara
alone. The two sides should put aside their differences so as to present a
united front against the Turks.[112]

Sigismondo admits that these remarks did not go over well. The doge
replied that there was no state or kingdom in the world more devoted to the
defense of the faith than the Venetian Republic. The Venetians had drawn
more Turkish blood than most in their many battles on land and sea (a fair
point, coming close on the end of Venice's sixteen-year-long war with the Ottomans). Moreover, the Venetians were a peace-loving people who had only
gone to war with Ferrara at the pope's urging: the entire campaign had been
the pope's idea! Now, after the loss of so much blood and treasure, it hardly
seemed fair to ask them to lay down their weapons when victory was almost
in their grasp. Nevertheless, the doge would consult with his councils and
draft a reply.

The task of composing this reply fell to the patrician Bernardo Giustiniani, who was instructed to explain the actions Venice had taken in its war against Ferrara and to show that the republic had been justified in pursuing it. While Giustiniani worked on his reply behind closed doors, Sigismondo continued face-to-face negotiations with the doge's secretary, Alvise Manenti, who tried to influence how Sigismondo should present the Venetian reply to the pope. Manenti gave Conti a complex set of instructions, veering between pleas and threats:

> [He asked that] I should soften the pope's heart, and setting out the many
> dangers of the situation, he urged me to dissuade the pope from declaring war
> or imposing censures, and to advise and beg him to use his good sense and
> loyalty, nor to let himself be persuaded by the advice of others to attempt any
> hostile action against the Venetians, who were eternally loyal to the Apostolic
> See; if he did otherwise, he would cast Italy into a perilous war whose end he
> would not see. . . . . All these things I faithfully reported back to Sixtus along
> with the letters.[113]

The two letters Giustiniani composed served as the official Venetian reply to the two briefs Sixtus had sent. Both were written in the name of Doge Mocenigo, one addressed to the pope and the other to the cardinals. In both texts, Giustiniani rejected the idea of peace and argued the justice of Venetian complaints against Ferrara and the justice of the war that Venice and Rome had undertaken together.[114]

So the exchange between Rome and Venice played out on two levels, written and oral, with official (but still, essentially private) letters drafted by high-level officials (Grifo and Giustiniani) providing the material around which face-to-face exchanges then took place—formal parleys like that between Sigismondo and the doge, and informal talks like those between Sigismondo and Manenti. Written texts, like Sigismondo's credentials, sought to set the climate in which oral remarks would be received, while oral performances reinforced the message of texts already written and influenced the tenor of texts yet to be composed. All of this happened in the relatively closed environment of the council chamber, preserved in the memoirs of elite observers who had access to those spaces. Their accounts make clear that the character of these communications, while official, was still private: whether the protagonists spoke to protest past actions, to excuse or justify the current state of affairs, or to promote or forestall certain consequences, their words were directed to the leadership of *another state*, not a wider public.

But they were not entirely confidential. As Sixtus sent Sigismondo to Venice with his briefs in December 1482, he also kept his new allies and other European states apprised of his maneuvers. Sigismondo himself had stopped in Ferrara on his way to Venice and shared a copy of the briefs with the ducal court.[115] In December and early January, the pope sent manuscript copies of his two briefs to the Venetians, along with further cover letters explaining his new peace policy, to more than twenty Christian princes: Frederick III, the duke of Saxony; the kings of France, Spain, Portugal, Hungary, Poland, Denmark, England, and Scotland; the dukes of Burgundy, Brittany, and Lorraine; Archduke Maximilian of Austria; and the seven imperial electors.[116] In his cover letters, the pope explained that his true goal had always been peace; moreover, by reconciling the Italian powers, he was also laying the ground for a new offensive against the Turks.[117]

Sixtus was bringing enormous pressure to bear on Venice, first by abandoning his alliance with the republic, then by making public peace with enemies they had only recently confronted together, and lastly by insinuating to just about every European court that his actions were a noble initiative aimed at the defeat of the Turks, which the Venetians—obstinately, inexplicably!—refused to join.

The Venetians saw things differently. Perhaps spurred by their sense of isolation, they resorted to the press to say so. Marin Sanudo reports that all four letters—the two papal briefs of December and Giustiniani's two replies—were put into print shortly after Sigismondo left Venice,[118] and indeed two quarto editions of the letters survive, issued by two separate presses in Padua and Venice.[119] Like the Florentine publications around the Pazzi War, these represent one of the first attempts by one state to rebut, via mass-produced text, the political claims of another—only here, instead of dueling statements issued in print by each side of a dispute, the Venetians went further by printing the pope's own words in order to rebut them.

Each of the editions, only six leaves long, was meant to circulate and be read quickly. In the Paduan edition, the Venetian side of the correspondence is attributed to Bernardo Giustiniani, while in the Venetian edition the letters appear without an identifying heading, and they appear in a different order.[120] In both, the Venetian letters come second, appearing after, and thus appearing to rebut, the papal briefs. Moreover, in both editions the entire correspondence is prefaced by an unsigned introductory text (*Argumentum*) that sets out the background to the conflict, including a list of Venetian grievances with Ferrara, an account of Venice's recent alliance with Rome, and the unjustness of the pope's *volte face*. The pamphlets do more than advertise a

diplomatic overture already made: they take control of the narrative by framing the entire debate in terms favorable to Venice.

The *Argumentum* is a very different text from any of the letters, Roman or
Venetian. The letters, written in the elevated, eloquent style of their humanist authors, assume a certain amount of prior knowledge of the quarrel. The
Argumentum, by contrast, speaks directly to a new audience, that of the reading public, still in Latin but in a much simpler style:

> The matter at hand: Ercole, duke of Ferrara, established in his duchy by Vene-
> tian arms, afterwards violates their ancient immunities. He enters an alliance
> with Ferrante king of Naples, the duke of Milan, and the Florentine republic,
> which, under the terms of his pact with Venice, he is not allowed to do. The
> Venetians demand their rights. He turns his back on them. Pope Sixtus IV,
> having abandoned his pact with Ferrante, allies himself with Venice. Ferrante
> is outraged; to draw the pope away, he attacks him with violence. The pope
> encourages the Venetians to declare war on Ferrara. They declare and attack
> by land and by the River Po.[121]

So it continues in its paratactic simplicity. Toward the end, the narrative
shifts from old campaigns to more recent history, including the origins of the
texts the reader is about to encounter:

> Then the pope writes to the Venetians and encourages them to join the peace
> of all Italy: they should put down their arms, give back what they have cap-
> tured, and raise the siege of Ferrara. The college of cardinals writes much the
> same thing. The Venetians—as fits the dignity of the pope and the indignity
> of their broken alliance—give the task of responding to Bernardo Giustiniani,
> soldier, ambassador, procurator of San Marco. He, being very eloquent and
> wise and also at that time president of the senate, is judged by all to be the best
> man for the job.[122]

This is followed immediately by the text of Sixtus's letters and then the
Venetian ones. The shift here is worth remarking on: from the printed reproduction of a single document to the creation of an entirely new compilation,
one that interleaves documents by different authors into an artificial dialogue
and frames the whole with introductory words aimed at the reader.

Rome followed suit. A month later, on February 13, Sixtus wrote yet again
to the Venetian doge, composing a longer brief, almost a short tract, setting
out his reasons for wanting the Venetians to join his newly formed league.
This was sent to Venice, but in addition, and very likely in response to the

Venetian publication campaign, it was also printed and made ready for distribution at a Roman press, possibly that of Stephan Plannck, another German recently arrived in the city who would have a long career printing materials for the papal court.[123]

The next month, Giustiniani composed a reply, dated March 15, which was sent to Rome. This letter was somewhat conciliatory: though Giustiniani protested the pope's harsh treatment of Venice, he reiterated that the republic had always been, and would remain, loyal to the Church. The Venetians trusted in God, who knew the truth of the situation, and hoped for a peaceful resolution to the crisis. This letter was not printed, for reasons I explore below.[124] Nor did it succeed in diverting Sixtus from his plan, for on May 23 the pope put the city under interdict. The bull was posted by heralds on the doors of St. Peter's the following day.[125] Once again, Sixtus copied the Christian princes: manuscript copies of the bull were sent to the emperor; the kings of France, Spain, Hungary, England, and Portugal; the electors; and others, with a request that they publish the bull in their own domains.[126] He had a copy of this text printed too: a folio volume of ten leaves by Johannes Philippus de Lignamine.[127] The text of the bull was both a legal document and a rhetorical appeal. The edition showed some attempt at reproducing the graphical elements of the original bull, including the subscriptions of the scribe and expeditor and a note about its promulgation "in valvis apostolorum" and the heralds who had seen to it. But printing the bull in Rome masked a weakness in Sixtus's position: he had been unable to get the actual bull of excommunication into Venetian hands.

Venice had already recalled its ambassador from Rome, and the agent staffing the embassy in his absence refused to accept the document. Sixtus then sent a sergeant-at-arms (*mazzier*) to deliver a copy to the patriarch in Venice. But the patriarch feigned illness and refused to receive the *mazzier*. Shortly afterward, the Venetians appealed the sentence to a future council, and they managed to get *their* appeal posted on the doors of the church of San Celso in Rome, a stunt that caused a minor sensation in Rome.[128] The asymmetry of the campaigns was galling: the pope, supreme head of the Church, was barred from publishing his documents in the Veneto or even getting them into the hands of his prelates, while the Venetians freely infiltrated the pope's own neighborhood to lay their countercharge at his feet.[129]

The Venetians took their arguments to the press again. Their moves must be reconstructed with guesswork. Sixtus had sent the Venetians his two manuscript briefs in December 1482; in January 1483, Giustiniani had composed two letters in reply. The whole correspondence then appeared in print in the Veneto, along with a new, tendentious preface, in two editions that can be traced to Padua and Venice. Back in Rome, Sixtus issued a third brief on February 13, which was sent to Venice in manuscript and also printed in Rome by Plannck. In reply to this, Giustiniani sent another, somewhat conciliatory letter on March 15, seeking to persuade the pope to step back from his threats of excommunication. But this letter was not printed in Venice or anywhere else.

Instead, someone, perhaps Giustiniani, composed another letter: a different, much sharper reply to the pope, unsigned and undated, which began to circulate with the rest of the collection of letters.[130] This letter, defiant in tone and bristling with threats, all but invited the pope to do his worst: "we will repel force with force, we will outspend your every expense, we will repay atrocity with atrocities if we have no other choice."

This (partly fictional) correspondence, now consisting of three letters from each side, was likely printed again, sometime in 1483, somewhere in the Veneto. There is no record of such an edition, but in 1484 an edition of all six letters (the three papal briefs with their three Venetian rebuttals, the first five of these genuine, the last never sent to Rome) and the tendentious *Argumentum* was printed in Westminster by William Caxton.[131]

The letters were edited by Pietro Carmeliano, a Venetian citizen resident in England at the time. We can imagine how the collection came to be: after Sixtus's third brief was sent to Venice (and printed in Rome), Giustiniani sent his official reply of March 15 to Rome through private channels, while this other letter, masquerading as an official Venetian reply but much more combative and triumphant in tone, was composed for public circulation. It was appended to Sixtus's brief of February 13, making it look like this was how the Venetians had answered the pope. This pair of letters was then added to the earlier collection of four letters (with the *Argumentum* preface), and the whole was printed in a single edition. It is possible that all six letters were put together in a manuscript that then made its way to England, or even that Carmeliano composed the sixth letter and put the edition together himself, but it seems more likely that an edition was printed in Venice that is now lost. A copy of this edition would have traveled to England, perhaps sent directly to Carmeliano, who then oversaw its republication by Caxton.

Carmeliano was an itinerant humanist scholar, a native of Brescia who, after a peripatetic early career, landed in England in 1482 and found work in the English civil service.[132] Hoping for royal patronage, he prepared elaborate presentation manuscripts of his Latin poetry and dedicated them to various figures at the court of Edward IV. In 1482, he presented the king himself with a printed copy of Cicero's *De oratore* in an edition printed at Venice in 1478, to which Carmeliano added a dedicatory poem in manuscript and additional marginal comments.[133] So Carmeliano could appropriate products of the Venetian book trade for his own purposes. In the case of the Venetian-papal letters, however, it is unlikely that Carmeliano took on the project as a way to win English patrons. More likely he acted as a Venetian agent, securing Caxton's services in order to get the Venetian version of events bruited abroad at a moment, in early 1484, when nearly every Italian power had aligned against the republic.[134] In later years Carmeliano worked quite openly as a Venetian spy in England. Marin Sanudo records that his letters on English affairs were frequently read out in the Venetian Senate.[135] In 1490, he offered similar information to the duke of Milan. Very likely he was acting on behalf of his native republic in 1484. If so, then the reach of the Venetians seems once again to have exceeded that of the pope.

The Caxton edition offered its readers not only a political argument but also a model for rhetorical imitation. The title of the edition ("Six very elegant letters") presents the correspondence as an example of fine Italian humanist prose. The brief poem Carmeliano adds at the end of the sixth letter invites the reader to enjoy its Ciceronian eloquence;[136] he also provides a short table explaining the abbreviations used by humanist chancery scribes for the names and titles of Italian princes and the stock formulas they employed in such documents.[137] The edition may have been intended—and read— as a formulary of high-level, humanist diplomatic correspondence.

Carmeliano's treatment of the text recalls that of other epistolary collections, like those of Aeneas Sylvius Piccolomini, Francesco Filelfo, and, indeed, Bernardo Giustiniani, which circulated in both manuscript and print in the fifteenth century, admired as much for their rhetorical quality as for their political or historical content. The Caxton edition, with its back-and-forth between pope and doge, resembles the fictitious collections of letters that fifteenth-century readers also enjoyed—for example, the one assembled by the Italian Laudivius Zacchia that circulated under the title *Epistolae Magni Turci*. In that collection, printed some nineteen times across Europe between 1473 and 1500, the Great Turk, rather like the pope in this one, issues a series of menacing messages to various Christian powers around the Mediterranean, each of whom (rather like "Doge Mocenigo" in Carmeliano's edition) then replies with a rejoinder that neutralizes or even undermines the sultan's position.[138] Zacchia's collection capitalized on a contemporary political crisis to offer examples of rhetorical prowess for the entertainment of the reader; Carmeliano's appropriation of the Venetian-papal correspondence may represent the same impulse, even as it was also intended to win support for Venetian policy at the English court.[139] The possibility of multiple motives for its publication underscores the multiplicity of ways that political texts could be read, repackaged, and marketed in this period.

Carmeliano's epistolary compilation had a distant editorial counterpart in Rome. In 1482, the Veronese jurist Johannes Nicolaus Faela composed a flattering history of the Maffei family of Rome, which he had printed at least twice in the city. Both editions appeared with an appendix of letters written by the apostolic abbreviator Benedetto Maffei, one of several Maffei who had carved out a successful career in the curia.[140] Maffei's second letter, to Cardinal Oliviero Carafa and dated the same day as the publication of the papal peace treaty of December 1482, is written *de laudibus pacis*. The third letter, dated August 13, 1483, is addressed to Antonello Petrucci, one of Ferrante's secretaries in Naples, and includes an encomium of another Neapolitan official, Aniello Arcamone, who had been Ferrante's ambassador in Rome during the War of Ferrara. Maffei identifies Arcamone as the key figure in the negotiations over the peace between Naples and Rome in the autumn of 1482.[141] The letters offer yet another example of political commentary put to the service of clientelistic projects. Maffei praises Arcamone as the architect of the peace of December 1482 and sent copies of this letter not only to Arcamone but also to Petrucci. Appearing in print, the letter illustrated Maffei's literary talents in the context of a panegyrical history of his family, which *its* author, Faela, dedicated to the pope's nephew, Girolamo Riario. The main purpose was surely to flatter all concerned, even as it also broadcast another message of support and praise for the peace of 1482. Carmeliano and Maffei's efforts remind us that the humanist commentariat examined in chapter two was still hard at work in the 1480s: in Rome and in Venice, in Naples and in distant Westminster, men of letters continued to publish texts on current events as a way of elevating their own profiles, serving their diplomatic masters as well as themselves.

Sixtus IV and his secretaries used the press to promote the temporal powers of the Church and assert the ecclesiastical supremacy of Rome. His rivals used the press to push back publicly against these assertions, rebutting, questioning, and even mocking papal claims to authority before a wide reading public. In these efforts we see the seeds of future papal practice in the high Renaissance—the use of bulls as a political weapon, intended to intimidate and embarrass their targets not only before God but also in the court of public opinion. The pope's press campaigns were still limited to the publication of official documents—bulls, briefs, and treaties—which often delivered or published a legal sentence. The point was to advertise that the declaration had been made; the gravity and formality of the documents impressed on the public the severity of a sentence already imposed at Rome. In the nimbler and arguably more persuasive responses of the Florentines and the Venetians, we have a foretaste of future papal failures, when opponents like Louis XII of France, Emperor Maximilian, or indeed Martin Luther would follow their Italian predecessors in finding creative and embarrassing ways to parry the thrust of the papal bull.

CHAPTER FOUR

Broadsides in Basel

If Sixtus's conflicts with Florence and Venice prompted the papacy's first pamphlet wars, what followed was its first-ever *affaire des placards*: a controversy waged in printed broadsides that were posted in public, passed from hand to hand, and circulated across southern Germany. Archbishop Andrija Jamometić's attempt to reconvene the Council of Basel in the summer of 1482 was seen by contemporaries as a major crisis in the papacy of Sixtus IV. In his printed pronouncements, the archbishop revived the conciliar claim that the power of the pope could be limited by an assembly of prelates representing the universal church. He also called some of Sixtus's signature policies into question: his nepotism, his extravagant spending on his court and the city of Rome, his financing of those enterprises via the indulgence trade, his interference in Italian politics, his willingness to go to war in Italy, and his failure to pursue the crusade against the Turks.

Jamometić's choice of Basel as the place to launch his attack put him not only at the heart of the old conciliar movement but also at the crossroads of Europe, in a welter of overlapping jurisdictions. Multiple authorities were drawn into his dispute with the pope, and most of them followed Jamometić in using print to articulate their positions. The result was a furious exchange of broadsides and pamphlets issued by presses in upper Germany and in Rome. Serious questions of ecclesiastical law and policy hung in the balance. But the exchanges also reveal how fraught the very idea of printed publication still was in these early years, as various parties debated who had the right to publish ecclesiastical decrees and what force and value they had when mechanically reproduced. In the end, it was not at all clear that there was a public interested in or ready to receive the publications these authorities issued.

On March 25, 1482, Andrija Jamometić, a Croatian-born, Italian-educated Dominican, arose after mass in Basel's cathedral and issued a call for a new church council. A university-trained theologian and diplomat, Jamometić was no crank, but his rhetoric was incendiary. He accused Sixtus IV of heresy, simony, murder, embezzlement, warmongering, defrauding the faithful, and conniving with the Turks. For these and other crimes, he summoned the pope to appear before a new Council of Basel that would soon convene in the Swiss city to judge the bishop of Rome and, if necessary, depose him.[1]

Within weeks, the text of his appeal had been printed and was circulating not only in Basel but also throughout southern Germany. It prompted at least a dozen printed rebuttals from the pope and his agents, including briefs and bulls of excommunication and interdict, while further texts were published by Jamometić and his allies, the Basel city council, and Emperor Frederick III. The controversy drew the interest of Lorenzo de' Medici, still smarting over the Pazzi affair, who sent emissaries to Basel to support Jamometić and intensify the embarrassment he was causing the pope.[2] It also unfolded in the same months as the War of Ferrara; contemporary chroniclers linked the two episodes as twin threats to papal fortunes in the winter of 1482–83.[3] But the printed exchanges that defined the Basel affair lasted longer than those associated with either the Pazzi War or the War of Ferrara, and the issues at stake were more complex.

Instead of two sides, at least four different parties were involved: Jamometić, the pope and his legates, the emperor, and the civic government of Basel. Moreover, the papal legation sent to silence the archbishop was rent with tensions, as different legates pursued different agendas, often on conflicting instructions from Sixtus himself. A mixed group of Franciscans, Dominicans, and diocesan clergy, some sent from Rome, others long resident in southern Germany and deeply invested in the local indulgence trade, at times they even used print against each other.

Moreover, the dispute unfolded in a city where printed broadsides were already a common medium for publicizing decrees. Papal agents quickly adopted these local practices, which were only just beginning to be used in Rome or elsewhere in Italy, but the new medium presented unexpected challenges. The papal legate Antonio Geraldini was astonished by the proliferation of printed texts in the affair; at times he despaired of ever controlling the flow of publications, even those coming from his own side. And despite their impressive volume, the status of many of these documents was unclear.

For example, Roman chroniclers knew in July 1482 that Sixtus had stripped Jamometić of his office and issued a bull of excommunication against him, but when Geraldini brought a copy of the bull to Swiss territory in September, his local counterparts ignored his requests to publish it; the bull was not fully promulgated till early November. Another papal commissioner, Peter von Kettenheim, placed Basel under interdict in September 1482, but by the end of December other clerics were printing notices that the interdict had been raised; furious arguments ensued over the next few months, as no one could agree on whether the ban was in force or not. Geraldini accused his fellow legates of forging and printing false notices of revocation, leaving him powerless in his negotiations with the city and allowing local clergy to ignore his decrees. In February 1483, Geraldini had another papal bull in his pocket, this one declaring a crusade against the city because it continued to harbor the heretical archbishop in defiance of papal commands. Geraldini told the pope he intended to publish this bull in phases, circulating a few printed copies in advance of an imperial conference but reserving the rest in case the meeting went well and the "crusade" could be called off. (The meeting did not go well, and Geraldini promptly released his stack of printed bulls.) Meanwhile the Basel city council printed statements protesting the pope's actions, which Geraldini was powerless to stop; the city paid armed guards to escort the papers through the streets and stand watch at their posting places, he said, and they were taken down in the night lest anyone try to destroy them.

These episodes of partial or incomplete publication, deliberate forgery, and conflicting legislation suggest how slippery the use of print as a means of official communication could be, even in a city used to its conventions. In addition, material evidence from the broadsides points to continuing anxiety over issues of the authority, authentication, and distribution of political texts in the Renaissance city—anxieties that were especially acute on the papal side, as the surviving documents with their wax seals, manuscript signatures, and notarial authentications suggest.

Finally, while Sixtus's disputes with Florence and Venice were primarily diplomatic, with print playing at best an adjunct role to negotiations going on behind closed doors, the conflict in Basel was largely a conflict *about* printing or, at least, a conflict about publication. The controversy began with Jamometić's public call for a new council, which he read aloud in the cathedral of Basel, had printed as a broadside, posted publicly in the city, and then recast as an open letter to the pope that circulated widely in manuscript (and possibly print) north and south of the Alps. As the conflict dragged on, drawing in different protagonists from Rome to Vienna, it continued to be prosecuted in print. Broadsides and pamphlets arguing both for and against the archbishop came off presses in Basel, Strassburg, Mainz, Nuremberg, and Rome. Although ambassadors and government officials spent many hours in private negotiations, each side was quick to use print to broadcast its case and secure support for its positions. Some texts were printed solely to counter previous publications by the other side. All told, a dozen different texts relating to the controversy were printed in at least twenty distinct editions, and there are archival references to still more texts printed in editions now lost.[4]

The participants in the dispute were acutely aware that printing had amplified the crisis; in many ways, print *was* the problem, even as it also offered means for achieving its resolution. The pope mentioned Jamometić's printing activities in his (printed) bull of excommunication, and Jamometić included the printing and posting of libels among the crimes he confessed to in a formal statement issued after his arrest—a document the city council then had printed and posted throughout Basel in hopes of defusing the papal sanctions still hanging over the city. The arguments advanced in these publications often turned on questions of publication—who had the right to publish, the legal standing of particular published texts, the risk of scandal to the faithful that publication posed, and the vexing problem of decrees that were printed, posted, and legally published but that the public chose to ignore.

Andrija Jamometić, educated in Padua and Florence,[5] rose quickly through the ranks of the Dominican Order, and in the 1470s he represented Frederick III on several embassies to Rome. In 1476, the emperor secured an archbishopric for him, the remote see of Krajina on the border between Albania and Macedonia. In Rome, Jamometić helped maintain the emperor's neutrality during the Pazzi War, and in return the emperor lobbied for a cardinal's hat for his envoy. Jamometić believed that Sixtus had agreed to an elevation: he would later style himself the "cardinal of San Sisto," in protest at what he called a broken promise on the part of the pope.

Feeling the sting of this disappointment, on his last embassy to Rome the Dominican envoy began to behave undiplomatically.[6] He voiced criticism of the pope, first privately and then in public, assailing him for nepotism, an aggressive foreign policy, the corruption of his court, and his neglect of the crusade. In June 1481, Sixtus had him detained in Castel Sant'Angelo, but he was released a few weeks later, after the emperor intervened on his behalf.[7] Jamometić fled north to Basel, stopping in Florence along the way, where he seems to have consulted with Lorenzo de' Medici or advisors in his circle about his future plans. When he arrived in Basel, he wasted no time setting these into motion: standing in the cathedral, he read out thirty-eight theses condemning Sixtus for everything from heresy to simony to abetting murder and deliberately undermining prospects for a new crusade. Much of this diatribe was sensationally exaggerated, but some of it derived from the first-hand knowledge of papal politics he had acquired as a legate in Rome, and quite a bit of it recapitulates the charges that Lorenzo's circle had published against Sixtus during the Pazzi War. When Jamometić left Florence, he may well have carried copies of Becchi's letters and *Florentina synodus* north with him; the idea of convening a church council to restrain Sistine excess—or at the least *saying* that a church council had been convened—was one both authors shared. Jamometić's appeal, which was also printed, inaugurated an ecclesiastical war of words that took more than two years to settle and was only resolved with the archbishop's suicide in prison in December 1484.

This was not the first time that the papacy had been drawn into ecclesiastical disputes in a Swiss city, nor even the first time that its agents resorted to print. In 1475, after the see of Constance fell vacant, a dispute arose over the election of the next bishop. Sixtus appointed one candidate, Ludwig von Freiberg, but his bull of appointment only reached Constance after the chapter elected their own candidate, Otto von Sonnenberg. There followed a flurry of broadsides and pamphlets, printed in either Constance or Ulm, issued by Otto, by Frederick III in Otto's support, and by Sixtus in support of Ludwig. The controversy dragged on through the summer of 1480, when Frederick issued yet more decrees denouncing Ludwig and supporting Otto.[8] The matter was only settled after Ludwig died in the fall of 1480.[9]

As that crisis subsided, a new controversy erupted in Basel over the wealthy Dominican convent of Klingental. In 1480, the Dominican Order made an attempt, supported by the pope, to introduce observant reforms into the community. But the nuns resisted; when visiting Dominicans read out the papal bull mandating the reforms, the sisters shouted them down and threatened them with violence. All but two of the nuns withdrew from the convent.[10] The campaign against them continued in print. Their recalcitrance was assailed in six broadsides published by the Basel printer Bernhard Richter between the spring of 1480 and the fall of 1481: Sixtus's reform bull of February 1480 appeared in both Latin and German versions,[11] as did his April bull excommunicating the nuns;[12] that same month, Frederick issued a German-language decree exhorting his subjects to lend support and protection to the reforming nuns who remained inside the convent;[13] eighteen months later, Guillaume de Perrier issued a judicial citation against the renegade sisters.[14]

The controversy drew political support on both sides: the bishop of Basel, a relative of a former prioress, opposed the reform, as did Sigismund of Austria, while the pope and emperor supported it. But when Jamometić, a Dominican, launched his council appeal, it became clear that he both supported the reform and was supported by Stephen Irmi, prior of the male Dominican house in Basel and leader of the campaign against the nuns. In May, Sixtus withdrew his support from the reform attempt, giving the recalcitrant nuns their old convent back and transferring them from Dominican obedience to supervision by the bishop of Constance; a few months later, he removed Irmi from office.[15]

As concerned as Sixtus was to support his various causes in the Swiss cities between 1475 and 1480, it is unlikely that he or anyone in Rome ordered the printing of decrees promoting his initiatives, neither his 1475 bull in the Constance dispute nor his 1480 bulls supporting observant reform in Basel.[16] The erstwhile bishop Ludwig and the Basel Dominicans surely commissioned these publications. The same can probably be said of the numerous papal decrees and notices printed in upper Germany between 1480 and 1482 supporting indulgence campaigns for the crusade after the Turkish occupation of Otranto. In the controversy over Jamometić's call for a new council, by contrast, the pope was kept minutely informed of the battle being waged in print by his legate, Angelo Geraldini, bishop of Sessa. The broadsides that Geraldini had printed are themselves full of details *about* their publication, and these, taken together with the letters he sent to Rome, offer one of the most detailed accounts of fifteenth-century ecclesiastical publication and its perils.

Jamometić's Appeal and Early Responses

"The holy Roman church, weakened, desolate, and ruined, seeks a council for the restoration and reformation of itself!"[17] So Jamometić opened his call for a new council in March 1482. He castigated the emperor, the Christian princes, and Christians everywhere for their negligence in allowing Sixtus to destroy the Catholic faith and holy Mother Church. Sixtus had bought his office and now profited from the sale of major and minor sees and even cardinals' hats. He took bribes, charged for spiritual dispensations, issued bulls for a price, retracted them, and then charged again for their reissue. The faithful could not see him without bribing his palace guards; he had surrounded himself with a troop of stablehands and illiterate boys with whom he squandered the treasure of the church on banquets and debauches. He had shown special favor to his brother Pietro and to Girolamo Riario, the papal nephew whom Jamometić claimed was really Sixtus's son; he had given both men fortresses and castles that were properly the patrimony of the church. In his mad desire to enrich "the boy," Girolamo, he had even resorted to murder.

After his support of the Pazzi led to Giuliano de' Medici's death, the pope had gone to war with the Florentines and unjustly placed their state under interdict. He had made secret pacts with the Mamluks and Ottomans, hoping to undermine European plans for a crusade and free himself for further war against Florence. He had even pimped the beautiful queen of Cyprus to the sultan in hopes of getting him to declare war on Venice.[18] He had made Girolamo lord of Forlì, ordering the murder of the legitimate lords of the city; Girolamo in turn had killed an innocent young man from Imola and thrown his body in the Tiber.

Sixtus, "father of pomp and vainglory," had everywhere kindled hostilities among the Christian powers and perversely sought peace with the infidel. He had forced Venice to sign a treaty with the Turks and sown division among the Italian powers for the sole purpose of preventing a crusade. He had stirred up the Swiss against Milan, fomented civil war in Genoa, and tried to start a war between Hungary and Venice. He had provoked hostilities between the king of Hungary and the emperor, between the emperor and his electors, between the king of France and the duke of Burgundy, between the king of Spain and the bishop of Toledo, between Poland and Russia, and in places even farther afield. Most recently he had encouraged the Ottoman sultan to invade the kingdom of Naples and oust Ferrante from his realm.

The wicked pope had turned Rome into a new Babylon, filled with blasphemy, murder, and rapine, a city where widows and orphans were oppressed and virgins defiled, adultery and fornication ran rampant, and sodomites and whores filled the streets. Most abhorrent of all, he had abused his spiritual power to forgive sins, seeking to turn a profit on the fears of the faithful. He sold indulgences, which Jamometić called "great extortions," and showered the proceeds on Girolamo. Donations collected for the crusade against the

Turks went instead to benefit the Turks, "in whose name Sixtus has deceived the entire world." It fell to Jamometić, and to the Christian princes he addressed, to compel the pope to appear before the council and, if necessary, depose him.

Jamometić's proclamation reveals the breadth of his diplomatic knowledge, as well as the urgency of his reforming spirit. His text is structured as a legal *citatio*, laying out charge after charge against the pope and summoning him to appear for judgment. But on closer inspection, internal inconsistencies point to the weakness of Jamometić's position: the archbishop had no legal authority to call a council, and in fact he did not presume to do so.[19] Instead, he begins by invoking the *Sacrosancta ecclesia Romana*; it is this body that cries out for a new council. He then calls on the emperor and Christian princes to examine the evidence that he will set out—implicitly investing them with the power to convene the council and judge the pope. Later, he says that it is "divine justice" that calls Sixtus to account for himself before the assembly that will soon convene. And yet, in his conclusion he declares that the council is already underway (*iam pronunciatum et inceptum*). By announcing that the council had already convened, Jamometić could sidestep the question of just who had convened it. The proclamation is also localized and dated like a formal decree even though it was not at all clear what body was issuing it. Like the *Florentina synodus* that circulated in print during the Pazzi War, Jamometić's appeal used the fiction of a council "already underway" to publicize charges against the pope that no legitimate church institution would or could prepare.

To broadcast this appeal, the archbishop turned to the Basel printer Johannes Besicken, who published the text sometime before the middle of May.[20] The long and densely typeset broadside survives in just two copies, both in Swiss archives.[21] Its rarity is due, in part, to the fact that it was an ephemeral publication, meant to be posted on doors, city gates, and walls, and in part because the bishops of Basel and other nearby cities soon condemned it, ordering all copies to be destroyed.

In April, Jamometić produced a new redaction of his council appeal,[22] framed not as a legal citation but as an open letter to the pope. This version repeats the thirty-eight charges of the original but puts the entire discourse into the second person ("We hereby inform your holiness that a council has convened . . . ") and expands the text in certain places, including longer, dramatic apostrophes addressed not just to Sixtus but also to the Christian princes, his brother bishops in the curia, and the Christian faithful. As provocative as the initial proclamation was, this second redaction was yet more incendiary.

It also seems to have circulated more widely. Three contemporary chroniclers—the Erfurt historian Konrad Stolle, the Nuremberg humanist Hartmann Schedel, and the Venetian diarist Marin Sanudo—copied the text into works of their own concerning events in the spring of 1482.[23] Each of these writers was an avid collector of news and copied all kinds of documents—letters, broadsides, even songs—into his work. The second-person version traveled farther than the original printed broadside, which is preserved only in Basel and Chur. Was this version also printed? It seems very possible, given its propagandistic tone, and it is tempting to imagine Stolle, Schedel, and Sanudo each acquiring a printed broadside and copying its contents into their notebooks. But no trace of a printed edition survives.

Jamometić still fretted about the status of his initiative. On May 14, he composed another quasi-legal document, another citation issued in the name of the *Sacrosancta generalis synodus*, an assembly he again described as already sitting in session and *legitime congregata* to boot. He named himself as its president, and in the name of the council he announced that Sixtus had six months to appear before the assembly in Basel or else be deposed.[24] A manuscript copy of this letter accompanied the printed broadside sent to the bishop of Chur, and it is likely that other copies circulated to other bishops and city councils in the region.[25]

Ecclesiastical Responses

There is no evidence that any assembly of clerics ever convened under Jamometić's leadership in Basel. But his textual appeals to convene such a body certainly made a stir. In the spring of 1482, Rudolf von Scheerenberg, bishop of Würzburg, sent a copy of Jamometić's council appeal to Cardinal Francesco Piccolomini in Siena, asking him to forward it to the pope. The cardinal's response was dramatic. On receiving the *libellus*, he said, "Outraged by the mendacious calumnies, the obscene words, and the shameful blasphemies it contained, I went out of my mind; I couldn't restrain myself; I didn't read to the end but simply ripped the text to pieces, using my hands and even my teeth, and then I threw it on the fire. And I would have done the same to the author himself, if I'd had the chance!"[26] Because of the sudden (possibly strategic) fury that overtook him, Piccolomini did not have a copy of the text to forward to the pope; instead, he sent a letter to another cardinal in Rome explaining how the text had been destroyed and asking him to let Sixtus know of its existence.

Piccolomini included two additional pieces of information: first, not only had the archbishop of Krajina "issued" (*ediderat*) the text, but he had also had it printed and had copies distributed throughout Germany.[27] Second, local officials were taking steps to contain its spread. The bishop of Würzburg had banned the text in his diocese, issuing an edict forbidding printers from printing it and also requiring them to get episcopal approval before printing any other texts like it.[28] This fairly early act of ecclesiastical censorship would soon be repeated in Basel.[29] On May 24, Bishop Caspar zu Rhein published a ban on the creation, circulation, publication, or possession of any defamatory statements against Sixtus IV anywhere in the diocese of Basel. Noting that various letters purporting to be official ecclesiastical documents were circulating both in manuscript and in print, the bishop declared that no one should "execute, make known, publish, or post" any such letters before he himself had seen and authenticated them, nor would he allow anyone to "read out, write, print, publish, or otherwise make known" any other, newer documents like them, on pain of excommunication. The faithful must surrender any copies of such letters, written or printed, to the bishop's vicar general within three days of their hearing his edict, again on pain of excommunication.[30] About the same time, the University of Heidelberg asked the bishop of Worms to ban publication of Jamometić's proclamation in his diocese.[31] These efforts proved to be about as effective as Venetian bans on the printing of texts and images about Simon of Trent.

Piccolomini's letter is undated, but it was probably sent in May 1482.[32] By June, the Roman diarist Jacopo Gherardi records that Jamometić's activities were the talk of Rome.[33] By then, Sixtus was well briefed on the situation and had already made his first moves against the archbishop. On April 27, the pope sent letters to the bishop, the cathedral chapter, and the Council of Basel condemning Jamometić and calling for his arrest; the following week, he sent letters to German princes seeking their help in silencing the archbishop.[34] The week after that, he withdrew his support for the Observant Dominicans at Klingental, now tainted by the support Jamometić had expressed for them.[35] He also sent two German clerics, Hugo von Landenberg and Johannes Ockel, north on a mission to detain Jamometić;[36] they left Rome carrying yet more letters to prominent Dominicans and Franciscans in southern Germany seeking further support.[37]

Sixtus's envoys could not secure the archbishop's arrest in the summer of 1482. Jamometić was moving from city to city in the Swiss cantons, trying to drum up support for his council, as he persisted with his attacks on the pope. On July 21, he issued another blistering attack, addressed directly to Sixtus, which was also put into print.[38] In this diatribe, at turns accusatory, self-pitying, sarcastic, and patronizing, Jamometić no longer addressed the pope as "Sixtus," but merely as "Franciscus of Savona." (In this, he echoed the *Litterae Florentinorum* of 1478, which hailed the pope by the insulting title "Frater Franciscus.") This so-called pontiff was in truth the son of the devil; he had brazenly bought his office and now profited from selling positions to clerics and graces to the faithful. He must answer for his crimes: for letting the church be ruled by lascivious young men, for squandering its treasure on earthly vanities, for neglecting the threat of the Hussites, for ignoring the chance to unite Eastern and Western churches, and most of all for disregarding that decree of Constance (*Frequens*) that required the pope to convene councils on a regular basis.

Jamometić then declared the grounds on which he was authorized to call a council. Invoking the Gospel instruction to seek resolution of disputes privately, and then before one or two others, and only finally before the whole assembly,[39] he reminded the pope of the several times he had admonished him privately in Rome. When Sixtus persisted in his ill behavior, he had rebuked him before the cardinals, but still the pope had turned a deaf ear. He had even put Jamometić in prison! And so he had come to Basel, the place where questions of church government had been so long debated in previous decades. He had summoned Sixtus to appear, invoking the grave dangers that threatened the church from within and without; and Sixtus could have come, as Peter came to meet James in the first council of Jerusalem. Did Sixtus think he was somehow better than Peter? He had no right to think so. He was not "lord of the bishops" but only first among equals. Jamometić himself held the same rank, and he had the right to summon him before the council assembly. History showed that a wicked pope could be deposed—here Jamometić recalled the infamous cases of Benedict IX and the antipopes of the Schism— but this was a risk Sixtus would have to take. The pope had had eleven years to call a council. Any further delay would be intolerable.

At this point, Jamometić's argument took a new turn: although he had plenty of authority to call a new council, in fact there was no need to do so, because the Council of Basel was still in session! Though attendance had dwindled and no more sessions were held after the 1440s, the council had

never been formally adjourned; all Jamometić had done was call its next session to order. This was why, when he came to Basel, he had received such a
warm welcome from the city fathers. They knew he had right on his side. The
pope had responded by sending a "tongue-tied boy" (Johannes Ockel?) with
a letter demanding his arrest. This had scandalized the faithful in Basel and
in neighboring districts too. All would rather obey God and support his Church
than submit to this pope, and if Sixtus tried to punish them, God would bless
them in their travails. Jamometić was prepared to become a martyr in pursuit
of his cause.

Finally, Jamometić touched again on the question of his council's legal
standing. His summons might not have legal standing at all, he was prepared
to admit. But "even if it is not issued with the lead seal of the council, or in
the usual, solemn form," he said, "even if I summon you with only my naked
voice," still divine justice would favor the enterprise, and any attempts by the
pope to resist would only result in yet more scandal to the church.[40]

This text, also printed by Besicken in Basel, ran about twice as long as the
original set of charges from March 25 and filled the rectos of two large folio
broadsides, with the versos of each sheet left blank. This is a curious way of
printing, for Besicken could have more economically printed the text on both
sides of a single sheet. It seems that the two sheets were meant to be butted
against each other at the short ends and glued together to make a single, long
broadside that could be posted on a wall or a door, with the whole text visible
at a single glance. The only copy of the invective to survive is preserved on two
separate sheets, but other broadsides printed in Basel and elsewhere in the
course of the controversy consist of separate printed sheets glued together
in this way.[41]

It is not known if or when this second text reached Rome, but Sixtus had
already decided to take further action against the rebellious archbishop. On
July 16, 1482, he drew up a bull of excommunication (*Grave gerimus*) that
condemned Jamometić not only for leveling false and blasphemous charges
against the pope but also for printing them and distributing the printed copies among clergy and laity alike.[42]

The pope also commissioned a new legate *de latere*, Angelo Geraldini,
bishop of Sessa, who was to travel north to Basel, publish the bull of excommunication, and secure Jamometić's arrest.[43] In Rome, Stefano Infessura noted
around July 27 that the pope had appointed Geraldini as his legate and stripped
Jamometić of his title of archbishop.[44] This suggests that the *Grave gerimus*
was published, or at least its contents were made public in some way, almost

immediately after its issue in Rome; getting it published in Basel, however, would prove more difficult.

Geraldini went north in August 1482. On September 22, he told the pope that he had fallen ill with a fever on his way over the Brenner Pass and was more dead than alive when he finally staggered into the Austrian town of Feldbach.[45] There, he met the Franciscan indulgence preacher Emeric von Kemel, who agreed to take the bull of excommunication on to Basel and publish it there, while Geraldini dispatched another colleague to take it to Constance.[46] Neither cleric made good on his promise, as the legate would later learn.

As soon as Geraldini was well enough to travel, he went to Constance, where he found the bishop little inclined to receive him. (The bishop was Otto von Sonnenburg, the same candidate for the episcopal see whom Sixtus had opposed for over five years and only recently agreed to recognize.) Geraldini managed a brief meeting at which he was able to *show* the bishop the bull of excommunication, but the bishop would not publish it on church doors or read it out at mass. Geraldini demanded that Otto come with him to Basel to secure Jamometić's arrest, but the bishop demurred, claiming he was expected at a meeting elsewhere. He then left town, leaving Geraldini with only vague promises of help against Jamometić.[47] Geraldini published the bull himself in the churches of Constance (*feci postea publicari per omnes ecclesias*), but he and others took steps to publish the bull several more times in October and November, suggesting that his September publication in Constance had little effect.

Matters were proceeding no better in Basel: Geraldini would later declare that not only had Friar Emeric not published *Grave gerimus* on the left bank of the Rhine, but he had also entered into secret talks with the Basel councilors, offering to blunt the most serious sanctions that Geraldini was threatening against the city, in exchange for Jamometić's arrest.[48] Sixtus himself may have known about—or even ordered—this double game, in which the unwitting Geraldini thundered grave threats from outside the city while Friar Emeric, a trusted local intermediary (and Franciscan), negotiated more quietly with the city council within.[49] Whether Emeric deliberately meant to undermine Geraldini, or whether Geraldini somehow misunderstood their mandates, the legate would complain bitterly of the Franciscan's behavior in the coming months.

Meanwhile, another Franciscan in papal service, Peter von Kettenheim, was proceeding on a different tack. Von Kettenheim, a collector of annates and tithes in the diocese of Basel, had traveled to Rome in May 1482 to represent the nuns of Klingental before the pope. They were still hoping to be excused from their obedience to the observant Dominicans. On May 4, the pope granted their appeal and restored them to their convent; on the same day he drew up letters condemning Jamometić and ordering his detention.[50] Von Kettenheim took both sets of orders with him back to Basel, and at the end of June Sixtus sent another letter north authorizing von Kettenheim himself to seek Jamometić's arrest. According to his instructions, he could impose ecclesiastical penalties on any city or locale that impeded his mission.[51] Since the Basel government would neither arrest nor hand over Jamometić, von Kettenheim placed the city under interdict on September 6.[52] The city council immediately began issuing formal appeals to the pope protesting these actions—six in all between September and November 1482.[53] Only one of them, dated October 4, survives in print,[54] but the council probably had all of them printed.

In their October 4 appeal, the councilors tactfully assured the pope they had no complaint against him but only wished to protest the actions of his agent, Peter von Kettenheim, whose unjust and disproportionate measures were threatening the honor of the city and the salvation of its citizens. The friar had gone about hanging up sentences of excommunication, interdict, and suspension on the doors of churches and other buildings, even on the doors of the cathedral of Besançon, seat of the archdiocese, which brought great shame on the pope and grave danger to the souls of the faithful.[55] What purpose could this interdict serve except to deprive innocent Christians of the sacraments? The councilors were sure that the Holy Father could not have approved these actions.

The councilors addressed their appeal to Sixtus, who presumably sent a copy to Rome, but they also chose to combat von Kettenheim more locally, on the doors and walls of their city. They paid Besicken to print up their appeal as a broadside, the text of the appeal buttressed by authentication formulas supplied by notaries, witnesses, and the city clerk.[56] The notary Johannes Struss then signed and notarized each printed copy by hand; only then were they posted around the city.

The ultimate cause of the impasse was the emperor's refusal to weigh in on either side of the dispute. Without an order against Jamometić from Frederick, the Basel government would not arrest him, but without a sign of imperial support for him, foreign powers like Milan and Florence, which watched the affair closely, would not send delegates to the proposed council. Their ambassadors remained in Basel offering Jamometić encouragement but nothing more.[57]

Other documents relating to the controversy may have been printed in this period. The city account books refer to separate payments for printing *appeals* in the plural. The next printed document in the controversy that survives is Sixtus's bull *Grave gerimus*. Confusingly, this document was "published" several times by different agents at different times and places. Geraldini had it published in the churches of Constance in early September, though he does not say whether it was merely read out in churches, or also posted, or printed. The problem was that Geraldini could not get anyone to *observe* his decree. The Basel clergy met in early September and agreed, with the approval of the bishop, that the interdict was illegitimate on the grounds that neither von Kettenheim nor Geraldini had the authority to impose it.[58]

Geraldini's cause got a small boost when the Dominicans in the city turned on Jamometić. The archbishop, sensing that their support for him was waning after their loss of Klingental, had published a stinging invective calling them "fake friars" and lamenting their slack commitment to orthodoxy. As a result, the Basel Dominicans began to preach openly against Jamometić and his council project.[59] On October 9, Salvo Casetta, minister general of the order, appeared before the city council with orders from the pope to publish the bull of excommunication unless Jamometić was arrested.[60] Later that month, he wrote to Sixtus confirming that he had indeed published *Grave gerimus* in Freiburg and in all its surrounding towns: he had had the text "read out in public, preached on, and made known," and on October 27, in the cathedral of Freiburg, he formally presented the bull to the rector, who read out its text before the clergy, city councilors, university faculty, and general public. The next day, Casetta himself read out the text in the city's Dominican church.[61] Soon after, Geraldini reported that *he* had had the bull published, again, too: on November 9, he sent copies throughout the region, including one to Besançon, where it was posted on the doors of the cathedral.[62] He also published a *citatio* against Jamometić's followers in the surrounding towns of Rheinfelden, Ensisheim, Mülhausen, and Neuenburg im Breisgau.[63] At some point during these multiple publications, one of the pope's agents— perhaps Geraldini, perhaps Casetta, perhaps von Kettenheim—also had *Grave gerimus* printed; it appeared as a broadside from the press of Georg Husner in Strassburg.[64]

From this welter of redactions, partial publications, repeated promulgations, and republications, we can draw a few conclusions about the ambiguous nature of ecclesiastical "publication" at this time. It is not surprising that Jamometić issued and reissued his council proclamation and citation; the documents had no legal standing anyway, and Jamometić was still trying to conjure his council into being. What is more surprising, perhaps, is the disorganization of the papal legation. Their lack of cooperation extended even to the publication of the grave sentences against Jamometić that they carried with them. Geraldini had to try three times before he could get *Grave gerimus* published even once (when he did it himself in Constance); it was published again, orally and in manuscript, in Freiburg at the end of October, and a third time by Geraldini at Besançon in early November; and at some point it was also printed. Earlier, Peter von Kettenheim had published his own decree of interdict against Basel, but it was not clear that his decree was valid since *Grave gerimus* had not yet been fully published. (The bull laid the foundation for an interdict by forbidding Christians from supporting Jamometić on pain of excommunication.) Ultimately, as Geraldini was beginning to realize, ecclesiastical decrees were hollow without the force of a secular power to back them up.

Like any political crisis, the conciliar controversy attracted commentators who sought to raise their own profiles by pronouncing on the affair. On August 10, in the nearby city of Schlettstadt, the Dominican Heinrich Institoris, inquisitor general for upper Germany, who had earlier involved himself in the Simon of Trent case and within a few years would help compile the infamous *Malleus maleficarum*, issued a strident letter denouncing Jamometić and his efforts to revive the council.[65] Institoris orchestrated a multipronged print campaign for his text: at Strassburg, the nearest major city to Schlettstadt, Heinrich Eggestein's press printed Institoris's letter as a (very long) broadside—two sheets glued together and nearly 90 centimeters long—as well as in a folio edition of six leaves.[66] The text also appeared in another pair of editions, again as a broadside and as a quarto, issued by an anonymous Nuremberg printer.[67] Both Nuremberg editions seem to be reprints of their respective Strassburg editions; Institoris likely sent the originals to Nuremberg for reprinting there. Copies of the broadside version were posted on public buildings and churches in Basel, on the public notice board in the middle of the Rheinbrücke, the main bridge across the Rhine, and elsewhere by critics of the supposed council.[68]

Finally, a copy of the letter made its way to Rome, where another quarto edition appeared from the press of Johannes Philippus de Lignamine.[69] Always ready to seize an opportunity for self-promotion, de Lignamine added a preface to Sixtus IV, recommending the text and excoriating the schismatic archbishop. Perhaps as a reward for this and other services to the papacy in print, de Lignamine was awarded a papal commission to sell crusade indulgences in Sicily.

De Lignamine was not the only observer in Rome to remark on the Basel controversy. On October 18, the humanist Lionello Chiericati addressed a letter to Georg Wilhelmi von Keppenbach, provost of the church of St. Peter's in Basel, who had come to Rome that same month to deliver the city's appeal against Peter von Kettenheim to the pope. Wilhelmi gained the pope's trust and returned to Basel in December with secret instructions of his own. During his time in Rome, he seems to have made the acquaintance of several humanist scholars. Chiericati, a tireless commentator on political affairs who was no stranger to the press himself, saw his letter published by Guldinbeck in Rome.[70]

The printing of an older conciliar tract from the original Council of Basel may also date to this time. The anonymous *Verteidigung des Konzils zu Basel*, written by a supporter of the earlier council in 1442, argued that the council represented the Church as a whole, it stood over the pope and could hold him accountable, and its deliberations in matters of faith were governed by the Holy Spirit and therefore infallible.[71] A quarto pamphlet of this text has been assigned to a Mainz printer whose dated editions range from 1479 to 1482. The edition itself is undated, but it is certainly possible that the reissue of this text was commissioned by Jamometić or his supporters.[72]

Geraldini recorded the publication of two other tracts in the controversy that do not survive. In June 1483, he told the pope that "a certain theologian" had composed a tract in Jamometić's name arguing that princes and bishops had the right to convene a council without papal approval; the council was superior to the pope; and, according to the decree *Frequens*, the Council of Basel continued to be in session. Worse still, this theologian had had the tract printed and distributed throughout Germany and France.[73] The good news was that a champion had stepped forward to rebut it: Geraldini had met a "certain Dominican inquisitor living near Strassburg" (almost certainly Institoris) who was working on a long refutation of its conciliar claims. The inquisitor had titled his work *De potestate papae*; Geraldini had seen a part of it already printed and estimated that it would fill three hundred pages when complete.[74] There is no trace of a printed edition of either theological tract, and it is not clear that Institoris ever completed the one that Geraldini says he had partially published. (Geraldini may have seen the folio or quarto edition of Institoris's letter and believed that this was the first installment of the longer tract.) At the time, Geraldini certainly hoped that it would help in the battle for public opinion.

Jamometić's Arrest and Retraction

As the stalemate dragged on, Sixtus grew impatient. In mid-December, he switched sides in the War of Ferrara, abandoning his Venetian allies; he also prepared a new document, a bull threatening military action—literally, a "crusade"—against the people of Basel, if they did not hand over Jamometić. This document, dated December 14, 1482, he sent north to Geraldini in the care of the Basel envoy, Georg Wilhelmi von Keppenbach. In a cover letter, however, he also advised Geraldini to allow Wilhelmi von Keppenbach to try his hand at negotiations with the council first, before publishing the dire penalties in the bull.[75] The pope may have begun to suspect that Geraldini was not up to the task of negotiating with the Swiss.

By the time this document reached Basel, events had already moved on. On December 18, a new papal legate, Antonio Gratiadei, another Franciscan, arrived from the imperial court. He brought news that Frederick had decided not to support Jamometić. The city council was relieved: now they could arrest the archbishop. No longer would they need to argue with the fractious papal legation.[76] The councilors agreed to the arrest, but they drew the line at handing Jamometić over to the pope. They had some grounds to refuse: in the course of their negotiations, different legates seem to have promised different things. Geraldini had his crusade bull, which he was ready to issue unless the city both arrested Jamometić and handed him over to papal custody. But Gratiadei, supported by Emeric von Kemel, took a gentler line; he seems to have promised the council that if they arrested Jamometić and put him in imperial custody, not only would the crusade bull stay unpublished but the interdict itself would also be raised. The ambiguity of their instructions opened the way to another six months of disputes.

On December 21, with Gratiadei observing, the city council arrested Jamometić and confined him to the prison over the city gate known as the Spalentor.[77] The fiery conciliarist crumbled quickly. On January 2, between six and seven in the evening, a group of city officials, clergy, and Gratiadei's secretary, the notary Johannes de la Woestine, gathered in Jamometić's cell and witnessed him formally retract his charges against the pope and his calls for a new council. Jamometić acknowledged that he had not only defamed the pope verbally but also committed his insults to writing and posted and distributed these statements widely.[78] He now wished his statement of retraction to be posted publicly on church doors and read out to the faithful so that all might know the vanity of his schemes.[79] The text of the retraction was carefully authenticated, witnessed, and notarized. De la Woestine then composed a long introduction (about three times as long as the retraction itself) vouching for Jamometić's sincerity: he had made his statement freely and without compulsion, for the benefit of those Christians scandalized by his false statements about the pope, whose honor and reputation he hoped to restore.[80]

Shortly afterward, this entire collection of texts—de la Woestine's long preface, Jamometić's retraction, and the notarial certification—was printed by Besicken as a broadside. The Basel council paid for the printing.[81] De la Woestine then notarized the individual copies a second time by hand.[82] The text was also sent to Mainz, where it was printed as a broadside by Peter Schoeffer.[83]

1483: A Contested Interdict

Neither Sixtus nor Geraldini was satisfied: they wanted the archbishop in papal custody. But the city council, backed by the emperor, maintained that the archbishop's crimes fell under imperial jurisdiction and should be dealt with in Basel; indeed, the papal commissioners had suggested that with Jamometić locked up in the Spalentor, the interdict would be raised. Geraldini was bewildered. In a letter to Sixtus in early February, he said he suspected that Emeric von Kemel was the source of the confusion. It was Emeric who, back in September, had promised to publish the bull *Grave gerimus* and then failed to do so.[84] Now he seemed more interested in selling indulgences than bringing Jamometić to justice. As long as the city was under interdict, the friar had no access to customers. At this very moment, he was sitting on some six thousand *confessionalia* that he had printed up at a cost of 25 florins; the financial pressure he faced to make a return on his investment was acute.[85] To clear the way for their sale, Geraldini said, von Kemel had resorted to forgery (giving yet more work to local printers): "Not only did he revoke the orders I had given, but he even revoked the interdict, and the people of Basel then printed and everywhere published that counterfeit 'revocation,'" he told the pope.[86]

Geraldini also charged Gratiadei with printing forgeries. He complained that after the interdict had been in force for two months, the people of Basel had been ready to yield to the pope's demands, but when Gratiadei arrived, he doctored the text of a papal letter (written to Gratiadei at the end of December) to make it appear that the pope was ready to deal more leniently with the city. Gratiadei had had this falsified text printed and distributed throughout Germany.[87] Moreover, a member of his staff—Johannes de le Woestine, the same notary who took down Jamometić's confession in January—had forged an even longer document containing even more concessions to the people of Basel, and the pair had had that printed and distributed throughout the neighboring cities and towns as well, "so that people everywhere believed that the sanctions I had leveled against the city had been lifted."[88]

This is all very difficult to trace. No printed documents survive that match Geraldini's descriptions. His frustration with his colleagues, who were clearly engaging in separate talks with the Basel government, is palpable, but it is not so clear that he understood exactly what they were doing or that he represented their activities accurately to the pope. Did von Kemel, Gratiadei, and de le Woestine *all* publish documents revoking the Basel interdict between January and March 1483? Or was there perhaps just one publication, a carefully edited version of some letter from the pope, which Geraldini at different times accused different officials of printing?[89] Whatever the case, Geraldini was clearly vexed by his fellow legates and believed they were both plotting and printing against him.

In this same letter of February 3, Geraldini told the pope he stood ready to publish his bull of crusade against Basel. The council had not surrendered Jamometić, so the time had come to level this most serious sentence against the city. Geraldini had a hundred copies of the bull printed.[90] He sent copies to the Dominican Salvo Casetta in Cologne, so that they could publish the bull simultaneously in their two cities. Each of them would enter the relevant data about place and date of publication by hand in spaces left blank for them on the printed sheets.[91]

Geraldini's plans for the publication of this bull were peculiar. He told Sixtus he would publish the bull in just a few places on February 7, in advance of a meeting of the German princes, for he was afraid that if he waited to publish the bull at or after the meeting, the princes might order him not to. If he released a few advance copies, he could declare in all honesty that the decree was issued and could not be revoked. But Geraldini did not want to be provocative; there was still a chance that the meeting could go his way and the princes would back the pope. Only if the meeting were unsatisfactory would he then release the remainder of his printed copies and distribute them as widely as possible. This was not a common way of "publishing" a document; presumably, before the advent of print, no legate would have paid for multiple copies of a document before knowing for sure they were to be distributed. Geraldini may have thought of this half-measure himself.

In the end, the meeting of the princes was inconclusive, and Geraldini released his printed bulls on February 9. Printed by Eggestein in Strassburg, the broadsides were remarkable objects: nearly a yard long, each was constructed from two folio sheets of printed paper glued together end to end and covered with notarial signatures and flourishes, with Geraldini's enormous red wax seal attached to the bottom (figure 4.1). The broadside reproduced the entire text of Sixtus's crusade bull of December 14, surrounded by Geraldini's lengthy commentary: paragraphs setting out his mandate and his authority to issue the bull, a history of his mission in Germany up until the end of January, the briefs he had received from the pope in October and November, and instructions for reading the text out in churches and further publishing its contents. At the bottom, Geraldini included a set of talking points that, he suggested, could be used by preachers in the public recitation of the bull: the entire city of Basel was now subject to not just the interdict but a crusade. Their goods were forfeit, their contracts were void, and the rest of Christendom was encouraged to come and attack them. Anyone who took up arms against Basel was promised the same spiritual benefits as a soldier going to war against the Turks.[92]

Geraldini's threat of armed violence against Basel roused the emperor, at last: a furious barrage of letters from Frederick now issued from Vienna. On March 10, the emperor wrote to the pope, blaming Geraldini for the impasse and pointing out that the rest of the papal legation disagreed with him. He, Frederick, had had an understanding with the other legates: if Jamometić was arrested, the ban would be lifted.[93] On March 20, he wrote directly to Geraldini demanding that the legate stop persecuting the people of Basel; the other papal commissioners all agreed with Frederick that the city had kept its half of the bargain. Now this outrageous "crusade" must be recalled and the interdict must be raised.[94] Frederick also issued twin proclamations on March 20 and 21, addressed to the estates of the empire, declaring that Basel lay under his protection and no one should attempt to harm its citizens. He called out Geraldini by name as the city's unjust persecutor and set out the reasons why he, Frederick, had satisfied the terms of his agreement with the

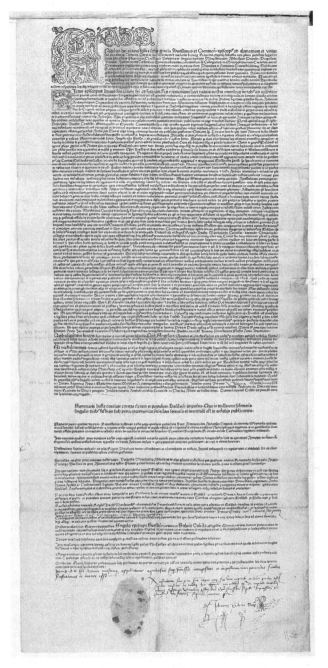

Figure 4.1. Angelo Geraldini, *Publication of a Bull of Crusade of Sixtus IV against Basel* ([Strassburg: Heinrich Eggestein, before January 31, 1483]). Staatsarchiv Basel-Stadt, Politisches H 1 II.

pope. All Sixtus had wanted was to have Jamometić arrested—and arrested he was! Calls to extradite him to a "foreign court" would be a violation of the man's rights as an imperial subject.[95] The city council had the German text of this decree printed, again by the redoubtable Besicken.[96] The council's account books record a payment for printing some 370 copies of the broadside, and yet more in wages to heralds who were to read out the contents of the decree and deliver copies to Constance.[97] Geraldini later complained to Sixtus that printed copies of the emperor's rebuke had been "posted everywhere."[98]

Geraldini's Final Moves

Geraldini found himself at odds with not just the city government of Basel but also the emperor and his fellow commissioners. Even worse, sometime in late March or early April he received a humiliating letter from Sixtus curtly instructing him to call off the "crusade."[99] The arrival of this letter prompted the last of Geraldini's moves in print. In April, he ordered up another gargantuan broadside from Eggestein in Strassburg, again made up of multiple sheets glued together to form a single poster, this one about 3 feet long. In a letter to Sixtus at the end of May, Geraldini said he had commissioned this publication as a countermeasure against the (false) documents his fellow legates had been printing and distributing. He wanted it known that the interdict was still in force, and he intended to punish those clergy who were not enforcing it.[100]

The broadside has been catalogued with the title *Publication of Two Briefs of Sixtus IV*,[101] but this hardly does justice to its remarkable contents. In fact, three papal documents appear on the poster, set in a large sea of Geraldini's plaintive prose. The legate's words take up more than half of the broadside and culminate in a legal citation charging just about every member of the Basel clergy with heresy and schism and summoning them to appear before an ecclesiastical court, chaired by Geraldini, where they might plead for mercy.

Geraldini's document provides yet another glimpse into the perils of fifteenth-century publication. In an introductory paragraph, he set out his credentials and attested to the authenticity of the documents he was publishing; he added the text of a brief from Sixtus to Geraldini of October 17, 1482, establishing his mandate, and then an abridged version of the crusade bull of December 14—the same text Geraldini had published in February, but this time edited so that only the bull's instructions for publication were included.[102] The legate explained that he had published this text widely in southern Germany and the Swiss cantons; several times he had had the bull posted in churches and in the *loca insignia* of the dioceses of Cologne, Mainz, Worms, Speyer, Strassburg, Constance, and Basel. He then noted, with perhaps a flash of irritation, that only after he arranged for the document's publication in so many places did a new letter from the pope reach him, telling him to hold off publishing the crusade bull until further notice. It seemed, the pope said, that it might still be possible to get Jamometić extradited without resorting to drastic measures. The letter had been delivered to Geraldini by a courier from Basel, where the pope had initially sent it. (That Geraldini chose to publish the complete text of this papal brush-off, which all but declares his irrelevance to the negotiations going on without him inside the city, suggests a faithfulness to his mandate so dogged as to border on the obtuse.) Geraldini stressed that the important thing about the letter was that it *only* suspended the crusade against the city; it left all other penalties, including the interdict, in force. Therefore, his earlier orders had not been overruled!

Geraldini had then drawn up a letter to the city council underscoring the fact that the interdict had not been suspended. But his words were disregarded. With a few pious exceptions, the city clergy continued to violate the interdict, kept the doors of their churches open, and celebrated mass in the presence of excommunicates; even worse, they did not blush to tell the faithful that they had nothing to fear from Geraldini's decrees. Worst of all were Gratiadei, who falsely claimed that the interdict had been raised, and Emeric von Kemel, who told the Franciscans in Basel the same thing; they had even forged documents in support of their position.

In response, Geraldini was now issuing a summons against the Basel clergy (he singled out about ten by name, including Gratiadei and Georg Wilhelmi, with the rest included in a general summons), who must appear before him in Mainz on May 25 to answer for their contumely and face charges of heresy and schism, possible imprisonment, and loss of their offices, benefices, and titles. Unfortunately, Geraldini admitted, the city had refused him access to publish this summons within its walls, so he was sending a preacher out with copies to be posted and published in churches in the surrounding districts: in Strassburg and Constance, as well as in Schlettstadt, Breisach, Zurich, Freiburg, Neuenburg, Rouffach, Rheinfelden, Colmar, and Ensisheim—and in the countryside around Basel too, even if the city remained closed to him. Given this widespread publication, no one in Basel could pretend ignorance of the summons; all whom it concerned must appear before his tribunal on the stated day, on pain of excommunication; all clerics who received a copy of the message must publish it in their churches within three days of receipt, again on pain of excommunication.

In a letter written at the end of the next month, Geraldini informed the pope that he had had a hundred copies of these "letters patent" printed up and that sixty-four of them were already in circulation.[103] One survives in the Basel city archive. Like its counterpart from February, it consists of two sheets glued together and is speckled with manuscript additions and notarial authentications, brightly stamped with Geraldini's red seal.[104] Geraldini intended to go on a tour of neighboring dioceses, where he would exhort the bishops in each place to uphold the ban.[105]

In this same letter to the pope, Geraldini vented his frustration by rehearsing the whole sorry tale of his legation, going all the way back to Jamometić's first proclamation of the new council back in March of the previous year. Now, thoroughly disgusted with the Basel city government, he accused the council itself of paying for printing, posting, and distributing Jamometić's original "articles" against the pope, both in Basel and abroad, in the spring of 1482.[106] The copies posted in the city, he said, had been "well-guarded by officers," while the city had sent no fewer than thirty-two messengers out onto the highways with printed copies to distribute to kings and princes, cities, towns, universities, and schools. Even more shocking, the printer of this sacrilegious screed had been rewarded with a seat on the city council.[107] Just as bad, the city government had endorsed Jamometić's "summons" directing the pope to appear before the council.[108] They had ordered armed guards to post a copy on the public notice board on the Rheinbrücke every morning. It was impossible to rip this text down from its hoarding on the bridge, he lamented, since every evening the same guards removed the paper for safekeeping overnight.[109]

Geraldini's account may not be completely reliable. He wrote well after Jamometić had been arrested, and after the council had refused for several months to hand him over for extradition to Rome. He may have decided to blame the council for actions that Jamometić himself had taken in the spring or summer of 1482. (The council's account books record payments for printing and distribution of *its own* publications on the affair in the fall of 1482 and spring of 1483, but none for Jamometić's texts from the spring of the year before.) Later in the letter, Geraldini included a startling menu of penalties he thought the pope ought to impose on the city: Basel should send Jamometić to Rome to be publicly executed in Campo de' Fiori; the city councilors should come with him so they might fall at the pope's feet and beg forgiveness, as the Florentines had done not long before; and the city should pay for two new chapels on either side of the Rheinbrücke where a team of priests would say masses for the pope; to atone for the city's transgressions, these chapels and their staff should be maintained in perpetuity at the city's expense.[110]

Nothing came of Geraldini's final press campaign. In August, Frederick returned to the issue, urging the pope to withdraw his sanctions on the city; the Basel council again paid to have his proclamations printed and distributed, including to the nearby towns of Delémont and Saint-Ursanne.[111] The controversy dragged on even after Sixtus's death in August 1484, but no more documents were printed. In December 1484, a thoroughly demoralized Jamometić committed suicide in his cell; his body was thrown in the Rhine, and the new pope, Innocent VIII, formally lifted the interdict.

By the early 1480s, the papacy and its agents were well aware of the possibilities of print, in a way they had not been in the early 1470s or even at the time of the Pazzi affair in 1478–79. Geraldini's campaign unfolded at the same time as the War of Ferrara examined in the previous chapter. Taken together, we can see the papal diplomatic corps beginning to exploit the press as they executed their mandates abroad, an innovation overlooked by historians of printing and historians of diplomacy alike.

As print multiplied the kinds and numbers of documents in circulation, the material evidence of surviving copies suggests a concern among all parties to establish their authenticity and validity. Was this anxiety occasioned by the fact that they were using print? Did Geraldini and his contemporaries fear that print was somehow less authoritative a medium for official publication than manuscript? Or was this concern occasioned by the massive contest *for* authority that the entire controversy represented? Jamometić knew he had no authority to call a council; everyone told Geraldini he had no authority to level an interdict; the Basel council had to assert its right to negotiate with the pope directly, via legates and appeals; the emperor wanted to prosecute his own subject without papal interference and to protect the citizens of an imperial city from unjust sanctions by the pope.

Whatever the source, the concern is manifest in the publications that survive from the controversy. The printed texts arguing in favor of the city—the council's appeal to the pope of October 4, 1482, and the five imperial decrees from March and August 1483, all printed by Besicken—are fairly similar. The October appeal concludes with a notary's statement, presumably taken from the printer's copy, verifying that the councilors were all present on the day and time specified and that the text was an accurate transcript of their deliberations. This is followed by a second, printed notarial statement that certifies that the text of the printed copy matches the original. This printed statement includes the name of the notary, Johannes Struss, and on the one copy that survives Struss has affixed his signature to the bottom of the sheet along with his notarial cipher.

We see the same formula for authentication in the imperial decrees. Each decree is composed as a letter from Frederick to princes, estates, clergy, or other groups of his subjects. At bottom, a printed notarial statement guarantees that the broadside is a true copy of the original, and on each copy that survives, a Basel notary—either the same Johannes Struss or his colleague Johannes Gerster—has added his signature and cipher. The printed copies must have been delivered by Besicken to the council, whose notaries then authenticated each copy before they were dispatched for posting or delivery. Not all were sent out: twenty-five copies of one of the imperial decrees from August 1483 remain in a tidy pile in the Basel city archive, each notarized by Johannes Gerster. Six copies of another August decree are filed with them, notarized by Johannes Struss. It is worth remembering that in March the city council paid for the printing of 370 copies of another imperial decree. Gerster or Struss probably sat and signed every one of them.

By contrast, Jamometić's first publication, his council proclamation from March 1482, was not certified or signed in this way, but his second one, dated July 21, 1482, contains a peculiar version of a notarial certification. At the bottom, a few printed lines assert that the printed text accurately reproduces the manuscript original, but this formula of authentication mentions no notary, nor did a notary sign it. Jamometić seems to have wanted to approximate the certifications the local government supplied its own documents, but the peculiarity of his situation meant he could only attach a fairly hollow formula to the bottom of his decree.

When we look at Geraldini's publications, the obsession with authentication extends from manuscript additions to the content of the printed text itself. His two publications of February and April 1483, printed by Heinrich Eggestein in Strassburg, are both giant broadsides constructed from printed sheets glued together.[112] The bottom of each of Geraldini's broadsides is covered in manuscript certifications, notarial signatures, and Geraldini's episcopal seal (see figure 4.1). Perhaps to further reinforce the local validity of the documents, Geraldini also left blank the spaces for the place and date of issue;

on the copies that survive, this information is filled in in manuscript. This is worth reflecting on: with his signatures, seals, and manuscript fill-in-the-blanks, Geraldini was presenting the printed object *not* as a diplomatic copy of a manuscript original that had been published at a prior time and place, but rather as an official object in and of itself, something that had legal weight and force on its own. My interpretation of this is that he hoped these material signs would combat the skepticism he encountered everywhere about the validity of his mandate and mission.

Geraldini's broadsides were meant to be posted on doors and walls. As large as they are, they could hardly have been read any other way. Their giant size would have further contributed to the impression of authority that Geraldini wanted to project. But his broadsides were also very long because Geraldini simply had so much to say. Both purport to publish his latest instructions from the pope: in February, the declaration of a crusade against Basel, and in April, the announcement that the crusade was suspended but the interdict was not. There were established conventions in southern Germany for the republication of instructions from a higher authority: a brief, printed notarial certification attached to civic and imperial decrees or, in the case of papal decrees, a headline above the papal text introducing it and establishing the authority of some local commissioner to republish it. What Geraldini did was very different: he published strings of different papal briefs and bulls; long narrative accounts of how and why he was publishing them; an abbreviated set of talking points, in the February broadside, for use by preachers; and, in the April one, his own sentences of excommunication and citations to appear before his tribunal. His broadsides are extraordinary confections of ecclesiastical legislation. It is not hard to sense, behind them, the impotence Geraldini felt in the face of his adversaries.

What was the audience for this printed controversy? All the parties printed up their texts in great numbers; we hear of dozens or even hundreds of copies of each edition. Jamometić admitted to posting and circulating his libels against the pope throughout Basel, and one of his proclamations may have been printed in a German translation. But ultimately, Jamometić's efforts were aimed at a relatively small audience of local lay and clerical authorities. In May 1482, he sent copies of his proclamation and citation to nearby city councils and local bishops. In June, he went riding circuit through the Swiss cantons, hoping to raise political support for his council. Above all, he wanted the emperor to endorse his project and for other states to follow suit. Jamometić seems not to have tried to connect directly with the general population—we hear little about him preaching, for example. His "reformation" was conducted on paper, directed at his peers and superiors. Three prominent humanists copied one of his diatribes against the pope into their diaries, but there is little evidence that his words resonated widely with the general public.

Geraldini and his colleagues were more concerned to broadcast their message widely. They wanted the people of Basel to feel the pinch of the interdict, for the clergy to observe the ban, and in the final stage of the conflict Geraldini seems to have thought he could summon an army of crusaders to attack the city. He tried to launch a genuine publicity campaign. Geraldini was certainly alert to the dangers the press could pose from the other side, as his frequent comments on the printing and distribution of pro-conciliar materials attest, but his greater interest in the possibilities of print did not translate into success in his mission. The Basel clergy ignored his interdict, and the pope himself repudiated his crusade. In the end, Jamometić and Geraldini both failed for the same reason: neither could generate credit in his project from the circumscribed public he had chosen to address. It is not clear that *any* printed publicity could have secured the support they needed. What both sides lacked was political backing: no authority wanted to endorse the two extremes they represented. Behind closed doors, a small group of moderates—Gratiadei, von Keppenbach, and the Basel city councilors— sorted the problem out between them; they threw their radical reformer into prison and politely ignored the legate and his thunderous bulls.

Ecclesiastical historians have long dismissed Jamometić's attempt to convene a new council as a failure: not only did the archbishop lack the standing to convene one, but he also lacked the political skills to persuade any other authority to do it for him. It is tempting to view his counterpart, Geraldini, as an equally ineffective figure, beset on all sides by nimbler characters who played a faster diplomatic game and had better command of the new technology of print to boot. Geraldini's struggles with his opponents, including his supposed partners on the papal legation, can read like farce. But for a brief moment, the controversy over a new Council of Basel seemed a serious matter: the Italian powers watched it closely, sending envoys to Basel who were poised to support Jamometić if a council seemed likely to convene. Frederick III tacitly supported his efforts and at one point was ready to defend the city by force of arms, and Sixtus himself was so alarmed by the prospect of an ecclesiastical rebellion that Sigismondo dei Conti, at least, thought that the crisis was what pushed the pope to desert his long-standing alliance with Venice and align with the Italian league in December 1482.[113]

The episode is also important because it offers such a detailed look at the processes and perils of official publication in late medieval Europe. All sides had forms and formats for conveying legal sentences (briefs, bulls, appeals, citations, decrees), and all sides had mechanisms for asserting the authenticity of these texts as they made their way into the public arena. Seals, signatures, and ciphers functioned as well on a printed broadside as they did on a manuscript parchment. But these conventions did not keep forgeries out of circulation; some agitators could even appropriate them to create the illusion of an official convention (a *Florentina synodus*, or a new Council of Basel) when none really existed.

In addition, those publications that were authentic (and authenticated) often failed to command assent from those they addressed. As Geraldini learned, it could be difficult even to get such a document properly "published." State power was needed to reinforce publication and make it effective. It took a long time for the notoriously indecisive Frederick III to determine where he wanted to deploy his authority. When he did, the crisis resolved quickly.

Finally, we can detect a certain sensitivity to the power of public opinion at work in these printed exchanges, however poorly the protagonists were able to harness it. Geraldini's constant reference to the efforts he had made to reproduce and broadcast his decrees, his tireless "publications" in towns and even tiny villages across upper Germany, his enthusiasm for Institoris's publicity campaign, and his dismay at the opposition's ability to forge their messages and distribute them broadly, sticking them on doors and bridges and *loca insignia* without compunction, all speak to his concern for managing the public response to Jamometić. Financial records show that the Basel council was equally willing to invest in the same sorts of activities: printing, notarizing, distributing, posting, and standing guard over their decrees in as many places as possible. Ultimately, the 1482 attempt to revive the council failed, but the circumstances under which the various parties contended with each other would persist in ecclesiastical disputes yet to come.[114]

CHAPTER FIVE

The Holy Face, Imprinted
and in Print

When Andrija Jamometić delivered his critique of Sixtus IV to the people of Basel in 1482, relatively few in the city proved receptive to his complaints. The long dispute over his arrest, while conducted largely in public and in print, was really a jurisdictional conflict. Representatives of the pope and emperor argued over who was obliged to detain the obstreperous archbishop, who could put him on trial, and what penalties could be leveled against those who protected him. Fewer were interested in the substance of his objections: while councilors and legates debated the pope's legal standing to intervene in the judicial processes of a free imperial city, hardly anyone picked up on Jamometić's tirades against papal nepotism, corruption, or the "great extortions" of indulgence sales. In the early 1480s, the faithful of upper Germany were still inclined to trust the spiritual authority of Rome and to seek pardons from its bishop.

Emeric von Kemel was not the only indulgence preacher to print and sell papal indulgences in upper Germany and beyond in these years. The Ottoman capture of Otranto in 1480 had unleashed an anti-Turkish panic across Catholic Europe reminiscent of the anxiety following the loss of Constantinople and Negroponte in previous decades, and Sixtus sought to capitalize on the sentiment in a series of highly organized indulgence-selling campaigns. More than 160 different editions of indulgence texts—both the actual slips of paper sold to the faithful and administrative publications announcing new indulgences, suspending others, or extending their terms—were printed in German-speaking lands between 1480 and Sixtus's death in 1484. The Turkish threat was a compelling concern to the faithful, and confidence in the pope's ability to issue spiritual benefits to his flock remained high.

Purchasing an indulgence was the most convenient way for a faithful Christian to acquire remission of penalties for his sins, but there were other ways to win papal grace. The Jubilee of 1475 drew thousands of pilgrims from across Europe to Rome, and in the following years devout visitors continued to make their way to the city to pray at the tombs of the apostles and before the relics their basilicas contained. Their interests, too, were reflected in print, both before and after Jamometić launched his campaign. In September 1481, a Bavarian printer issued a German translation of a popular Latin history of Rome (the *Historia vel descriptio urbis Romae*) that told the story of the city from Romulus to Constantine, alongside a survey of its principal churches, their most important relics, and the indulgences a pilgrim could gain by visiting them (usually titled *Indulgentiae ecclesiarum principalium urbis Romae*, this text was also printed here in German translation).[1] A Munich printer brought out a reprint of these two texts in June of 1482,[2] and a year after that, just as the conflict between Geraldini and the city fathers of Basel over Jamometić was coming to a head, a Basel printer issued a much-expanded redaction of this same text, in Latin, setting out the spiritual attractions of the papal city in minute detail.[3] Meanwhile, two editions of a prayer invoking the most sacred relic in Rome, the veil of Veronica, appeared in Ulm in 1482, one in Latin and one in German translation.[4] Popular belief held that reciting this metrical prayer, *Salve sancta facies*, while gazing at the face of Christ imprinted on the veil could earn an indulgence. The Ulm editions, both broadsides, included a woodcut illustration of the Veronica cloth; in the Latin version, it is held aloft by the flanking figures of Peter and Paul, surmounted by crossed keys and a papal tiara (figure 5.1).

The Jamometić affair was a significant chapter in the history of conciliarism, and in attempting to suppress the archbishop's revolt, the papal legate was correct to note the worrying potential of the press as a tool for critics of the papacy to broadcast their complaints to the German faithful. But, as the above examples suggest, at this early date German printers saw less of a market in fomenting unrest against Rome and more in catering to the vast majority of faithful Christians who remained devoted to and fascinated by the pope and his city. German print stimulated the trade in indulgences, promoted the cult of Roman relics, and advertised the pope as proprietor of both.

As in Germany, the press in Rome played a major role in advancing papal initiatives like the sale of indulgences, the Jubilee, devotion to the Veronica, and other new traditions. These efforts are the subject of the next section of this book. Where previous chapters explored political events of the late fifteenth century and the ways in which individual actors—popes, princes, am-

Salue fanda facies noftri redemptozis.Jn qua nitet fpeces diuini amozis.Jmpffa
pāniculo niuei colozis.Dataqz veronice fignū ob amozif.Salue deus fectū fpeculū
fanctozum.Ad videre cupiūt fpiritus celozum.Nos ab omni macula purga vidoz
Atqz nos confoztio ziunge beatozum.Salue noftra glo.ia in hac vita dura.Labili
ac fragili dzoqz trāfitozia.Perduc nos ad patriam o felir figura. Ad videndū fadē
q eft rpi ? ura.Efto nobis qfumus tutum adiuuamē.Dulce refrigeriū atqz ?folamē
Vt nobis non noceat hoftile grauamen.Sed fruamur requie omnis dicat amen
Verficl's.Signatum eft fuper nos lumē vultus tui dñe.Dedifti leticā in cozde meo
Ozatio   Deus qui nobis fignatis lumine vultus tui memoziale tuū ad inftantiā
veronice imaginez tuam fudario imprefſam relinquere voluifti prefta fuplicibz tuis
vt ita in terris per fpeculum in enigmate venerari et honozare valeamus vt te tunc
fade ad fade venientez fup nos iudicio fecure z ioanze videre valeamus per eundē
dñm ūſm ihm rpm qui tecū viuit z regnat deus per omnia fecula feculog Amen.
Oū cū deo pre ꝛꝛ

bassadors, scholars, poets, editors, and printers—used the press to publicize their responses, the next three chapters follow a different pattern. In each, instead of events, I explore how particular ritual, devotional, and diplomatic traditions, some rooted in the medieval past and some quite newly created, further developed in fifteenth-century Rome. These traditions both supplied material for and were greatly amplified by the new medium of print.

In each case, I examine a body of texts that grew out of and together with one of these traditions, collections of literature that developed organically, often without any direction from the papal court, but which still reflected contemporary images of papal authority to their readers. These texts were all mainstays of the Roman printing industry in the late fifteenth and early sixteenth centuries: pilgrim guides to the major relics of Rome and the indulgences they offered; accounts of new relics brought to papal lands, above all the pilgrim shrine of Loreto in the northern papal states; and obedience orations, diplomatic speeches delivered by foreign ambassadors to the papal court after the election of a new pope.

Relics and pilgrims, on the one hand, and ambassadors, on the other, may seem to have little in common beyond the fact that all three were on the move in the fifteenth century, traveling to Rome in unprecedented numbers. But several strands link them. All three were attracted to Rome by the newly resurgent authority of the Renaissance popes, who certified relics, dispensed graces to pilgrims, and deployed both spiritual and temporal powers to arbitrate the diplomatic affairs of Europe. All three engaged in a dynamic relationship with the papacy: they needed what the pope had to offer, but their presence in Rome brought notable benefits to the pope as well. And all three acquired a remarkable presence in print. Devotional and diplomatic texts alike were published in Rome in enormous numbers, both describing and prescribing the parameters of a visit to the papal city. Here, too, the relationships between authors, readers, and texts were dynamic: while popes and their propagandists might project one message to the world from the Eternal City, pilgrims and ambassadors often came away with a very different one. Roman print could direct experiences and shape visitors' priorities, but printed works just as often reflected the ideals and served the needs that visitors brought with them to the pope's domain.

Pilgrims, Relics, and Books

Since late antiquity, Christian pilgrims had come to Rome to visit the tombs of Peter and Paul and their treasuries of relics: apostles' bones and saintly skulls, instruments of the Passion, and miraculous images of Christ and the Virgin. These holy objects offered beholders an immediate connection to the heroes of the early Church.[5] Yet, despite the antiquity of items like the True Cross or the heads of Peter and Paul, the stock of relics in medieval Rome could be quite fluid. Ancient relics might emerge, miraculously, from church crypts or other hiding places; new objects, rescued from infidel hands in the East, could be brought to the city with great fanfare; and new devotions to certain objects might develop organically, as the interests and tastes of the faithful shifted over time.

The Veronica, the most famous late medieval object of pilgrim devotion, only emerged into view at the end of the twelfth century. Over the course of the next two centuries, it gradually triumphed over other miraculous images in the city, in a process shaped equally by changes in devotion and by the same municipal tensions between commune and curia, between barons and popes, that defined so much of life in late medieval Rome (chapter one). Initially promoted by Innocent III, the popularity of the Veronica was linked to the gradual triumph of St. Peter's over the Lateran as the most sacred and most visited church in the city. The Renaissance popes, once ensconced in the Vatican, co-opted the Veronica for their own purposes, deploying it in diplomatic as well as liturgical contexts, as its sanctity added luster to the daily business of the papal court. At the same time, the holy image of the face of Christ attracted an international following. From the start, the Veronica had been used as an emblem of the Jubilee. It featured on documents that Boniface VIII issued for the first of these yearlong pilgrimage events, in 1300. It dominated the souvenirs that Jubilee pilgrims took home with them after visiting Rome; silver and tin badges and, later, single-sheet prints and printed guides to the city all gave pride of place to the Holy Face. In the late Quattrocento, and with the help of print, the fusion of the Veronica with the Vatican, and the assimilation of both to the figure of the pope, became complete.

How did pilgrim guides to the relics of Rome, its indulgences, and its historical lore become vehicles for advertising papal authority and charisma? The three pilgrim texts I examine in this chapter—the *Historia vel descriptio urbis Romae*, the *Indulgentiae ecclesiarum principalium urbis Romae*, and the *Oratio de Sancta Veronica*—circulated widely in both manuscript and print, produced in great numbers in Germany (and elsewhere north of the Alps) as well as in Rome. The papacy looms large in these works. Various popes are praised in the brief history of Rome and acknowledged as dispensers of grace in the catalogue of indulgences. The papal basilica of St. Peter's is celebrated as the home of the Veronica, the relic itself is lauded in the *Oratio*, and images of the Veronica appear next to papal portraits and devices on title pages and before and after individual texts. Pilgrim handbooks thus projected a vision of papal authority that was both temporal and spiritual, rooted in the histories of pagan Rome and its empire and in the triumphs of the early Church, poised in a liminal position between God and man, dispensing grace and commanding obedience.

Surprisingly, there is little evidence that the popes had much to do with the production of this literary *corpus*. Unlike the political bulls and broadsides examined in previous chapters, neither the popes nor their delegates had a hand in the composition or publication of pilgrim guides. Rather than examining these guides as instances of direct papal propaganda, then, it is more useful to think of them as reflective texts, offering pilgrims a vision of Rome, its ruler, and its spiritual economy that they wanted to know and responded to enthusiastically, which also aligned with the images projected by the Quattrocento popes themselves. It was a vision of Rome as the papal metropolis, the site of Peter's tomb, a treasure house of relics over which the popes had presided since the earliest days of the Church, the place where, thanks to Constantine, the authority of imperial Rome had passed seamlessly to its papal heirs. The pope also possessed, from Christ himself, the power of the keys, to loose and to bind the souls of his flock, a power he dispensed through the mechanism of indulgence. Papal charisma, tightly bound to the fascination of the ancient city and its monuments, pagan and Christian, was something pilgrims were invested in and actively sought. Print codified and solidified these notions because they were ideas that popes and pilgrims shared.

The Holy Heads of Rome

Printed indulgence guides give pride of place to the Veronica, but the relic was actually a fairly recent arrival on the Roman devotional scene. The scrap of cloth, supposedly imprinted with the image of Christ's face as he was on his way to Calvary, only emerged as an object of ritual focus in the late twelfth century.[6] There were other relics the popes had deployed in both public and political contexts for far longer, whether as pilgrim attractions, in ordinary and extraordinary liturgies, for rallying popular support in times of crisis, or as tokens of papal authority.[7] The Veronica's success in the fifteenth century was due, in part, to the way it reproduced and then supplanted the role played by earlier Roman relics.

For many centuries, the Salvatore, the icon of Christ held in the Sancta

Sanctorum, the pope's private chapel at the Lateran Palace, said to have been painted by angels or by St. Luke, was regarded as the premier relic of the city. The icon first appears in the documentary record in the eighth century, when Pope Stephen II (752–57) carried it in procession seeking divine help for the city against a Lombard siege. By the middle of the ninth century, the pope took the icon on procession every year, in an overnight ritual that involved an extensive tour of the city. On the night of August 14, the eve of the Feast of the Assumption, the pope would carry the Salvatore from the Lateran down the via Maggiore, past the Colosseum and through the Arch of Titus to the Forum, and finally back up the Esquiline Hill to the basilica of Santa Maria Maggiore.[8] Here the icon was brought face-to-face with the icon of the Virgin and Child known as the *Salus Populi Romani*.[9] The procession was staged as a reunion of mother and child, as the pope escorted the image of Christ on a visit to his mother on the eve of her feast.[10] The procession was also staged, in its early centuries, as a pageant of papal authority, with the pope cast as proprietor of the holy icon, of its keeping place in the Lateran, and of the monumental ancient ruins through which the icon processed.[11]

Like the Salvatore, the Veronica, too, was at first exposed and carried in an annual procession through the streets of Rome. Miraculously created, both images offered viewers an experience of virtual contact with the face of Christ, and the presence of both in Rome was taken as proof of the city's sanctity and the supernatural protection it enjoyed.[12] By the fourteenth century, each was venerated as though it had been in Rome from time immemorial, but in fact there was much that was unstable in the history of both. In the twelfth century, the pope began to lose his grip on the Lateran icon—perhaps as papal patronage at the Vatican increased, perhaps as the popular commune and its baronial allies emerged on the other side of the river. In an account of the ceremony composed in 1143—about the same time as the first communal statutes asserting the rights of the Roman people—the pope is described as going to the Sancta Sanctorum to prepare the Salvatore for its nighttime journey, but cardinals and deacons, followed by the Roman people, carry the icon through the streets; the pope only meets it again at Santa Maria Maggiore. A century later, the laity had taken over the procession completely, making it a civic pageant at which clergy, including the pope, only intervened at minor points and on the fringes of the action.

In the thirteenth century, a number of new hospitals were built in Rome, where the poor, the sick, orphans, and pilgrims could all take shelter.[13] In 1288, Cardinal Pietro Colonna rebuilt and expanded a small hospice on the

Figure 5.2. Stone plaque depicting the icon of the Salvatore flanked by the arms of the papacy and the commune. Rome, Ospedale San Giovanni Addolorata. Author's photo.

Lateran square and founded a lay confraternity, dedicated to the Salvatore, to manage it. This group rebuilt the hospital in the 1340s; gradually, they also took possession of the icon for which it was named.[14] In the Assumption festivities, the pope may still have received the icon at Santa Maria Maggiore and participated in mass there at the end of the night, but the confraternity, supported by city guilds, took over the job of escorting the icon through the streets.[15] The procession marked a high point in the city's civic calendar; it was at the climax of the event in 1347, just after the arrival of the Salvatore at Santa Maria Maggiore, that Cola di Rienzo had himself crowned tribune of the people and declared he would restore the liberty of the ancient republic.[16]

Technically, the Salvatore remained papal property, lodged in the Sancta Sanctorum for most of the year, but its *image* became the emblem of the confraternity, one of the most powerful institutions of the medieval municipality.[17] When they marched in procession, members carried banners painted with the image of the Savior's head. As the confraternity acquired land around the Lateran through purchases and bequests, members put up stone plaques of the head to mark their holdings (figure 5.2).[18] Though the Lateran was Rome's cathedral, the district around it was branded by the Salvatore as the territory of the confraternity and its noble patrons—Roman, baronial, and municipal, not papal.[19]

Meanwhile, at the Vatican, the cult of the Veronica was newer but no less subject to change.[20] There are records of a holy image of Christ's face kept in

St. Peter's going back to the eighth century, and twelfth-century documents mention that the basilica possessed a *sudarium* (towel) of Christ, though they do not mention its bearing a sacred image.[21] The Veronica proper, the cloth miraculously imprinted with the Savior's face, only emerged as a focus of ritual attention in the late twelfth and early thirteenth centuries.[22] The relic was kept in a special ciborium inside the Oratory of John VII, at the back of the (liturgical) south aisle of the basilica. One of the first references to it concerns a visit to Rome by Philip Augustus of France in 1193, when Celestine III honored him with a private showing.[23] Its fame grew when Innocent III co-opted it for his own political and urbanistic projects.[24] In 1203, Innocent refounded the ancient "Saxon" hospital of Santo Spirito in the Borgo and assigned a new religious order, the Order of the Holy Spirit, to administer it.[25] In 1208, he established a lay confraternity to support the hospital and issued an indulgence for pilgrims who visited it. Once a year, the pope declared, the confraternity and the canons of St. Peter's were to take the *effigies Christi* in procession from St. Peter's to Santa Maria in Sassia, next to the hospital, and then back again. In his bull establishing the new station, Innocent specified that the procession should be understood as a reunion of Christ with his mother, echoing the ritual logic of the Salvatore procession.[26]

The whole complex of building, relic, and rite echoed that which obtained at the Lateran: another basilica, another miraculous image of Christ, another procession to a Marian church.[27] Innocent's foundation of the new hospital also prompted the baronial patrons of the Lateran to promote their own hospital and confraternity in response. The stage was set for centuries of rivalry between the two basilicas and their charitable precincts. Innocent's patronage of Santo Spirito and promotion of the Veronica were part of a larger campaign to establish the Vatican and Borgo as a new papal enclave, one that would match if not outstrip the Lateran in sanctity and prestige.[28] In this, Innocent anticipated Nicholas V's more permanent move to the district in the 1450s.

Innocent's promotion of the relic would also resonate far beyond the walls of Rome. As a miraculous image to rival the Salvatore, the Veronica was a remarkable choice. A contact relic of Christ, one of the few traces of the Lord's bodily presence on earth, it was even more sacred than a miraculously painted portrait, and, as Gerhard Wolf has argued, it gained yet more stature *because* it was a reproduction, one that lent itself to copying, distribution, and veneration from a distance.[29] In the thirteenth century, a prayer attributed to Innocent (the "Office of the Holy Face") began to circulate, which anyone could say, whether in the presence of the image itself or a copy of it, and thereby earn an indulgence.[30] The prayer was usually illustrated with hand-drawn or painted images of the Holy Face (or, perhaps, the image was "captioned" by the prayer). Either way, the combination of text and image extended the reach of the Veronica even further, allowing virtual pilgrims in distant lands to participate in its cult.[31] In fourteenth- and fifteenth-century manuscripts, the office was sometimes supplemented or even replaced by one of two anonymous metrical prayers or hymns in praise of the icon, *Ave facies praeclara* or *Salve sancta facies*.[32] The latter text would have a long afterlife in fifteenth- and sixteenth-century printing; this is the text printed, for example, on the Veronica broadside printed in Ulm in 1482 (figure 5.1).

As the icon of the Salvatore became localized within the confines of baronial or communal Rome, associated less with the pope and more with the civic laity and their annual procession, the Veronica became a more exclusively clerical, indeed *papal*, but also universal image of Christ. It traveled less. More often, it was shown in situ to pilgrims inside St. Peter's. In the Jubilee of 1300 these expositions were weekly events at which throngs of pilgrims gathered, as both Dante and Giovanni Villani attest.[33] In the century that followed, the canons of the basilica gave private expositions to visiting dignitaries. For much of this time, the popes were ruling in Avignon, not Rome.[34] The hospital of Santo Spirito fell into disrepair, and the annual procession of the relic through the Borgo may no longer have occurred.[35]

The Veronica retreated from the streets, giving up its place in urban ceremony to become, instead, the emblem of the international pilgrim, reproduced and broadcast on the many paper, cloth, and metal souvenirs that pilgrims bought in Rome, wore on their hats or clothing, and carried with them across the continent.[36] While the Salvatore continued to beat its annual path across the *disabitato* of baronial Rome, the Veronica no longer moved at all. It sat at the center of a vast circle of devotion encompassing the whole of Europe.

The Veronica had a dual character—it was *the* image that pilgrims to Rome longed to see, to pray before, to buy in reproduction, and to carry home with them.[37] But the image also served as a potent emblem of papal sanctity and authority, one that enhanced the status of St. Peter's and the Vatican as the international seat of the Church, in contrast with the Lateran, now "merely" the cathedral of Rome.[38] Already in 1300, a parchment announcement of the first Jubilee was headed by a painting of the Holy Face flanked by Peter and Paul;[39] a fourteenth-century seal belonging to one of the canons of St. Peter's shows the same saints lifting the Veronica in its frame between them;[40] and after the Jubilee of 1350, the Roman Senate minted a new gold coin, modeled on the Venetian ducat, that included various insignia of the city: the crossed keys, the SPQR monogram, the words *Roma Caput Mundi*, and a tiny image of the Holy Face.[41] The Veronica functioned as both relic and emblem, an object of pilgrim devotion and a sign of papal authority, associated with the pope, his city, and its place at the head of the universal Church.[42]

The Veronica and Salvatore in Pilgrim Guides

The changing fortunes of the Veronica and Salvatore can be traced in the pilgrim guides to Rome's ancient and sacred attractions that proliferated over the fourteenth and fifteenth centuries. The pilgrim experience of Rome overlapped only partially with the civic religion of the city; pilgrim guides pointed visitors to relics and liturgies where one might earn indulgences, which not all civic processions could offer. Nevertheless, the minor role that the Salvatore plays in these texts is striking. The calendar of stational liturgies known as the *Stationes Romae*, for example, mentions no special events at all for the Feast of the Assumption. The text does list other holidays when relics were exposed: a period in Lent when the icon of the Virgin at Santa Maria del Popolo was displayed; St. Mark's Day, when "all the relics of the city" were gathered and carried in procession to St. Peter's; St. John's Day, when the head of John the Baptist was exposed at San Silvestro in Capite; and the Feast of the Ascension, when the *vultus sanctus* or Holy Face would be exposed at St. Peter's.[43] But amidst all this, there is no mention of the Salvatore at all.

Both relics make an appearance in the *Indulgentiae ecclesiarum principalium urbis Romae*, the text German printers were publishing in translation in the early 1480s.[44] This extraordinarily popular text, first compiled in the late fourteenth century, sets out the attractions of the great churches of Rome: their shrines, relics, and festivals and the indulgences that could be earned by visiting them. Different redactions survive, some longer and some shorter, describing different churches and relics and the indulgences they offered. But all follow the same pattern, starting with the seven major basilicas of the city and moving on to lesser Roman churches.

Probably composed during the Great Schism, and well before Nicholas V moved the papal court from the Lateran to the Vatican, the text gives pride of place to the Lateran and its relics. The cathedral of Rome comes first in the text, and its description fills more than a third of the section on the seven basilicas, while the description of St. Peter's that follows is just over half as long.[45] The author extols the attractions of Rome's cathedral as a pilgrim destination, but his account suggests that already in the fourteenth century the Lateran *needed* promotion. Were fewer pilgrims making their way there? Every day at the Lateran, the author says, indulgences are granted amounting to many years of remission—by one reckoning, more than fifty years off, by another, so many "that God alone can count them."[46] If people only knew how many indulgences the Lateran offered, he remarks, they would see no need to go to Jerusalem or Compostela.[47]

Pope Sylvester himself founded this church; its treasures included the Ark of the Covenant; the rod of Aaron; the Virgin's veil, as well as her milk, hair, and clothes; Christ's foreskin; the water and blood that spilled from his side; the table from the Last Supper; and the heads, bodies, and contact relics of countless saints. As the author describes the most famous of these, the two great silver reliquaries over the high altar containing the heads of Peter and Paul, he drops another telling remark: by viewing these relics, the pilgrim may receive as many years of indulgence as can be had from an exposition of the Veronica.[48] The sense of competition is palpable: pilgrims neglecting the Lateran in favor of the Holy Land or Spain or the Veronica across town were missing out on something special.

Given this, the author has every reason to draw the reader's attention to particularly attractive relics at the Lateran, and yet he barely mentions the Salvatore at all. It comes at the end of the entire section and is treated very briefly. The nearby Scala Sancta is touted as the more worthwhile pilgrim attraction:

> In the chapel of San Lorenzo . . . which is called *Sancta Sanctorum*, there is remission of all sins; in this same chapel there is an image of Christ when he had lived twelve years as God and man, and around it are the steps down which He fell and on which he spilled blood, and where a mark appeared as each drop of blood fell; these were the steps in front of Pilate's house in Jerusalem, and whoever climbs these steps will receive, for each step, nine years and nine times forty days' indulgence and remission of one third of all his sins.[49]

The Salvatore seems to have counted for very little. It is not clear that the author had even seen the icon, which does not depict a twelve-year-old youth but a bearded and regal adult. The author notes that the *chapel* of the Sancta Sanctorum offers a plenary indulgence, but this seems to depend on entering it (or perhaps hearing mass within), not viewing the icon. The Scala Sancta is the more interesting relic, and the *really* high-value object is the double reliquary of Peter and Paul over the high altar of the basilica—the object the author casts as the Lateran's main attraction, the only one to offer as much grace as the Veronica.

The next section of the *Indulgentiae* treats St. Peter's. The church is presented in almost identical terms as the Lateran: it, too, was founded by Pope Sylvester, it offers the same plenary indulgence, and the pilgrim can gain additional indulgences from climbing another set of sacred steps.[50] The church has multiple altars, each offering further indulgences, and the relics of numerous saints (Simon and Jude, Petronilla, Chrysostom, and of course Peter and Paul—or part of them, their heads being at the Lateran and the other half of each body being held at San Paolo fuori le Mura). Most potent of all is the Veronica, "the holy face of our Lord Jesus Christ." The author enumerates the several feast days on which it is displayed and the extraordinary indulgences that can be had from witnessing an exposition: three thousand years of remission for Romans, six thousand for those coming from the outlying countryside, and twelve thousand for those who come from abroad, "over mountains, hills, or valleys."[51] In other words, the Veronica was the relic for the long-distance pilgrim—of far greater value to the ultramontane than the Roman.

The *Reliquie rhomane urbis atque indulgentie* (first printed in Basel around 1483 but dating to earlier in the fifteenth century and derived from the original *Indulgentiae*) likewise reflects the changing status of the two icons and adds interesting details about their respective settings. In the oratory of St. Lawrence, known as the Sancta Sanctorum, one can find at least a dozen important relics—Christ's umbilical cord, pieces of the True Cross, the bones of various saints, and so on. Among these is a painting of Christ started by St. Luke and finished by angels. Having mentioned it, the author turns almost immediately to describing the sacred stones and relics embedded in the altar beneath it. A few lines later, he devotes twice as many words to another icon in the chapel, a "face of the savior, painted on a wooden panel," which miraculously survived an attack by hostile Jews and saved a faithful Christian who owed them money.[52]

In the section on St. Peter's, by contrast, the author presents the Veronica as one of the main attractions. The "holy face of Christ, which appeared from his sweat when he was on his way to his Passion," is kept behind one of the seven major altars of the basilica and shown every Sunday that the hymn *Omnis terra* (i.e., *Jubilate deo*) is sung, as well as every day in Holy Week.[53]

The author then repeats the same schedule of indulgences as is found in the *Indulgentiae*.

The relative fortunes of the Veronica and Salvatore were all but fixed once the Schism was settled and the popes returned to Rome for good. Giovanni Rucellai, in his description of Rome during the Jubilee of 1450, mentions the Salvatore as a great treasure of the church,[54] but he also reveals how little the Sancta Sanctorum now figured in the ritual life of the popes: "It is said that in this chapel [the Sancta Sanctorum], no one but the pope may say mass, and that it has been more than a hundred years since mass has been said there, neither by the pope nor by anyone else, except that the previous pope Nicholas V had a chaplain say mass there once in 1448."[55] The Veronica, by contrast, drew great crowds of pilgrims for the intense emotional experience of an exposition. The German pilgrim Nicholas Muffel was overcome during his visit in 1453: "One's heart is struck with terror the moment one catches sight of it."[56] Basilica and relic alike together attract high praise in Flavio Biondo's *Roma instaurata* (ca. 1448–58). Biondo praises the Vatican as the seat of a newly triumphant papacy and the Veronica as its most prized possession. Rome was and is the greatest city the world has ever known, Biondo maintains. The ancient empire may have been lost, and most of the city's monuments may now lie in ruins, but Rome still rules over the world—now through the power of religion, not arms. The pope is the new dictator, the cardinals are his senate, and all the world pays tribute to the church of Rome. And what a splendid city it is. Built on sacred ground, watered by the blood of martyrs, Rome possesses not only the tombs of the apostles but also "the image of the face of the Lord, which is preserved on the cloth of Veronica and presents his very features, and which is kept in the basilica of St. Peter, the first and mother of all the churches in the world."[57]

The Veronica in Print

The descriptions in the *Stationes*, *Indulgentiae*, and *Reliquie Rhome* and the accounts by Rucellai, Biondo, and other fifteenth-century visitors to Rome reflect the steady rise of the Veronica as a focus of pilgrim attention. The icon was to achieve even greater prominence with the arrival of print. The text of the *Indulgentiae* was first printed in Rome in the early 1470s; during and after the 1475 Jubilee, both visual representations of the Veronica and textual accounts of its exposition would dominate the printed pilgrim guides to the city.

The printing history of the *Indulgentiae* is complex and closely linked to that of several other texts popular with pilgrims and printed in Rome in great numbers: the *Mirabilia Romae* (an account of the ancient monuments of the city—buildings, arches, bridges, and statues—and the legends that attended them), the short historical survey *Historia et descriptio urbis Romae* (hereafter *Historia*), the *Stationes Romae*, and the metrical prayer or hymn *Salve sancta facies* (which circulated in print under the title *Oratio de Sancta Veronica*), as well as other devotional texts like Andreas Escobar's *Modus confitendi*, the *Orationes* of St. Bridget, and the account of the Holy House of Loreto known as the *Translatio miraculosa* (discussed in chapter six). These are all quite different texts, but they tended to travel together, and multiple Roman printers published and republished them in varying combinations over the last three decades of the fifteenth century.[58]

Bibliographies have tended to obscure the more precise contours of this corpus of literature, often subsuming editions of the *Mirabilia*, the *Historia*, the *Indulgentiae*, and the *Oratio* under the general heading of *Mirabilia Romae*. The *Historia*, like the *Mirabilia*, is mostly concerned with pagan Rome, and, adding to the confusion, both texts were often printed with the same title-page woodcut depicting the ancient foundation myth of Rome: the Vestal Virgin Rhea Silvia praying in a temple of Mars while the she-wolf suckles her twins, Romulus and Remus, nearby. However, the two texts were quite distinct. In the entire incunabular period, the *Historia* was almost never printed in an edition together with the *Mirabilia*. And the *Historia* was, in fact, printed more frequently.[59]

But it was not exactly an independent text. In every surviving edition but one, the *Historia* was printed as a preface to the longer text of the *Indulgentiae*, not on its own. And the *Indulgentiae* actually appeared in print first, published some nine times on its own starting in the early 1470s; it was only in the mid-1480s that the *Historia* began to be paired with it in movable type editions.[60] More than fifty further editions of the *Indulgentiae* then appeared, in composite volumes together with either the *Historia* or the *Oratio* or both. The point of all this is that the *Indulgentiae*, not the *Mirabilia* or the *Historia*, was by far the most popular of "Mirabilia" texts in late fifteenth-century Rome; it was printed more often than any other pilgrim text, and it tended to be the constant, appearing either on its own or in varying combinations with other texts that hardly ever appeared on *their* own.

Finally, while the *Mirabilia* and the *Historia* had their single illustration of

Rhea Silvia, the *Indulgentiae* gradually acquired an entire suite of woodcut illustrations (copied by one printer from another, in several different runs, each distinct from the next but iconographically consistent). The relationship of the *Indulgentiae* to its attendant texts and the development of their iconography will be my focus here. My point is to stress, first, the ubiquity of the Veronica in both text and image and, second, the assimilation of the Veronica to a body of historical and political images (and texts) that underscored its identity as a particularly papal icon and an icon of Rome.

The printing history of the *Indulgentiae* can be divided into several phases:

1(a). Between the introduction of printing with movable type to Rome in the late 1460s and the Jubilee of 1475: a block book edition of the *Indulgentiae* in German was produced, paired with the *Historia* in German, with woodcut illustrations.

1(b). The same period as above: quite independently, short editions of the Latin *Indulgentiae* were published in print, almost always on their own and always without illustrations.

2.   1478–84: the Latin *Indulgentiae* began to be published together with the *Oratio de Sancta Veronica*, still without illustrations.

3.   1485–87: these various traditions fused to produce two composite editions: a Latin *Historia, Oratio,* and *Indulgentiae* with woodcut illustrations; and a German *Historia* and *Indulgentiae,* without the *Oratio,* but with the same suite of woodcuts. This model was established practically simultaneously in the mid-1480s by the Roman printers Andreas Freitag, Bartholomaeus Guldinbeck, and Stephan Plannck and then copied by other Roman printers till the end of the century.

4.   1488–1500: production of composite, illustrated editions in Latin, German, and Italian steadily increased until the start of the 1500 Jubilee, at which point the numbers soared, with dozens of individual editions issued and reissued on an almost monthly basis.

Printing was in its infancy in Rome in the years leading up to the 1475 Jubilee. Sometime before or early in that year, a block book was printed, unsigned but probably printed in Rome, with a German translation of the *Indulgentiae* prefaced by a German translation of the *Historia*.[61] This book was once thought to predate the arrival of print in Rome, possibly as early as the Jubilee of 1450. But the book, some ninety-two pages long, is printed on both

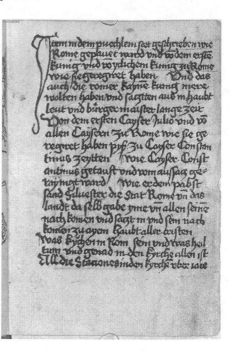

Figure 5.3. *Historia et descriptio urbis Romae + Indulgentiae ecclesiarum principalium urbis Romae* (German). Munich, BSB, Xylogr. 50, fols. 1v–2r.

sides of the leaf, a technique that only became possible after the development of the movable type press.[62] It has five woodcut illustrations, of which two depict the Veronica.

Seven copies of this book survive. Thanks to the flexible system of imposition that block book production allowed, some of the woodcuts appear in different places in different copies. The variations are significant enough that it is impossible to say precisely where the maker of the book intended the cuts to appear. The Munich copy examined here is representative but need not be considered the standard.

The first recto of the book is blank, but on the verso of the leaf a woodcut shows the Veronica as a fluttering cloth imprinted with the face of Christ, held aloft by two angels who hover over a set of three shields containing, from left to right, the crossed keys of the papacy, the della Rovere oak, and the SPQR monogram (figure 5.3). The central shield is surmounted by the papal tiara and pallium. (In the Gotha copy, this image appears not at the start but at the very end of the book as a sort of visual colophon.)[63] Opposite, on the recto of the second leaf, is the preface to the *Historia* ("Item in dem

Figure 5.4. *Historia et descriptio urbis Romae* + *Indulgentiae ecclesiarum principalium urbis Romae* (German). Munich, BSB, Xylogr. 50, fols. 2v–3r.

puechlein stet geschrieben wie Rome gepauet ward . . . ”); this is followed on the next verso by a full-page woodcut image of Rhea Silvia praying in a temple, with a castle representing Rome in the background and her infant twins suckling from the she-wolf in the foreground (figure 5.4).

The main text of the *Historia*, which starts with the births of Romulus and Remus, follows on the recto of the third leaf. The opening lines are surmounted by another ensemble of shields emblazoned with the imperial double eagle and the title “Imperium” one on side, the SPQR monogram with the title “Roma” on the other, and the crossed keys of the papacy, with papal tiara and pallium dominating the visual space of the center. The *Historia* concludes some twenty-five leaves later with the life of Constantine and his donation of Rome and the western empire to Pope Sylvester. Thus, the text, while ostensibly a history of pagan Rome, both starts and ends with indications of papal supremacy over temporal powers, ancient and modern.

At the next opening, before the start of the *Indulgentiae*, a full-page woodcut on fol. 27v presents another image of the Veronica, this time a scene depicting the moment of exposition in St. Peter’s, when the relic was held aloft

Figure 5.5. *Historia et descriptio urbis Romae + Indulgentiae ecclesiarum principalium urbis Romae* (German). Munich, BSB, Xylogr. 50, fols. 27v–28r.

before crowds of the faithful (figure 5.5). In a high, freestanding pulpit, a priest, flanked by acolytes carrying burning tapers, presents the icon in its frame for veneration to a group of standing pilgrims below. The text of the *Indulgentiae* begins on 28r. Thus, an image on the verso of one leaf again serves as a visual introduction to the text that follows on the recto of the next. As Rhea Silvia and her twins were to the classical Rome of the *Historia*, so the Veronica is to the Christian city of churches and shrines: its oldest relic, its emblem, and a sign of the divine favor the city enjoys.[64]

The *Indulgentiae* begins with another reference to Pope Sylvester—specifically, his catalogue of the churches of the city; accordingly, the text begins with a historiated initial containing an image of that pope. The *Indulgentiae* occupies the rest of the book; in some copies, the collection concludes with the woodcut of the Veronica with angels and shields (figure 5.3) that, in the Munich copy, occupies fol. 1v.[65] In these copies, the cut of the Veronica exposition is moved to fol. 1v to serve as the frontispiece to the book, and the break between the two texts at fol. 27 is left without an illustration.

Wherever it is placed, the Veronica serves as the emblem of this Jubilee book. As the supreme goal of the pilgrim, it is the one relic out of the hundreds described that is illustrated, and not just once but twice. Both times it appears with papal images—with the crossed keys, the insignia of the pope, in one instance, and with a portrait of Sylvester in the other. It is presented as the property of the papacy, just as the Jubilee itself is a papal enterprise. The Jubilee pilgrim made his way not just toward the Veronica but to the city of the pope, who presided over the ancient city, decreed the Jubilee, and was the author of the indulgences the pilgrim sought. As Sylvester founded the first great churches of Rome, now Sixtus would welcome the Jubilee crowds inside them, presiding over city and church alike as Christ's vicar on earth.

The block book set out an iconographic program for the *Indulgentiae* that was enormously influential over the next half century, although, curiously enough, after their publication around 1475, the images disappeared from circulation for more than a decade. Block books were never widely produced in Italy, and the format was already giving way to books printed with movable type by 1470. In these early years, though, printers working with movable type rarely incorporated illustrations into their products. Around the same time as the block book was produced, in the years leading up to the 1475 Jubilee, several Roman printers published editions of an abbreviated Latin text of the *Indulgentiae*, but none had illustrations.[66] These were short, single-quire quarto or octavo pamphlets, six to eight leaves long, and many seem to have been printed with matching editions of the *Mirabilia Romae*, though never in a single edition *with* that text.[67] As soon as the Jubilee of 1475 was over, this run of production stopped.

Then, starting in 1478 or so, the *Indulgentiae* began to appear again, this time prefaced by the *Oratio de S. Veronica*. It was an obvious pairing to make, given the centrality of the Veronica to the pilgrim experience. In addition, the year 1478 saw Sixtus IV refound the Ospedale di Santo Spirito in the Borgo; an imposing new brick building, surmounted by an octagonal lantern, rose on the bank of the Tiber between Ponte Sisto and Ponte Sant'Angelo. The pope issued a bull granting the hospital certain privileges and issued indulgences to those who participated in the celebrations every year to commemorate the anniversary of its refounding. The Veronica would be displayed in St. Peter's while other relics would be carried in procession to the hospital church. The streets of the Borgo would have echoed with the words of the *Oratio*, and Sixtus's initiative may account for the appearance of the *Oratio* in print starting in 1478.

The prayer, or hymn, consists of forty-eight rhyming verses praising the image of Christ on the cloth and giving thanks to Christ for leaving this sign of himself for the world to adore:

> *Salve sancta facies nostri redemptoris*
> *In qua nitet species divini splendoris*
> *Impressa panniculo nivei candoris*
> *Dataque Veronice signum ob amoris . . .*

> Hail our Redeemer's Holy Face,
> All aglow with heaven's rays,
> On snow-white cloth it left its trace,
> Veronica's sweet sign of grace . . .

The verses are followed by an abbreviated version of the Office of the Holy Face attributed to Innocent III, asking God's blessing on the viewer of the relic. In the printed editions, the prayer is marked with its liturgical headings: "versus" for the Psalm text (4:7) and "collecta" for the collect.

The *Oratio* has been variously attributed to Giles of Rome (d. 1316) or John XXII (d. 1334). By the fifteenth century, pilgrims believed that either this hymn or the original office could be recited in front of an image of the Veronica in order to gain an indulgence. Text and image tended to travel together, in manuscript at least, as can be seen in Petrus Christus's *Portrait of a Young Man* (ca. 1450–60) in the National Gallery, London, where an illuminated manuscript copy of the *Oratio*, illustrated by a copy of the Holy Face, is tacked to a wall behind the figure of a young man at prayer.[68] By contrast, the first four Roman printed editions to contain both the *Oratio* and the *Indulgentiae* still present the texts without illustration.[69] This was a less than ideal arrangement: the point of reciting the prayer was to intensify the experience of viewing the image, and it was believed that the indulgence could only be earned by the combination of the two. North of the Alps, as we have seen, German printers would soon combine the elements in just this way: the two broadsides produced in Ulm in 1482, mentioned above, included the text of the prayer (in Latin and German, respectively) surmounted by a woodcut image of the Holy Face (figures 5.2 and 5.6).[70] Printers in Rome soon supplied the prayer text with an image as well. Between 1485 and 1487, a new group of printers began to issue illustrated versions of the *Indulgentiae* in an expanded redaction, one that included indulgences for many other churches in Rome beyond the seven major basilicas, printed together with the *Oratio* and with the *Historia*, a text that had appeared in German translation in the block book

Grieſt heſtu hailiges antlit vnſers behalters · Jn dē
da ſchinet dʒe geſtalt des götlichen glanczes. Gedru
ket in ain ſchne wiſſes dieʒblin Vñ gegebē weronice
cʒü ai...nem ʒaichen der liebe. Grieſt heſtu geczierd der
woelte ain ſpiegel der hailigen · Den da begerend cʒü
ſchowen die hymelſchē gaiſte Künige vns von allē
hindē · Vnd ſieg vns zü der ſeligē geſelſchafft. Grieſt
heſtu vnſer gloſi in diſem hertten hinflieſſenden vnd
ſchwachem leben · Fier vns cʒü dem vatterland o du
ſelige figure · Zü ſehend das wöneuglich antlit criſti
vnſers herren · Bis vns ain ſichere hilff ain ſieſſe erkie
long troſt vnd ain ſchirme · Das vns nit ſchadē müg
die beſchwerong vnſer künde Sonder das wir niel;
ſend die ewige rüo amen

So fil ſind gegeben tag applas vnd karen diſem
gebet das ich ſy hie nit künd wol begriffen

Figure 5.6. *Oratio de S. Veronica, Salve Sancta Facies* (German) ([Ulm: Conrad
Dinckmut, ca. 1482]). Munich, BSB, Einblatt. VII,4.

of ca. 1475 but had yet to be printed in movable type in Rome. (The German text of the *Historia* had already been printed with type, as we have seen, twice in southern Germany in 1481 and 1482, but as of yet there had been no editions of it printed in Rome). Roman printers now produced illustrated editions of the *Historia, Oratio,* and *Indulgentiae* in Latin, as well as illustrated editions of the German-language *Historia* and *Indulgentiae.*[71]

The pioneers of these composite editions were a new group of German printers who had arrived in Rome in the late 1480s: Andreas Freitag, Bartolomaeus Guldinbeck, and Stephan Plannck, followed by Petrus de Turre, Eucharius Silber, and Johannes Besicken, who worked with a changing array of partners and in the early 1490s further refined the model by adding a new series of woodcuts to illustrate each of the seven churches described in the *Indulgentiae.*

It is not immediately clear who came first. At some point, Freitag and Guldinbeck each acquired the woodcuts from the block book edition published before 1475. Each reused some (but not all) of the cuts in new editions of the text printed with type. Freitag's earliest edition is undated. Guldinbeck's is signed "28 August 1487." But the redaction of the *Indulgentiae* that every one of these printers used had a number of updated entries, including information on relics recently introduced to the city, new indulgences granted by Pius II and Sixtus IV, and recent miraculous events. The latest of these refers to an image of the Virgin in the church of Sant'Agostino that performed miracles in August 1485.[72] It seems likely that some edition of this updated text would have appeared in print earlier than August 1487. The most plausible scenario is that Andreas Freitag came into possession of the woodcuts from the block book and used them to illustrate the newly updated Latin text of the *Indulgentiae,* together with the *Oratio* and *Historia.*[73] His edition may be the first. The cuts would have then passed to Guldinbeck, who used two of them to illustrate his dated 1487 edition, which presented the German translation of the *Historia* and *Indulgentiae.*[74] At some point, Plannck commissioned an entirely new set of woodcuts, modeled closely on the block book ones but with some important differences (discussed below). Plannck's first illustrated German edition has been assigned a date of sometime between September 1484 and 1487, a range that can be narrowed given the reference in the *Indulgentiae* to events of August 1485;[75] his first two illustrated Latin editions, each including the *Oratio,* are assigned to 1485–87,[76] as well as ca. 1486.[77] Plannck's editions should probably be dated after Freitag's, but not necessarily after Guldinbeck's. To summarize a complex set of relationships:

LATIN EDITIONS	GERMAN EDITIONS
Indulgentiae	*Historia + Indulgentiae*
[1471–75] Various printers	+ woodcuts (I)
(eight editions)	[1475?] Block book
Indulgentiae + Oratio	
[1478–87] Various printers	
(four editions)	
Historia + Oratio + Indulgentiae	
+ woodcuts (I)	*Historia + Indulgentiae*
[1485–87] Freitag?	+ woodcuts (I)
	1487 Guldinbeck
Historia + Oratio + Indulgentiae	*Historia + Indulgentiae*
+ woodcuts (II)	+ woodcuts (II)
[1485–87] Plannck	[1485–87] Plannck

Woodcut Illustrations in the Printed Guides

In his first edition of the composite text (*Historia + Oratio + Indiulgentiae*), Andreas Freitag recycled all four of the woodcuts from the block book edition: Rhea Silvia opposite the start of the preface to the *Historia*; the three shields directly above the start of the text; the Veronica exposition later, opposite the start of the *Oratio*; and the Veronica held by angels, above another set of shields, facing the description of Santa Maria in Trastevere, the first church after the seven major basilicas in the long redaction of the *Indulgentiae*. In this last case, Freitag made some changes to the shields, removing the della Rovere oak from the central shield since Sixtus IV was no longer pope and removing the papal tiara and pallium from above it as well. (Guldinbeck simply dropped the shields altogether from his German-language edition and included just the cuts of Rhea Silvia and the Veronica exposition.)

Once the basic pattern for the new composite edition was established, it quickly became the standard. The combination of texts, the particular set of woodcuts, and their placement in the edition were all fixed and then repeated dozens of times over the next fifteen years, first by Plannck and Freitag, and then by Silber, Petrus de Turre, and Besicken. Each printer commissioned his own set (sometimes, more than one set) of woodcuts to serve as illustrations. In all, more than fifty illustrated, composite editions of the

Historia and *Indulgentiae* in Latin, German, and Italian appeared between 1485 and the end of the Jubilee year of 1500.

Freitag's was almost certainly the earlier Latin edition, but it was Plannck who commissioned new and slightly different woodcuts that became the model for later printers. Plannck's new woodcut of the Veronica exposition appeared on the verso facing the *Oratio* (figure 5.7).[78]

Compositionally, there are some important differences between Plannck's new cut (figure 5.7) and its original (figure 5.5). In the block book version, the cleric, flanked by two acolytes holding burning tapers, displays the relic from a freestanding pulpit to a crowd of pilgrims in the foreground below. In Plannck's version, the clergy are more firmly located in an architectural setting, a balcony or pulpit set into a wall, with a perspectival ceiling receding above them, while the burning tapers are now held in twin brackets fixed to the structure or the wall behind it. Below, the crowd of pilgrims now kneels, though a single pilgrim at the back of the crowd remains standing, perhaps to get a better view of the exposition.

The scene is more detailed and more firmly fixed in architectural space than the one in the block book. This marks the end point of a curious shift: manuscript and woodcut images of the Holy Face tend to show it in complete isolation, either on its cloth or in a frame surrounded by a halo, a model that German prints (e.g., figures 5.2 and 5.6) also follow. Their Veronica purports *to be* the Veronica of Rome: a faithful reproduction of the original image. What the Roman block book offered instead was a visual souvenir of the moment of exposition as it was orchestrated inside St. Peter's. Presumably the tiny framed image of the Veronica at the center of the scene was a close enough reproduction of the original to secure the indulgence that came from reciting the *Oratio* before a copy of the *sacra effigies*. But the pilgrim purchaser of one of these new, printed editions also took away an almost photographic tableau of the intense, perhaps even ecstatic, moment of her encounter with the relic. Plannck's further revision of the scene sets it even more firmly in its architectural setting, the nave of St. Peter's.

In the block book, the exposition scene accompanied the text of the *Indulgentiae* with its general survey of the churches and relics of Rome. In the new composite edition, the scene instead accompanied the text of the *Oratio* alongside the liturgical texts of the Office of the Holy Face. Here, response and collect were typographically marked out for the reader, further grounding the text and image in a particular moment: the prayer was a text the faithful reader could say anywhere, but the pilgrim also took away a transcript of

Figure 5.7. *Historia vel descriptio urbis Romae + Oratio de S. Veronica + Indulgentiae ecclesiarum principalium urbis Romae* ([Rome: Stephan Plannck, ca. 1485–87]). Munich, BSB, Inc.s.a. 195, fol. 19v.

the liturgy she herself may have participated in.[79] The book thus fixed the Veronica in time as well as space—at the climax of the pilgrim's journey, at the moment of exposition and benediction, inside the papal basilica, in the sacred city of Rome. The inclusion of one, standing pilgrim at the back further serves as a proxy for the viewer—the one farthest from the relic who stands and strains to get a better view.

Following Plannck, Andreas Freitag in the 1490s commissioned new woodcuts for his own composite editions of the *Historia*, *Oratio*, and *Indulgentiae*. In his 1492 edition of the collection, Freitag included a new woodcut of the Veronica, printed on the same page as the start of the *Oratio*, modeled on the frontispiece image of the Veronica with shields that appeared in the original block book (figure 5.3), which he himself had reused in his first edition in the 1480s: two hovering angels holding the cloth, imprinted with the Savior's face, aloft in space. (This is cut down from a larger woodcut that also showed the shields below; a few stray lines from the shields can be seen at the bottom of the image.)[80] In his 1494 edition, however, Freitag presented a new suite of illustrations modeled closely on Plannck's own (figure 5.8). Later, in the 1490s, Plannck produced another woodcut of the exposition scene, which quickly passed to Besicken. In this cut, the exposition scene is even busier, with more pilgrims in the foreground, a geometric band representing stonework on the front of the pulpit or some kind of fabric antependent or banner hanging over it, and a triangular roof, pediment, or canopy sheltering the clerics as they present the icon. The entire scene is framed by a thick border of white vines, vegetation, and the emblems of the four Gospels in the corners. This woodcut was first used by Plannck and later passed to the printing partners Johannes Besicken and Sigismundus Mayer. Later, a close copy of the cut appeared in books printed by Besicken in partnership with Martinus de Amsterdam (figure 5.9). Two other printers, Petrus de Turre and Eucharius Silber, included their own versions (figures 5.10 and 5.11) of the scene in editions printed around 1499.[81]

Thus, the Veronica maintained its place at the center of the pilgrim books (just as it occupied the pinnacle of the pilgrim experience in Rome); more than that, the books presented *the experience* of viewing the relic, not just the relic itself, as a scene the pilgrim could take home.

The final stage of development, in the early 1490s, saw Besicken add a new series of woodcuts to illustrate each of the seven major basilicas of Rome described in the *Indulgentiae*. These were images not of the buildings but of their patron saints (John, Peter, Paul, Mary, Lawrence, Sebastian, and, for

Figure 5.8. *Historia et descriptio urbis Romae + Oratio de S. Veronica + Indulgentiae* (Rome: Andreas Freitag, December 2, 1494). Rome, Biblioteca Nazionale Centrale, 70.1.F.13.1.

Santa Croce, a scene of the Crucifixion), and other printers followed suit.[82] Now every church in the first part of the indulgence catalogue had an illustration of its patron saint, including Peter. But the Veronica did not go away: it still illustrated the *Oratio*, and even in those editions that do not include the *Oratio* (whether vernacular translations of the two other texts or Latin editions that simply omit the *Oratio* for some reason), the Veronica scene was

Figure 5.9. *Historia et descriptio urbis Romae + Oratio de S. Veronica + Indulgentiae* (Rome: Johannes Besicken and Martinus de Amsterdam, April, 30, 1500). Munich, BSB, Inc.c.a. 356, fol. 16v.

Figure 5.10. *Historia et descriptio urbis Romae + Oratio de S. Veronica + Indulgentiae* ([Rome: Petrus de Turre, ca. 1499]). Florence, Biblioteca Nazionale Centrale, Magl. G.6.2, sig. d8v.

Figure 5.11. *Historia et descriptio urbis Romae + Oratio de S. Veronica + Indulgentiae* (Rome: Eucharius Silber, September 12, 1499). Munich, BSB, Inc.c.a. 325, fol. 17r.

still included, as a second illustration for the section on St. Peter's. The Vatican basilica was the only church to be illustrated with two images, and the Veronica was the only relic in the entire city to be depicted.

The other text that Plannck and his followers included in their composite editions was the *Historia vel descriptio urbis Romae*, a rapid survey of ancient Roman history from the foundation of the city to Constantine. The *Historia* covers the story of Romulus and Remus, legendary tales of the archaic republic, the careers of the triumvirs and dictators, Augustus and the foundation of the empire, and the reigns of the more important emperors, culminating in the life of Constantine, his conversion, and his Donation to Sylvester. The program of illustration worked out for this text in the block book was preserved more or less without changes throughout the early printed editions. The preface to the *Historia* (which begins, "In isto opuscolo dicitur quomodo Romulus et Remus nati sunt . . . ") usually fills fol. 1r, followed by the woodcut of Rhea Silvia with Romulus and Remus on fol. 1v (figures 5.12 and 5.13).

The *Historia* proper then starts on fol. 2r, now surmounted by a half-page woodcut of three shields. Where the block book, Freitag, and Guldenbeck all used a central shield with the generic crossed keys of the papacy, in the mid-1480s Plannck introduced the arms of the current pope, Innocent VIII, and added crossed keys and a pallium above the central shield and behind the papal mitre (figure 5.14). Later printers followed him, substituting the arms of Alexander VI after his coronation in August 1492 (figure 5.15). The shields appear in countless editions; printers were willing to invest in multiple versions of the cut. It is worth reflecting on the fact that hardly *any* official, papal documents had yet been printed in Rome with woodcut insignia of this sort, despite the fact that these same printers published several bulls and briefs for the papal chancery in these years. The absence of papal *stemme* or insignia in printed papal documents of the time underscores how unusual it was that such strikingly political imagery should appear in pilgrim texts. J. R. Hulbert notes that few other medieval shrines produced publicity for their relics and indulgences at all, but in Rome the city's spiritual treasures were not only celebrated in print but also associated with the pagan history of the city, its particular *loci* of authority, and the dynastic profile of the current pope.[83]

The arrangement of the three shields can be read in two distinct ways, contemporary and historical. The current pope, whose family arms occupy the central position in the image, takes precedence over both the contemporary

Figure 5.12. *Historia vel descriptio urbis Romae + Oratio de S. Veronica + Indulgentiae ecclesiarum principalium urbis Romae* ([Rome: Stephan Plannck, ca. 1485–87]). Munich, BSB, Inc.s.a. 195, fol. 1v.

Figure 5.13. *Historia et descriptio urbis Romae + Indulgentiae ecclesiarum principalium urbis Romae* (Italian) ([Rome: Petrus de Turre, ca. 1499]). Florence, Biblioteca Nazionale Centrale, Magl. G.6.2, fol. 1v.

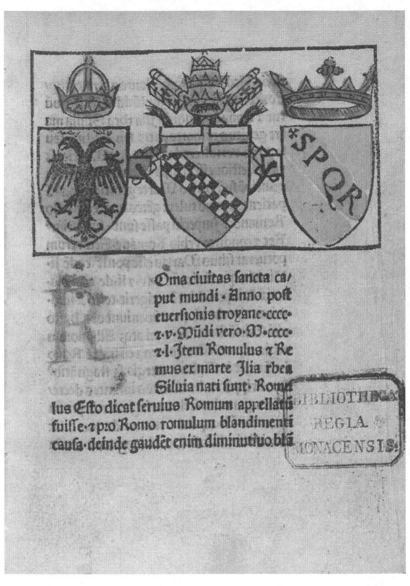

BIBLIOTHECA REGIA MONACENSIS

Figure 5.14. *Historia vel descriptio urbis Romae + Oratio de S. Veronica + Indulgentiae ecclesiarum principalium urbis Romae* ([Rome: Stephan Plannck, ca. 1485–87]). Munich, BSB, Inc.s.a. 195, fol. 2r.

Figure 5.15. Papal, imperial, and Roman insignia on fol. 2r of the *Historia*.

Roman commune and the Holy Roman Emperor, whose emblems occupy subordinate positions on either side. The pope is the supreme authority in Rome, within the Church, perhaps over the whole world. Second—and a corollary to the first—the papacy *as an office* is heir to the two older institutions, the Roman Republic (SPQR) and the ancient Roman Empire, whose stories the *Historia* recounts. In this sense, the *Historia* is not just a narrative of Roman history from Rhea Silvia to Constantine and Sylvester; it also explains the foundations of the current pope's power: Rome, founded with divine sanction, grew to become a mighty empire, which Constantine first made Christian and then entrusted to the popes. Fittingly, the *Historia* closes with a fairly long account of the life of Constantine; the story of how Sylvester cured him of leprosy, secured his conversion, and baptized him; and, last, an abbreviated version of the text of the *Donation of Constantine*.[84]

The *Indulgentiae* then starts with a further discussion of Pope Sylvester. The text begins with a reference to the pope and his interest in cataloguing the city's churches and goes on to identify Sylvester as the founder of several major basilicas, including the Lateran and St. Peter's. It is striking that the text identifies the pope, not Constantine, as the founder of these basilicas; this aligns with a larger theme of the book, that the pope is the proprietor of the city, custodian of its holy places, and dispenser of its spiritual graces.

Thus, in addition to inspiring devotion, the composite volume invites the pilgrim—or any reader—to wonder at Rome as a timeless seat of authority, both imperial and apostolic. The *Historia* traces Rome's history from its legendary beginnings. The conclusion of that narrative, the Constantine and Sylvester legend and the *Donation of Constantine*, connects nicely with the start of the *Indulgentiae* and its foundational emphasis on the reign of Sylvester. The pilgrim guide thus tells two tales at once: the graces available in Roman churches, and the ancient authority of the pope who offered them. The Veronica—whose arrival in the city is never documented or mentioned in the text—sits at the center of the two narratives, in the middle of the book, gazing directly out at the reader. A timeless relic, it was still fixed in a particular moment, the encounter that was the high point of the pilgrim's stay in Rome. The Veronica was the centerpiece of the book, just as its viewing was the climax of the Jubilee experience.

By the late 1490s, as Rome prepared for the Jubilee of 1500, the printers of Rome were reissuing the *Historia-Oratio-Indulgentiae* collection in new editions almost monthly, often swapping woodcuts and types as they coordinated their output: in 1499–1500, Plannck produced six editions (including one German and one Italian); Besicken, seven (also including one German and one Italian); and Silber, three (of which one was Italian). All told, about twenty editions appeared in these two years, all of them illustrated with the central exposition scene.

A final innovation saw, for the first time, some divergence in practice among the different printers. In addition to the standard suite of woodcuts, most added yet one more image to their editions, either at the start as another frontispiece or at the point in the *Indulgentiae* where the original catalogue of seven churches ends and the supplementary catalogue begins. Since the first church in the supplement was Santa Maria in Trastevere, the image added here was often one of the Virgin.[85] But sometimes it was Petrine or papal. Plannck added an image of Peter and Paul in a landscape, with a tower (representing Rome?) in the background (figure 5.16), either at the start of the book or at the start of the supplementary catalogue.[86]

Besicken used a smaller cut of Christ giving the keys to St. Peter, which appears in almost all his editions of the 1490s and the Jubilee year of 1500. Petrus de Turre, in an Italian translation of the collection printed in 1499, included a full-page illustration containing four scenes from the life of Pope Sylvester, including him baptizing Constantine and receiving the Donation.[87] (In far-off Gouda, meanwhile, a Dutch printer of the *Indulgentiae* included, in the section on St. Peter's, an image of the pope standing in the door of the basilica, holding the Veronica in his own outstretched hands.)[88] These images in their different ways further reinforced the connection that the whole collection asserted: the relics of Rome were important not only because of their inherent sanctity but also because they were lodged in the sacred basilicas of the city, watched over by the pope, and endowed with indulgences by his authority. Popes from Peter to Sylvester to Gregory the Great had together amassed this treasury of grace, and popes from Boniface VIII to Alexander VI had put them on offer to the pilgrim. To come to Rome was to seek communion with Peter, the prince of the apostles, and to earn indulgence from his modern successor.

The composite editions of the *Indulgentiae* provided inspiration for a variety of other printed publications—books, broadsides, and images—where the relics of Rome were branded as papal property and the pilgrimage to Rome as an enterprise overseen by the pope. In 1492 or 1493, the poet (and penitentiary of the Lateran) Giuliano Dati published an *ottava rima* guide to the indulgences of Rome based loosely on the texts of the *Stationes* and *Indulgentiae*.[89] Dati recapitulates much of the information on the Salvatore and Veronica found in the *Indulgentiae*: at the Lateran, the Salvatore is lit by perpetual lamps and positioned near the Scala Santa, where the drops of Christ's blood will make all who see them swoon. At St. Peter's meanwhile, the Santo Volto awaits the pilgrim, the first thing to be seen on entering the basilica.[90] Freitag's first edition of the text carries a full-page woodcut on the first page depicting Pope Gregory the Great, wearing the triple tiara and seated on an ornate throne, delivering a document with one hand and a blessing with the

Figure 5.16. *Historia vel descriptio urbis Romae + Oratio de S. Veronica + Indulgentiae ecclesiarum principalium urbis Romae* ([Rome: Stephan Plannck, ca. 1485–87]). Munich, BSB, Inc.s.a. 195, fol. 32v.

other, to a group of cardinals, bishops, and other clerics; presumably, he is issuing an indulgence or indulgences to the group. The Florentine edition by Lorenzo Morgiani and Johannes Petri uses a different cut of the same scene: Gregory enthroned, wearing the tiara, delivering a blessing from a book to a group of seated cardinals.

Other printed publications follow the *Indulgentiae* in wedding the Veronica to particularly Roman, Petrine, or papal imagery. Zainer's broadside of ca. 1482 had already shown the Veronica held up by SS. Peter and Paul, an iconographic tradition that, as we have seen, dates back to at least the 1300 Jubilee. In 1498, Alexander VI issued a printed bull suspending other indulgences in advance of the 1500 Jubilee (figure 5.17). The pamphlet, printed by Eucharius Silber, begins with a striking woodcut version of the same scene—the first illustrated papal bull to be printed in Rome.[91]

A silver badge created for that Jubilee, now in the British Museum, is imprinted with the Veronica, held up by Peter and Paul, surmounted by the papal tiara and crossed keys, with letters spelling out the name of Alexander VI surrounding the two saints and his arms occupying the space at their feet.[92] A German-language guide for travelers to Rome, published at Erfurt in the Jubilee year of 1500, has the same arrangement of figures—Peter, Paul, the tiara, the keys, and the Veronica—on its title page (figure 5.18).[93]

In the early sixteenth century, the late twelfth-century ciborium that held the Veronica within the Oratory of John VII was provided with a panel painting by Ugo da Carpi showing Veronica holding her cloth imprinted with the face of Christ, flanked by Peter with his keys and Paul with his sword. A drawing in the Uffizi, attributed to Parmigianino, shows the same three figures in the same arrangement,[94] as does a print by Dürer from 1510 (figure 5.19).

Finally, two undated prints by the Augsburg printmaker Hans Burgkmair, roughly contemporaneous with Dürer, borrow even more directly from the iconography of the *Historia/Oratio/Indulgentiae* editions. One shows Peter and Paul together holding the Veronica cloth over an arcaded balcony; in the spandrels of the arches are three shields containing the arms of Pope Julius II, the imperial double eagle, and the SPQR monogram.[95] The other shows Veronica herself, holding her cloth in an aedicule decorated by the twin shields of pope and emperor (figure 5.20). Beneath is the legend *Salve effigies Christi sacerrima*—which recalls, though inexactly, the opening lines of the *Oratio* (*Salve sancta facies nostri redemptoris*).[96] The popes did not commission these prints, any more than they commissioned the dozens of

Figure 5.17. Alexander VI, Bulla, April 12, 1498, "Consueverunt Romani pontifices praedecessores nostri appropinquante anno iubilaei" ([Rome: Eucharius Silber, not before April 12, 1498]). London, British Library, IA.19030, fol. 1r.

Figure 5.18. *Weg und Meilen von Erfurt bis Rom* (Erfurt: [Hans Sporer], 1500). Staatsbibliothek Bamberg, Inc.typ.Ic.I.74, fol. 1r.

fifteenth-century editions of the *Indulgentiae*, nor did they direct their iconography.[97] But prints and printed books alike reflect papal policy in a more indirect way: it was popes who endowed the churches of Rome and their relics with indulgences, and popes who declared Jubilees. The *Indulgentiae* certainly echo images and themes that were central to the cultural and devotional programs of the Renaissance papacy, even if the popes had no part in their production. Indeed, the fact that the many printers who published editions of the *Indulgentiae* added papal coats of arms to their works is all the more striking given that these editions were *not* produced on commission. The same printers were, in these years, publishing papal documents like bulls, constitutions, and chancery regulations as pure text, without any graphical signs of papal authority. Roman printers pioneered the use of papal arms in printed books essentially "on spec," in a context that was quite unofficial. Their presence must have added prestige or authority or charisma to the product they were trying to sell, and pilgrims seem to have responded positively to the presence of the papal arms as a sort of seal of apostolic ap-

Figure 5.19. Albrecht Dürer, *St. Veronica between SS. Peter and Paul*, 1510. New York, The Metropolitan Museum of Art, 19.73.192.

proval on their books, on their indulgences, and on their entire pilgrim journey. The visual elements branded the Veronica as a papal icon, the Jubilee as a papal enterprise, and Rome as a papal city, a fusion of elements and ideas that Boniface VIII had first promoted some two hundred years before.

Figure 5.20. Hans Burgkmair, *The Sudarium of Saint Veronica*. New York, The Metropolitan Museum of Art, 37.43.6.

The Renaissance popes could deploy relics in other ways too—not just as pilgrim lures. In moments of crisis, relics could serve, as they had in earlier times, as protective or apotropaic totems. In July 1470, as rumors reached Rome that Negroponte was under siege by the Turks, Paul II took the Salvatore out of the Sancta Sanctorum and led it in procession around the Lateran. Accompanied by the cardinals, he took the icon around the city together with the head of John the Baptist and the *Madonna del Popolo*, an icon also thought to be painted by St. Luke and usually housed in Santa Maria del Popolo; on the same day, Infessura says, a Turk was baptized at the basilica and a golden chalice, a gift to the Lateran from the king of France, was displayed to the crowd. In the face of a threat to the safety of Rome—indeed, a threat to all Christendom—the Lateran regained its ritual importance as the city's cathedral, and the Salvatore served as Rome's palladium once again.[98] This is a useful reminder that print can only show a fraction of what actually happened in a Renaissance city: unlike pilgrims, the local faithful had no need of guidebooks or souvenirs of their religious experiences; their devotion to the Salvatore (and countless other relics as well) went uncommemorated in print.

Yet more relics and miraculous images were pressed into service against the Turkish threat in these same years, though not all of them had such long pedigrees as the Salvatore. The next chapter considers how the popes promoted certain "refugee" relics as weapons against the Turkish threat, imagining them as protective totems for all of Christendom. As they integrated these into the ceremonial, architectural, and devotional fabric of the papal domain, some—but not all—were also memorialized in print.

CHAPTER SIX

Refugee Relics

The sanctity of the Veronica and the Salvatore derived largely from the sense of timeless permanence that surrounded each. Both images had been created by a direct encounter with Christ, to be sure; they also seemed to have been lodged in Rome, in or near their Constantinian basilicas, for as long as those buildings had stood. Texts like the *Indulgentiae* further fixed the images in the sacred landscape of the city, and the printing of those pilgrim texts projected that sense of stability to a wider world.

Elsewhere in medieval Europe, relics moved frequently, and the story of their peregrinations was often integral to their sanctity.[1] Mobile relics were not unknown in Rome, either: in the Middle Ages and in the Renaissance, certain holy objects were venerated precisely because they had come from somewhere else. Impelled by divine providence, they had chosen Rome as their new home.

The paradigmatic such relic was the True Cross, brought to Rome in 328 by St. Helen, the mother of Constantine, after a pilgrimage to Jerusalem. Helen had given the relic to Pope Sylvester, later legend held, who consecrated the basilica of Santa Croce in Gerusalemme to hold it. Soil from Jerusalem was scattered into the footings of the church, integrating particles of the Holy Land itself into the foundation of the newly Christian city. Relic and dirt together sealed Rome's claim to be the new Jerusalem. These events were recorded in printed indulgence guides. The *Reliquie rhomane* notes Helen's discovery of the True Cross, its transfer to Rome, the dedication of its basilica, and the indulgences available to pilgrims who visited it.[2] In his *ottava rima* indulgence guide, Giuliano Dati devotes six stanzas to Santa Croce, describing how St. Helen gave the church the True Cross and other relics of the

Crucifixion—a nail, the sponge, a piece of the good thief's cross—and how Sylvester consecrated the new basilica in their honor.[3]

Another part of the True Cross remained in Jerusalem, where it had other adventures. In the *Golden Legend*, Jacobus de Voragine tells how the Jerusalem relic of the Cross fell into infidel hands when Chosroes II captured the city in 614. The Roman emperor, Heraclius, led a Christian army against the Persians, recovered the Cross, and brought it back in triumph to the city in 628. Here a miracle occurred: Heraclius could not pass through the city gates until he stripped himself of his regalia; only then, barefoot and humble, did he pass through the walls and restore the relic to its rightful home.[4]

In Jacopo's account, the True Cross is an index of the health of the Christian polity: threatened by infidel hostility, the cross can only be recovered by a Christian king of proven virtue, whose devotion lets him triumph over the forces of unbelief. The relic itself issues a call to arms, both urging and enabling Heraclius to restore Jerusalem to Christian possession. In Italian fresco cycles of these scenes (e.g., by Agnolo Gaddi in Santa Croce in Florence and Piero della Francesca in San Francesco in Arezzo), the implications of this narrative for late medieval Christendom became clear: Heraclius was a Christian hero whose victory over the Persians recapitulated Constantine's victory over Maxentius at the Milvian Bridge. The victory of the Cross was not just a triumph of faith but also the triumph of an imperial Christendom. Moreover, in turning the power of the Cross against pagan armies in the East, Heraclius played the role of a crusader *avant la lettre*.[5]

The Roman relic of the True Cross missed out on all this action, but in the 1490s Antoniazzo Romano and Melozzo da Forlì decorated the apse of Santa Croce with a massive fresco illustrating Heraclius's recovery of the True Cross. The scene introduced to the Roman relic the same associations with crusading and recovery from infidel captivity as had long accrued to the Jerusalem portion, even though the Roman one had been safely lodged in its basilica by the Aurelian Walls since the time of Constantine.[6]

In the fifteenth century, new relics came to Rome from the East. Like the True Cross, they were celebrated for having escaped from infidel clutches. With divine help, they had made their way west to find sanctuary with the popes, the heirs of Constantine and Heraclius, rulers of an imperial Christendom poised to lead a new crusade against unbelief in the East. After the fall of Constantinople in 1453, for example, a Byzantine icon of the Virgin miraculously appeared in the church of Sant'Agostino, near Piazza Navona, having fled the Turkish-occupied city for a safer refuge in the West.[7] In 1499, another

Marian icon appeared in Rome, this one with a scrap of paper attached to it declaring that it had been stolen from its original home in Crete and now wished to be housed in a Roman church.[8]

This chapter surveys some other refugee relics that made their way from Muslim to Christian possession in the later fifteenth century. Once introduced into papal territory, they were incorporated into papal ceremonial, commemorated in papal building projects, deployed in papal crusade propaganda, and recorded in Roman print. St. Andrew's head and the Holy Lance were both brought to Rome by fifteenth-century popes hoping to rally the faithful to a new crusade. Both traversed the city in ritual processions that mimicked the annual progress of the Salvatore; both found new homes in St. Peter's and were incorporated into papal chapels or tombs. Both were also commemorated in printed pilgrim guides like the *Indulgentiae*, though with varying degrees of attention.

These same years saw the growth of a new devotional tradition in central Italy focused on the Holy House of Loreto. Legend held that this ancient structure, commanding a remote hillside near the Adriatic Sea, was actually the house of the Virgin Mary; it had been flown by angels to Italy after the Holy Land fell into Muslim hands. Accounts differed as to whether the house fled after the Arab invasions in the seventh century or after the fall of Acre in 1291, but the story that it did so only began to circulate in the late 1460s, as anxieties about the Turkish advance grew keen. Long a popular pilgrimage site, Loreto now attracted papal patronage: popes from Paul II to Leo X showered the shrine with gifts, built a new basilica around it, fortified it against Turkish attacks, and eventually brought its cult to Rome. The legend of the house's miraculous flight from Nazareth to Italy also became one of the most popular pilgrim texts of the Italian Renaissance, printed over forty times between the 1480s and 1520s in Latin and vernacular versions.

Most Loreto editions were printed by the same Roman printers who published dozens of editions of the *Mirabilia* and *Indulgentiae* in these same years. Small in format and invariably illustrated with a woodcut on the title page, the Loreto texts resemble these other Roman pilgrim texts and are frequently bound together. One might think, then, that Loreto presents another case of Roman print infusing an object of popular religious enthusiasm with papal prestige, another corpus of pilgrim texts with political as well as devotional resonances. But the story is more complicated, as this chapter will show.

In the *Mirabilia* and *Indulgentiae*, print brought together a collection of late medieval texts concerning relics and miraculous images and fixed them into a single, stable corpus; print also helped fix an iconography for the collection that remained equally stable for many decades. That iconography fused devotional and political imagery in a way that set Rome at the heart of Christendom and the pope at the head of it, a set of associations that was not only acceptable to pilgrims but also something they looked for when they came to Rome, an idea they would invest in and take home with them. By contrast, the relics examined in this chapter were appropriated by popes for various propagandistic purposes, but printed texts suggest that pilgrims responded to these efforts fairly indifferently, showing little regard for papal promotion.

Negative arguments are difficult to prove, but they are central to this chapter. The first part treats St. Andrew's head and the Holy Lance, two relics brought to Rome from Turkish possession and promoted in Rome as political totems, with ceremonial and architectural honors equal to that of the Veronica, but which found little resonance in Roman print. The significance of this is not immediately clear: the previous chapter argued that the Salvatore never figured prominently in printed indulgence guides because it was not an object of *pilgrim* devotion. Beloved and jealously protected by local guildsmen and their baronial patrons, the icon needed no further publicity. In that case, print functioned as an index not of popularity but rather of focus. The Veronica, an icon sought by pilgrims, was ubiquitous in books printed for that international market, but the Salvatore's cult was intensely local and thus almost invisible in print.

Here, print's indifference to St. Andrew's head and the Holy Lance reveals something simpler: despite intense promotion by the popes, few people, Roman or foreign, developed a devotion to either relic. Since they do not figure much in printed pilgrim guides, we can conclude, as a corollary, that influential figures in the papal curia made little effort to shape the contents of those books—or if they did, that they were unsuccessful. The *Indulgentiae* evolved in response to market demands, not curial pressure.

The Holy House of Loreto, by contrast, enjoyed a remarkable *fortuna* in print. But the cult proved surprisingly resistant to papal appropriation. Even as popes fortified the Madonna's domestic shrine against infidel attack, pilgrims showed little interest in reimagining the object of their devotion as a political relic, the palladium of the papal states. The Madonna di Loreto transcended the narratives that were composed for her, and print served her cult in curious ways.

St. Andrew's Head

St. Andrew's head came to Rome in a pageant staged by Pope Pius II in the spring of 1462. Like the Veronica and the Salvatore, the relic played various roles as it moved about the city: an object of popular devotion but also an emblem of papal supremacy; it was also drafted into the pope's campaign for a new crusade against the Turks.

The Renaissance popes imagined the crusade as a pan-European enterprise with themselves at its helm, harking back to the heyday of the medieval papacy when (as they imagined) Urban II had launched crusader armies eastward and Latin Christendom looked to Rome for its instructions. Pius had campaigned for a new crusade since the start of his pontificate, just five years after the fall of Constantinople. In 1459, he convened the Congress of Mantua to secure commitments from the European powers, although little came of their meeting. In early 1462, hoping to inspire (or shame) the Christian princes into joining his project, he declared that he would lead the expedition himself. To mark this new initiative, he brought the relic of St. Andrew to Rome, casting it as a representative of the entire Greek world come to seek asylum and help from St. Peter and his successor.

Pius describes the ceremony at the start of the eighth book of his *Commentaries*.[9] After the Turks conquered mainland Greece in 1460, the despot Thomas Paleologus went into exile, taking the head of the saint with him. The sudden availability of the relic sent the princes of Christian Europe into an acquisitive frenzy, but the pope prevailed over rival contenders and secured possession, bringing the head first to Ancona on the Adriatic coast, then to a castle in Narni for safekeeping, and finally to Rome in April 1462. In Pius's telling, every aspect of the head's progress was miraculous: it chose to come to him rather than to any other prince, it survived gales at sea, it settled the stormy weather plaguing central Italy that spring, and in Rome the city's warring factions set aside their differences once they heard that the head was on its way.

Pius offered plenary indulgences to any pilgrim who came to Rome to witness the relic's entry. The reception he organized for its arrival stretched over several days, starting far outside the city and culminating in its deposit on the high altar of St. Peter's, at the heart of papal Rome. The event was not only a reunion between Andrew (or rather, his head) and the bones of his brother, Peter, but also a pageant of papal authority, confirming Rome as the universal home for all Christians, Latin and Greek, and Pius as the head of the church and its defender against infidel aggression.[10]

Pius wanted to bring two other relics out to welcome the saint: the heads of Peter and Paul, encased in their massive silver reliquary over the high altar of St. John Lateran. The gesture would have mimicked the procession of the Salvatore to meet his mother at Santa Maria Maggiore: in this case, brother would go out to meet brother. But the reliquary proved impossible to detach from its iron case, so Pius decided that living people should stand in for the long-dead apostles. Cardinal Bessarion, titular patriarch of Constantinople, a passionate advocate for the crusade, and head of the Greek émigré community in Rome, would speak for Andrew, while Pius himself would play the part of Peter.

Escorted by a party of cardinals, the head reached Rome on Palm Sunday, a feast commemorating another triumphal entry into another holy city. The following day dawned bright after weeks of pouring rain. Pius, the cardinals, and the entire curia processed out of the Porta del Popolo, carrying palm branches from the previous day, and met the head at the Milvian Bridge, a site with multiple associations: the conversion of Constantine, the tradition of his "donation" of imperial authority to the pope, and the triumph of soldiers signed with the cross over armies of unbelievers.

The pope welcomed the head with a speech that exalted it but also pointedly described it as a victim of Turkish aggression, a refugee seeking asylum after defeat.[11] The head was not just the relic of Andrew but a contact relic of Christ (rather like the Veronica). Here were eyes that saw God in the flesh, a mouth that conversed with Him, and cheeks that Jesus surely kissed. By coming to Rome, the head brought the city another powerful patron in heaven and greater protection against the Turks. But the head was not omnipotent: it had not protected its country against the Turks; it had fled before their advance. Andrew came to Rome seeking his brother's protection, reduced to client status before Peter, prince of the apostles, and Pius, his successor as vicar of Christ.

With the Greeks put in their place, Pius's narrative turns to rivals closer to home. That night, the head was lodged at Santa Maria del Popolo, just inside the city gate; overnight, it rained again, and in the morning the streets were slick with mud. Some cardinals wanted to keep clear of the filthy streets and asked permission to ride on horseback, but the pope refused. All must show the head reverence by walking through the muck. Only Pius, disabled by a (well-timed?) attack of gout, would ride on his *sedia gestatoria*: "It was a fine and impressive sight to see those aged men walking through the slippery mud with palms in their hands and miters on their white hair. . . . Some who had been reared in luxury and had scarcely been able to go a hundred feet except on horseback, on that day, weighed down as they were in their sacred robes, easily covered two miles through mud and water."[12] Fourteen muddy cardinals led the way through the city, followed by Roman clergy, conservators, *rioni* captains and magistrates, ambassadors of kings and princes, and Roman barons, some thirty thousand in all. The pope brought up the rear with Andrew's head. Thus, the entire civic structure of the city was organized: foreigners, laity, and communal officials bookended by clergy, with pope and relic in the place of honor at the back.[13] The procession was so long that the start of it reached St. Peter's before the tail left Piazza del Popolo. The length of the parade may account for the route that it took through the city, but stagecraft, too, played a role.[14] Eschewing the direct route to the Vatican, the procession took a special detour past the Pantheon, "which the heathen once consecrated to all the gods, that is, demons," but which the early Church had exorcised and reclaimed for Christian worship. Like the Milvian Bridge, here was another site of Christian triumph over the forces of unbelief. Taking the head past the ancient rotunda also mimicked the annual procession of the Salvatore through the Forum: a triumphal process through the heart of the old pagan empire, and one that portended Christian victory over new empires in the East.

The procession finally arrived at St. Peter's, where Pius staged another fraternal reunion, this time between the head and Arnolfo di Cambio's statue of St. Peter, then located in the atrium outside the basilica. Here, too, was an echo of the Salvatore's journey to visit his mother: the reliquary of Andrew met the statue of Peter, but each object also *was* the person it represented: "Imagining that the statue wept with joy at the coming of his brother, the pope burst into tears as he reflected on the meeting and the embrace of two brothers so long kept apart." Pius took the relic to the high altar for another round of orations and benedictions. Bessarion was again made to speak for the Greek head and again reiterated its—and his nation's—submission to Rome: "I have fled from the clutches of the pagans and come to you, most holy brother, teacher and master, appointed by God the universal shepherd of Christ's flock."[15]

Pius kept the head in the Vatican palace for the rest of Holy Week. On Easter Sunday he brought it back to St. Peter's and took it to the Veronica chapel at the back of the basilica. Now another reunion was staged, as the Veronica was exposed to Andrew's head as though to a supplicant pilgrim. And so the apostle came face-to-face with his lord. "A marvelous and awesome thing to behold: on one hand the holy face of the Saviour and on the other the precious relic of His apostle," Pius reports.[16]

Several important themes were at work in this procession—or rather, in the way that Pius describes it. Like many medieval relics, the head of St. Andrew seemed to choose where it wanted to go; supernatural events cleared its way. It put various elements in their place (rival princes, fractious barons, stormy weather, haughty cardinals, pagan ruins), while it and the pope who received it were singularly exalted. And as it was reunited with its apostolic companions, the holy heads and faces of Rome, it was also made to speak for its contemporary compatriots, the Greeks, emphasizing Greek submission to Rome and the supremacy of the pontiff who gave them shelter.

This tension, between the power of the relic and the weakness of the Eastern world it represented, had deep roots in Latin Christian thinking. Already in the legend of the True Cross there was a suggestion that Christian princes must make themselves worthy to possess a powerful relic. Heraclius, after reclaiming the Cross, found the gates of Jerusalem closed to him until he approached as a humble pilgrim. Fifteenth-century humanists who studied this episode were less sympathetic, recasting the figure of Heraclius as a grotesque incompetent who had lost the holy places—first to the Persians and then, more permanently, to the Arabs—through political ineptitude and personal vice.[17] In this interpretation, Greek weakness and sexual immorality allowed both the True Cross and Jerusalem to fall into infidel hands. It would take a better, stronger (Latin) champion to turn the tide. With the staged submission of Andrew to Peter, Pius drew the same contrast between a weak Greece and a powerful Rome, the refugee relic and its papal protector.

Pius also hoped to promote the head as one of the premier relics of the city, on an equal footing with the Veronica. He built a new chapel for the head at the back of the south aisle of St. Peter's, directly opposite the Veronica chapel.[18] Inside was a ciborium housing the relic of Andrew, patterned on the one that held the Veronica, while three semicircular marble reliefs depicted the head's providential transfer: in each, a pair of hovering angels holds the reliquary head aloft between them (figure 6.1). The flanking angels recall those carrying the Veronica in both woodcut and printed editions of the *Indulgentiae* and in early single-sheet prints. (They also foreshadow the angelic iconography that would soon develop for the holy house of Loreto.) From its chapel in St. Peter's, Pius imagined, Andrew's head would be exposed regu-

Figure 6.1. Isaiah da Pisa, St. Andrew's Head, ca. 1462. Photo courtesy of author.

larly like the Veronica. He ordered indulgences for those who viewed its display; he also intended the new chapel to serve as his tomb. And he erected a tabernacle at the Milvian Bridge, where every pilgrim coming to Rome from the north must pass, with a marble statue of Andrew and a long inscription commemorating the events of April 1462.

Despite Pius's efforts, the head of St. Andrew did not develop a popular following. The diaries of the papal masters of ceremonies mention occasional expositions of the relic in St. Peter's in the later fifteenth century, patterned on the more frequent exposure of the Veronica, but the head never matched the holy face in popularity.[19] The *Indulgentiae* notes that it was exposed on the feast of St. Andrew and on the anniversary of its transfer to Rome but lists no indulgences attached to it, nor processions or other signs of devotion.[20] Distinguished visitors rarely requested a showing. There is little evidence that pilgrims bought badges or prints of the head or sought it out during their visits. Writing in 1470, William Brewyn mentions the head briefly in a list of relics of the basilica but without mentioning its indulgences.[21]

Soon, its story was only hazily recalled: the *Reliquie rhomane* says that the head was brought to Rome from Constantinople.[22] After his visit to Rome in 1496, Arnold von Harff said that the head was installed over the tomb of St. Gregory (!) and occasionally displayed, but he made no mention of Pius II or its recent transfer to Rome.[23] Giuliano Dati listed the relic as one of many saintly heads preserved in St. Peter's: the sacred skulls of Andrew, Sebastian, Processus, Martinianus, Luke, and James were all there.[24] Andrew's relic was one of dozens, barely distinguished from the rest. Papal endorsement, by itself, was not enough to capture the attention of the faithful.

The Holy Lance and the Baptist's Arm

The same might be said of the Holy Lance, which came to Rome in the reign of Innocent VIII as a diplomatic gift from Sultan Bayezid II. Innocent spun his receipt of the lance as an act of recovery—a Christian triumph over Turkish perfidy and a sign of greater victories to come. Spin was certainly needed. A party of Ottoman diplomats brought the lance to Rome in 1492 as thanks for the pope's willingness to keep the sultan's brother (and rival) Cem Sultan as a secure "guest" of the papal court.[25]

The lance followed the same route to Rome as St. Andrew's head: it sailed across the Adriatic to Ancona, cardinals fetched it from there to Narni and from Narni to Rome, it was received in a ceremony outside the Porta del Popolo, it spent the night at Santa Maria del Popolo, plans were hatched to have other relics come out to meet it,[26] and pope and clergy escorted it through crowded streets to St. Peter's. Innocent hoped to use the lance as an object to rally the Christian princes to a new crusade (ignoring the shabby transactions that had brought him the relic in the first place). He had a chapel constructed inside St. Peter's to house the relic, midway along the south aisle, where he also built a ciborium patterned on that of the Veronica. He intended chapel and ciborium together to serve as his funeral monument, just as Andrew's chapel did for Pius.

Even before the lance arrived in the city, however, there were doubts about its authenticity. The Holy Roman Emperors had long claimed to possess the Holy Lance, which they treasured as one of the most precious in their large collection of relics. Records of their lance dated back to the reign of Otto I, and since 1424 it had been kept in a special treasury in Nuremberg where it was displayed to pilgrims every year. Pilgrim badges from Nuremberg featured the Holy Lance, and woodcut prints celebrating the German relic had been circulating for decades. In 1487 a pamphlet commemorating the annual festival was published in Nuremberg, with a full-page woodcut of the moment of exposition and text explaining its provenance and indulgences.[27]

In Rome, Burchard fretted as planning for the reception of the sultan's lance got underway. Was the relic even real? Was it embarrassing to receive it from a sultan? Would the Roman people turn out to welcome it?[28] Burchard records a fractious committee meeting where court officials debated its authenticity: "Some were of the opinion that the gift should be received with all solemnity and reverence, and in the same manner as the head of St. Andrew in the time of Pius II . . . others asserted, on the contrary, that they had seen the point of this spear in Nuremberg, where it is exposed each year on the Feast of the Spear."[29] There was debate, too, over whether the lance (if genuine) was an honorable gift, considering that the giver was "the archenemy of our faith." Some thought it better to accept the lance quietly and in private.[30] In the end, the faction in favor of a public celebration won out. Debating how to assemble a crowd for the procession, Innocent said he would issue indulgences. Burchard suggested they set up fountains of wine at strategic points along the route, arguing that this would be a more effective way to get Roman crowds to appear.[31]

Writing after the event, Stefano Infessura confessed himself impressed by the spectacle. "The lance was received by the pope and all the clergy who went in procession from Santa Maria del Popolo to St. Peter's. It was locked inside a kind of tabernacle, all of crystal and very beautiful, with the base and other decorations made of pure gold; it was a thing of great value."[32] Sigismondo dei Conti also praised the reliquary and the crowds who came to see it. On its progress to Rome, "remarkable numbers of people came out of the fields and towns to line the roads along which the relic was conveyed."[33] It restored the deaf and mute to their senses as it passed, and once inside the city, the Romans scattered flowers on the streets it traversed. The grandeur of the ceremony certainly quelled any doubts Sigismondo had about the relic's authenticity. Although Nuremberg claimed also to have the lance, he was sure the one he had seen in Rome was the real thing.[34] Giuliano Dati, writing within a year or two of the event, recorded that the lance was given to Innocent by the "Gran Turco" and praised the pope for delivering the relic to the Vatican "with his own hand."[35]

Within a few years, though, both the provenance of the relic and the pope who had procured it were forgotten. Arnold von Harff, visiting Rome in 1496,

Figure 6.2. Guillaume Caoursin, *Historia Rhodiorum* (Ulm: Johann Reger, October 24, 1496). Rare Books & Manuscripts Department, Boston Public Library, Q.403.97, sig. f7v.

mentioned seeing the lance in the Veronica chapel, not in Innocent's ciborium. He noted only that it had "recently come there,"[36] not where it had
come from or who had acquired it.

The Ottoman ambassadors who brought Innocent the lance in 1492 had
stopped in Rhodes and presented another relic, the hand of John the Baptist,
to the Knights of St. John.[37] This transfer was recorded in a nearly contemporaneous printed book: in 1496, Guillaume Caoursin's history of the order
included an account and a large woodcut of the scene (figure 6.2).

According to Caoursin, the Turkish ambassadors had tried to renege on
their gift, but angels appeared and pried the reliquary out of their hands,
flying it across the audience hall to the Master of the Knights. The relic later
ended up in Siena. This story had little resonance in Rome, but it underscores
how prevalent was the theme of divine intervention in the transfer of relics:
angels might appear anywhere, at any time, to fly a relic from Muslim captivity to the custody of a Christian protector.

The Holy House of Loreto

Angels bearing relics from the East bring us, finally, to Loreto.[38] This hilltop
shrine near Ancona had long housed a sacred image of the Virgin that attracted pilgrims seeking miraculous cures. Medieval legend held that the cult
image had been brought to the site by angels, but in the 1460s a variant story
began to circulate, claiming that it was not the image of the Virgin but the
little church housing it that angels had delivered to the site. The church was
in fact the Virgin's own house, brought to Loreto from Nazareth, the place
where she had grown up, received the Annunciation from the Angel Gabriel,
and raised the Christ Child. The apostles later converted the house into a
church, but after the locals embraced heretical errors, angels had come and
removed the church from its foundations, carrying it first to Dalmatia and
then to Italy, where it began to perform miracles.

The earliest textual witness to this tradition, a tract composed between
1455 and the late 1460s by the rector of the church, Pietro di Giorgio Tolomei,
dates the departure of the little church from Nazareth to the time of the Arab
invasions.[39] The church had been preserved with great devotion "for as long
as the people in those parts were Christian; but after the people rejected their
Christian faith and accepted the faith of Mahomet, then the angels of God
carried the church away."[40]

The story of the house's miraculous flight spread quickly through Italy,
carried to distant parts by pilgrims who had visited the shrine. The mystic

Caterina de' Vigri of Bologna (d. 1463) described how the *mansio* of the Virgin had been spirited away by angels on account of the "idolatry" of an invading race.[41] Writing after a visit to Loreto in the late 1460s, Giacomo Ricci of Brescia explained that the house departed Palestine after the Jews of that country were attacked and defeated by "infidel barbarians."[42] In a poem composed ca. 1475–80, Luigi Lazzarelli had the Virgin herself declare she would no longer dwell "among my enemies."[43] The Carmelite poet Battista Spagnoli, writing ca. 1489, linked the flight of the house directly to the Islamic conquest of the Holy Land: as orthodoxy faded in the East, the Virgin's house fled west to settle in more faithful precincts.[44] In a later poem, Spagnoli returned to this theme, describing how the Virgin had been offended by the invasion of "the tyrants of Arabia" and decided to remove her house to Christian parts.[45]

Like St. Andrew's head and the Holy Lance, the Virgin's house at Loreto was a sacred remnant of the apostolic age that fled Islamic aggression in the East, another relic connected to the life of Christ that sought asylum in Italy with his vicar on earth. Almost as soon as the story of the house's angelic transfer began to circulate, Loreto attracted papal attention. Various popes endowed the shrine with gifts, issued indulgences to attract pilgrims, took over administration of the site, fortified the town against Turkish attack, rebuilt the shrine on a massive scale, and sponsored ambitious artistic works inside. An imposing basilica was built around and over the little house-church; its octagonal dome, designed by Giuliano da Sangallo, was completed in the Jubilee year of 1500.[46] In 1507, Julius II elevated the shrine to a pontifical chapel, began work on a papal palace next to the church, and sent architects and sculptors to encase the primitive house in a marble screen, with sculpted reliefs of events in the Virgin's life, including the flight of her house to Loreto.[47] Julius and Leo after him confirmed that pilgrims making a journey to Loreto could earn the same spiritual merits as they might from the journey to Jerusalem, thus ratifying the claim that the house was a genuine piece of the Holy Land, miraculously rescued from the Turks and settled on Italian soil. Papal inscriptions and *stemme* proliferated across the walls of the new basilica, on its artistic works, and on the fortifications built around the town. The popes presented themselves as proprietors and protectors of the little house and the house itself as an emblem of Christian triumph over the Turks.

In these same years—and rather unlike St. Andrew's head or the Holy Lance—Loreto grew extraordinarily popular. The shrine attracted throngs of pilgrims who professed their devotion to the Madonna di Loreto with offerings and prayers and bought printed copies of her legend text. These were sold both at the site and in cities and towns across Italy and farther afield. Tolomei's account, entitled *Translatio miraculosa*, was printed more than forty times between about 1485 and 1525, both in the original Latin and in Italian and German translations; editions appeared in Rome and Ancona, as well as in Venice and Florence and in Augsburg north of the Alps. After the *Mirabilia* and the *Indulgentiae*, it was probably the most popular pilgrim text printed in Renaissance Italy. Other authors retold the story in poetry and prose versions that also found their way into print, while single-sheet prints of the Madonna di Loreto and her house circulated widely in Italy and the north.

Despite the emphasis in these accounts on the house's escape from Islamic peril, the *meaning* of the Loreto cult was not entirely stable in these early years. Even as popes left their mark on Loreto, and poets celebrated the holy house as a protagonist in Christianity's struggle against Islam, pilgrims continued to focus their devotions on what the house contained: the Madonna di Loreto and her ability to effect miraculous cures.

Early printed texts on Loreto thus tell a slightly different story than printed pilgrim texts like the *Indulgentiae* or the *Historia*. Pilgrims bought copies of the Loreto legend in great numbers, and like the *Indulgentiae*, the editions were almost invariably illustrated with woodcut images. But while the iconography of the *Indulgentiae* and *Historia* was fixed early on in its printing history and remained stable for decades, the Loreto cult could be illustrated by a wide variety of images. The Virgin and Child by themselves or in a pavilion carried by angels, the Virgin and Child atop the holy house in flight, the Virgin in the apocalyptic garb of the woman clothed with the sun, or the Virgin Annunciate—all these could be the Madonna di Loreto. The early iconography of Loreto was flexible, unstable even, reflecting the newness of the cult and the many meanings it could hold for its devotees.[48]

The Legend of the Holy House

Little is known about Pietro di Giorgio Tolomei, rector of the church at Loreto from 1455 to 1473, the reasons he wrote, or the date of his composition. Around 1467–69, Giacomo Ricci of Brescia made a pilgrimage to Loreto, and sometime after becoming a canon of Brescia Cathedral in 1473, he composed a version of the Loreto legend that depends heavily on Tolomei's.[49] This suggests that Tolomei's text was available at the shrine when Ricci visited in the later 1460s. There is evidence that some version of the story, if not Tolomei's full account, was already in circulation by about 1462.

According to Tolomei, the house in Nazareth where Mary was born and raised the Christ Child was also the place to which she retired after his death, resurrection, and ascension. The apostles used to visit her there. After her death, they consecrated the house as a church; St. Luke provided a portrait of the Virgin for them to remember her by. For generations, the house was a sacred shrine in Nazareth, a site of popular devotion, but eventually the surrounding population gave up their Christian beliefs to follow the errors of Muhammad. (Exactly when this happened remained vague: Tolomei suggests that the angels came to spirit the house away soon after the Arab invasions in the seventh century, but he also says that the house only landed in Italy in the 1290s.)

The angels flew the house first to a site in Croatia, but the people there proved irreverent, so they lifted it up again and flew it across the Adriatic to Italy, to a grove of trees just inland from Recanati. Pilgrims began to visit the house and leave offerings, but brigands hid in the coastal scrub and preyed on the pious visitors, so the angels came a third time and flew the house to a nearby hill owned by two brothers. When they began to fight each other for the pilgrims' gifts, the angels moved the house one last time, to a spot in the middle of the public road, where no private individual could claim it. And there the holy house and its miraculous image of the Virgin had remained to the present day.

Much ink has been spilled over what folk memories, crusader legends, or ancient connections between medieval Croatia and the Marche this legend may reflect. Tolomei himself claims that the story had been known for a long time and was verified by medieval investigations. When the church first alighted near Recanati, no one knew what it was or where it had come from. But in 1296, the Virgin appeared to a local man in a dream and revealed that the building was her house. The man told the city fathers of Recanati, who sent a delegation to Nazareth to see if the Virgin's house was still there. On arrival, they discovered the foundations of a building that matched the dimensions of the church in Loreto precisely; an inscription on a nearby wall explained that a church had once occupied the spot but had mysteriously disappeared.[50] Tolomei also mentions a man whose great-great-grandfather had seen the angels carrying the structure across the waves to the grove, as well as another whose 120-year-old grandfather had worshiped at the church when it stood there.[51]

Tolomei invoked these authorities to lend antiquity and authenticity to a tale that was in fact not at all well known before he wrote it down. Later authors who repeated his story would report that they had taken their information from a "very old" and dusty tablet posted on the wall of the church at Loreto. (Bartolomeo da Vallombrosa, author of an Italian translation of Tolomei's account from around 1483, Battista Spagnoli, whose 1489 account repeats much of Tolomei's tale, and Giuliano Dati, author of the Italian verse rendition of Tolomei's text published around 1492, all refer to this tablet or sign.)[52] Tolomei likely had his text written out on a wooden or parchment plaque and posted at the shrine for the information of pilgrims. Later references to this "ancient" tablet, taken together with the thirteenth- and fourteenth-century witnesses Tolomei invokes, have allowed apologists for the legend to argue that it dates back decades if not centuries before the 1460s. But before Tolomei, it seems, no other written source repeated its particulars.[53]

Origins of the Cult

This much the legend told. What of the real origins of the cult? Loreto, in a remote area of the Marche near the Adriatic coast, had long housed an image of the Virgin that could effect miraculous cures. Records of a shrine date back to the twelfth century, and in the late fourteenth century Urban VI and Boniface IX issued indulgences to pilgrims who visited the site.[54] In his *Italia Illustrata* (1447–53), Flavio Biondo described a little church, famous for healing the sick, its walls festooned with votive offerings left by grateful pilgrims.[55] In the fourteenth century, the cult image was probably a Byzantine icon,[56] but by the early fifteenth century the *imago* was a wooden statue, believed to have been made by the hand of St. Luke. In these years, it was thought that the cult image (*not* the church that housed it) had been miraculously conveyed to the site from foreign parts. As late as 1471, Paul II mentioned the belief that the Loretan image was delivered to the church by a crowd of angels.[57]

Fifteenth-century woodcuts, engravings, and paintings suggest that this notion was widespread. The earliest known depictions of the Madonna di Loreto depict her standing or seated beneath a baldachin held aloft by angels. A single-sheet woodcut of the Virgin of Loreto, accompanied by a long xylographic caption, probably created in the Veneto and dated to the middle of the fifteenth century, shows the Virgin nursing the Child on her lap under a domed structure whose slender columns are gripped by angels, while others lower a crown onto her head. The poem beneath celebrates the Virgin as queen of heaven, spouse and mother of God, advocate and true hope of sinners.[58]

A pair of engravings, both probably Florentine, present the same composition in a more sophisticated style.[59] In the first, dated to the 1460s, the

Figure 6.3. Fifteenth-century engraving of the Madonna di Loreto. Whereabouts unknown.

Virgin stands beneath a small octagonal dome while the platform beneath her feet carries the caption "SCA MARIA DELORETO."[60] The second is dated to 1490–1500 and bears a longer inscription: "SCA MARIA DELLO HORETA HORA PRO NOBIS" (figures 6.3 and 6.4).[61]

Figure 6.4. Fifteenth-century engraving of the Madonna di
Loreto. Whereabouts unknown.

In the title woodcut of the *Miracoli della Vergine Maria*, a book of verses
in praise of the Virgin printed in Brescia in 1490, the Virgin appears in this
Loretan guise: seated on a throne beneath a domed structure whose columns
are gripped by four angels (figure 6.5).[62] A related woodcut appears in a
printed calendar of the early sixteenth century.[63]

Panel paintings and frescoes of the fifteenth century also present the Madonna di Loreto in this way. More than a century ago, Corrado Ricci showed
that this iconography proliferated across central Italy for decades before any
artist introduced the figure of a flying house or church.[64] Ricci recorded ten

Irgo beata genitrix maria
Matre filiola del tuo pre e filio:
Fons facer ftella : rachel & helia
Celeſte humanita diuin confilio
Regina celi mater alma pia
Da cui principio in ogni cofa piglio
Aue gratia plena ſpira nel mio core
Che itoi miracoli ſcriua al tuo honore.

Figure 6.5. Miracoli della Vergine Maria (Brescia: Baptista Farfengus, March 2, 1490), fol. 1v. Florence, Biblioteca nazionale centrale, P.6.3, fol. 1v.

paintings of the Virgin under a baldachin held up by angels in the Marche, Umbria, Abruzzo, and Lazio.[65] (At least eight more paintings in remote churches of the Marche have since come to light.)[66] Few of these are dated: the earliest was painted in 1403 and the latest in the first or second decade of the sixteenth century. Ricci suggested that the baldachin held by angels was a stylized representation of the original house-shrine or its entrance facade, and thus that these images illustrated the legend as Tolomei told it. But it seems more likely that they represent a canopy or pavilion of honor for the cult image or statue as *it* moved with angelic help from east to west. When the iconography of the house on the move emerged (a development that, like Tolomei's text, only dates to the 1460s), it featured a real brick house, represented in perspective, with windows and a door, a solid roof, and a bell tower to indicate the church that it had become. There seem to have been no intermediate steps in the development of this image: pictures from the period feature either the Virgin under a stylized canopy or the Virgin atop a house-church.

The activities of one Italian painter help to fix the development of this iconography rather precisely. Between 1445 and 1450, Andrea Delizio was in Atri, near the Adriatic coast, where he painted a fresco of the old-style Madonna di Loreto on the wall of the church of San Nicola (figure 6.6).[67] The Virgin stands beneath a vaulted structure held aloft by angels, flanked by SS. Rocco and Sebastian. (The presence of the two plague saints beside the Virgin underscores her intercessory role in the healing of disease, the main reason pilgrims of the mid-fifteenth century visited Loreto.)

In 1460, Delizio returned to Atri and painted two more frescoes of the Madonna di Loreto, both in the cathedral. One was a traditional Virgin under a baldachin supported by angels. The second, known as the Madonna d'Alto Mare, or Madonna of the Deep Sea, depicts the Virgin floating in a curiously disembodied way over the tiled roof of a plastered building with a bell tower. The building is supported by angels who carry it not through the air but (as the traditional title of the fresco suggests) over the sea (figure 6.7).

The awkward composition suggests some confusion as to how to represent the miracle of the house's flight, as does the placement of the angels on the water rather than high in the air. The idea of the house sailing to Italy, rather than flying, would persist into the sixteenth century.[68]

One other piece of evidence helps to pinpoint the moment when the legend began to take shape: in 1462, an English pilgrim to Loreto recorded what he knew of the site: the church was not the Virgin's house but a chapel built

Figure 6.6. Andrea Delizio, Madonna di Loreto with SS. Rocco and Sebastian, San Nicola ad Atri, ca. 1450. San Nicola ad Atri.

by St. Helen during her visit to the Holy Land in the fourth century; angels had later transported the structure from Nazareth to Loreto.[69] Thus, for some time, important details of the legend remained unclear. Had the house flown to Loreto, or had it sailed there? Was it Mary's house, or Helen's chapel? When exactly had it arrived? At some point in this decade, Tolomei wrote his version of the story down, doing a good deal to fix the narrative, but further variants were still possible.

Papal Patronage

In 1464, two years after receiving St. Andrew's head in Rome, Pius II visited Loreto. On his way to Ancona, where he hoped to launch his long-planned crusade, the ailing pope prayed before the image of the Virgin, asking for relief

Figure 6.7. Andrea Delizio, Madonna d'Alto Mare, Atri Cathedral, ca. 1460. Atri Cathedral.

from his fever and cough and the pains in his limbs. He left a golden chalice as an offering, hoping for a cure for himself and success for his crusade.[70] Neither was granted: Pius died soon after reaching Ancona, awaiting a crusader fleet that never arrived. As the cardinals and courtiers who accompanied him started their journey back to Rome, Cardinal Pietro Barbo fell ill with a pain in his groin that left him unable to sit on his horse. He, too, called in at Loreto to pray to the Virgin. He received not only a cure for his embarrassing condition but also the news, delivered in a vision, that he would soon become pope.[71]

As Paul II, Barbo showered Loreto with attention, issuing a partial indulgence in 1464 to pilgrims who visited the site, "on account of the great and marvelous and almost innumerable miracles . . . which we most plainly have experienced in our own person."[72] Five years later, the bishop of Recanati began construction of a new basilica at the site. When the bishop died that same year, the pope took over the project. In 1471, Paul issued a plenary indulgence for contributions toward construction; it was then that he referred to the popular belief that the image of the Virgin had been brought to the site by angels.

Just how the new basilica would preserve or memorialize the original shrine is unclear. Ultimately, the new church was built over the earlier structure, which remained intact at the center of the crossing. This arrangement may have been integral to the design from the start, on the model of other sacred structures preserved inside larger sanctuaries like the Holy Sepulcher in Jerusalem.[73] On the other hand, the identification of the original church as a special building, the house of the Virgin, emerged only a little before the new construction began. Paul II's plans for a new basilica at Loreto may have included demolishing the existing chapel, prompting Tolomei, as guardian of the shrine, to compose a defense of the structure: did he codify, or even invent, the tradition that angels had brought to Loreto not just the image of the Virgin inside but also the building that housed it?

Tolomei's legend resonated deeply with contemporary anxieties about the Turkish threat. The decade in which he wrote saw Pius's reception of St. Andrew's head in Rome, the failure of his crusade at Ancona, and the fall of Negroponte in 1470. In 1474 and again in 1478, Turkish forces attacked Venetian Scutari on the Dalmatian coast. In 1480, they invaded Italy itself, occupying Otranto in the far south of Puglia for over a year. Italy's Adriatic shores seemed particularly vulnerable to Ottoman attack, and it was possible to imagine that the little church, with its rich treasury of pilgrim gifts, might be a target for Turkish greed or sacrilegious fury.[74] The project of building a new and more secure basilica at Loreto, begun by the local bishop and taken over by Paul II, was continued by Sixtus IV after his election in 1471.[75] Sixtus established the church in its own parish, granted new indulgences to pilgrims, and appointed his nephew Girolamo Basso della Rovere bishop of Recanati in 1476.[76] In 1482, Girolamo, now a cardinal, called in the architect Giuliano da Maiano to supervise construction; he also entrusted administration of the site to the Carmelite Order, weakening the hold of the local clergy. Girolamo continued building after Sixtus's death in 1484. In March of the next year, after hearing of possible Turkish raids on the Adriatic coast, he had Bacio Pontelli complete a ring of battlements above the cornice of the basilica, the so-called *camminamento di ronda*, creating its distinctive fortified appearance.[77]

Della Rovere patronage at Loreto was part of Sixtus's larger project to assert control over the Marche and establish his relations in local lordships. In addition to appointing Girolamo Basso della Rovere bishop of Recanati, Sixtus made his nephew Giovanni della Rovere lord of nearby Senigallia and installed another nephew, Girolamo Riario, as lord of Imola and Forlì farther north. Girolamo's designs on Forlì led to the Pazzi War between the papacy and Florence and, indirectly, the War of Ferrara in 1482–84. Whether for the security of the papacy or that of his family, Sixtus and his relations poured resources into the line of cities running across the northern papal states. Baccio Pontelli built the Rocca of Senigallia on the orders of Giovanni della Rovere, the Rocca di Ravaldino for Girolamo Riario in Imola, and further fortifications at Iesi, Osimo, and Acquaviva Picena. Alongside the papal fortresses Pontelli designed at Ostia, Tivoli, and Civitavecchia, the fortified basilica at Loreto projects the *roveresca* style of military architecture imposed across the papal states in the later fifteenth century.

When Julius II came to the throne in 1503, he continued his family's transformation of Loreto: both a Marian shrine and one in a chain of military installations securing the northeast frontier of the papal state. After his conquest of Bologna in 1506, Julius sent Bramante to design a new papal palace at Loreto. Plans were begun for the marble screen around the holy house, which would only be completed decades later. Julius had a medal cast celebrating his building works at Loreto, part of a series commemorating his architectural projects. In 1510, he came to Loreto as a pilgrim himself.[78] His successor Leo X fortified the entire village around the new basilica; the circuit of walls was begun in 1517 and completed in 1521.[79]

Loreto in Print

The popularity of Loreto was an overdetermined phenomenon. The Madonna di Loreto was already the focus of a thriving, late medieval thaumaturgical cult when Tolomei created a legend for her building that set it in the most pressing geopolitical drama of the day. And the story of the house's flight from Islamic persecution had already begun to spread through Italy when Sixtus IV and his relations chose to fortify Loreto as a strategic bulwark for the defense of the northeastern papal states.

Taken together, these factors are more than enough to explain the massive publishing success of the Loreto legend in these same decades. Unlike St. Andrew's head or the Holy Lance, here was a "refugee relic" that the popes tried to promote and the public embraced. As Rome issued calls for a new crusade, Loreto offered a very concrete indication that the Virgin favored the cause. Having no desire to dwell among infidels, she had brought a small but exceptionally sacred part of the Holy Land to Italy, where the popes now did their best to honor and defend it.

And yet, the printed literature doesn't quite bear this interpretation out. It is difficult to prove negatives and harder still to establish what readers or purchasers of printed books may or may not have wanted from their texts. But early printed editions of Tolomei's legend and other Loreto texts are conspicuous for certain absences that require comment if not explanation. For one thing, they contain no trace of papal imagery. While the popes rebuilt Loreto, covered the basilica with their emblems, and claimed the shrine as their own, and while most editions of Tolomei's legend were published in Rome, the books feature none of the papal portraits or insignia that feature in contemporary editions of the *Historia urbis Romae* and *Indulgentiae* (most of them printed in Rome, by the same printers). For printers and pilgrims alike, Rome was branded a papal city, but Loreto, thus far, was not.

The texts present other puzzles. In the 1480s and 1490s, a handful of lively literary accounts of the Loretan story were published, each suggesting dramatic ways the devout might respond to the site. But these works, once printed, were rarely republished. By contrast, Tolomei's original and strangely inert exposition of the legend, which offered the faithful little in the way of inspiration or material for reflection, was reprinted dozens of times in its original Latin and in Italian and German translations. For some reason, the text that said less was published more.

Beyond this, there were further curiosities. The origin story of the holy house, which only took shape in the 1460s, was informed by contemporary concerns about the Turkish menace, but printed texts make relatively little reference to the Muslim conquests that prompted the house's removal. Instead, they focus on the Virgin and the miracle of the Incarnation that took place in her house, concentrating on the building's role in salvation history. The texts argue that the structure now in Loreto really was the Virgin's house and that it really had come from Nazareth; they fix its arrival in time and produce witnesses to confirm it. But the argument about authenticity centered on establishing the house's association with the Virgin. The fact that it had fled *Muslims* was almost circumstantial.

Thus, important as it was to show that the house was genuine and that it had flown to Loreto with angelic help, the woodcuts illustrating printed editions of Tolomei's text only sometimes show the house in flight. Just as often, they present the Virgin in some other Marian guise, for example, as the Virgin Immaculate, the Virgin Annunciate, or the apocalyptic woman clothed in the sun. What was important about the house is that it was the Virgin's; the cult—and the legend—focused on her.

Loreto was a site of religious devotion whose most popular printed legend aligned only obliquely with the political and artistic initiatives that ecclesiastical patrons provided for its growing cult. Devotion to the Madonna di Loreto seems to have advanced with little regard for papal attempts to appropriate or direct its focus; the printing history of Loreto texts reflects this organic development. It is unlikely that the popes had anything to do with their publication: even as they issued indulgences to encourage pilgrims to visit the shrine, they made no effort to shape the kind of message pilgrims might encounter there.

A final point: if printed literature on Loreto was primarily devotional, untouched by propagandistic motives, print still functioned rather differently than it is supposed to have done even in religious contexts in early modern Europe. The market ignored lively, ingenious works of devotion and demanded, instead, a text of almost talismanic inscrutability, one that did little to stimulate the imaginative faculties of its readers. Print also helped *perpetuate* a situation of iconographic instability, as multiple visions of the Madonna di Loreto proliferated across the decades.

Spagnoli and Dati

The Loreto legend did not appear in print till the late 1480s. Although Tolomei (d. 1473) first wrote the legend down and served as a key source for later authors, his *Translatio miraculosa* was not the first text to be printed: the earliest surviving edition dates to 1486–90. Two other Loreto texts, printed earlier, bear examining first.

Battista Spagnoli, vicar general of the Mantuan congregation of the Carmelite Order and an accomplished humanist poet, gave his Loreto text the title *Redemptoris mundi matris lauretanae historia* and dedicated it to Cardinal Girolamo Basso della Rovere, bishop of Recanati and nephew of Sixtus IV, who brought the Carmelites to Loreto to administer the site. Spagnoli's text is dated September 22, 1489, and was printed in Bologna, a city where Spagnoli once studied and taught;[80] within a year, the Bolognese humanist Giovanni Sabbadino degli Arienti produced an Italian translation, also printed in Bologna.[81] Like Giacomo Ricci some twenty years before, Spagnoli drew heavily on Tolomei's text, recasting the rector's simple prose into a sophisticated literary composition, which he then dedicated to an important patron.[82]

Spagnoli sets his borrowed text in a larger frame meant to advance his own interests and those of his order. In his dedicatory letter to the cardinal, he explains that on a recent visit to Loreto, he had been looking at the walls of the shrine, hung with its many *ex-votos*, when his eye fell on an ancient tablet, covered in dust and grime, on which the story of the shrine ("from where and in what way the place derived its power") was written in faded letters.[83] To rescue this narrative from its obscurity, Spagnoli would now retell it as a service to history and as the pious act of a Carmelite with a special devotion to the Virgin.[84]

Spagnoli then rehearses Tolomei's tale, polishing the language and inserting additional details: he notes that the house's original site in Nazareth lay close to Mount Carmel, the most sacred site for Carmelites in the Holy Land.[85] He offers a more precise date for the removal of the house than Tolomei, who only says that it arrived in Loreto within living memory, and that its story was revealed to a local hermit in 1296. Spagnoli, by contrast, is precise about the house's first removal from Nazareth. The chapel was revered by Christians in the Holy Land right up to the time when the "Hagarenes" invaded. Then, "during the reign of Emperor Heraclius, the terrible Chosroes, king of Persia, invaded the Promised Land, sacked Jerusalem,. and carried away the wood of the Holy Cross as plunder. Rampaging far and wide in his hatred for Christianity, he inflicted grave wounds on the Eastern church. Next, the Mohammedan error began to spread through the region, and the worship of God and orthodox faith began to migrate from East to West; it was then that this little chamber was picked up with the help of angels and, leaving its foundations behind, was carried to Illyria."[86]

Spagnoli's narrative for the holy house parallels that of the True Cross. Just as the wood of the Cross miraculously rescued itself from the impious Persians in the reign of Heraclius, so the Virgin's house orchestrated its miraculous removal from the clutches of Arab infidels. The two relics were linked by their place of origin (the Holy Land), their moment of crisis (the reign of Heraclius), and their subsequent fortunes: both sought and found asylum in the West under the protection of the Roman Church.

Spagnoli then explains how the popes began to patronize Loreto; here he recounts Paul II's vision and miraculous cure at the shrine, just before his election to the papacy, which prompted him to build a new basilica to honor the Virgin.[87] After his election, Sixtus IV appointed his nephew Girolamo bishop of Recanati, who wisely continued the construction project at Loreto, investing so much in the massive new basilica "that it seems to rival those ancient monuments of Roman magnificence whose ruins even now are a source of amazement."[88] He had also fortified the new construction in order to protect it from Turkish attacks.[89]

Girolamo's next wise move had been to bring the Carmelites to Loreto. Here Spagnoli advances some extraordinary claims. The Carmelite Order had a longer history in the Holy Land than any other, and its members had always played a role in the custody of the Holy Places.[90] In fact, the Carmelites had looked after the Virgin's house in Nazareth until the judgment of God allowed the Saracens to drive the monks from the East.[91] Since the Carmelites had tended the house in the Holy Land, it was only fitting, now that it was in Italy, that they should look after it again. Spagnoli praised the cardinal for restoring the monks to their *antiquam possessionem*.

Among other supports for this claim, Spagnoli describes a dramatic exorcism he witnessed at Loreto.[92] A French pilgrim came to the shrine seeking relief from a demon that was possessing her. The Carmelites summoned it forth and subjected it to an interrogation: was the house at Loreto truly the house of the Virgin? At first, the demon refused to say, but eventually it was compelled to admit that it was. And who, the Carmelites asked, had guarded the shrine when it was in the Holy Land? The demon made the poor Frenchwoman roll her eyes and loll her tongue; she raged and she spat, but at last the demon replied, through her, that the original and true custodians of the house were the Carmelites.

For Spagnoli, it was important to establish the origins of the house in the Holy Land, not only to verify its miraculous character but also to justify the Carmelites' recent installation. One can imagine tensions at the site after the strict ascetic order took over its administration. Local clergy and laypeople may have bristled at their claims to the place. Spagnoli's text asserts a chronology for the house, including a relationship with the Carmelites, predating its arrival in the Marche. Flattering Cardinal Girolamo and his late uncle for their prudent management and patronage of the shrine—a monument to rival that of the ancient Romans, and one that would also protect agains the Turks—also helped protect his order's tenure at the shrine.

If the thrust of the tract thus far was pragmatic, even tendentious, in the final section Spagnoli shifts direction entirely. Here he waxes poetic, offering a series of meditations on the sanctity of the little house of Loreto that are quite moving in their fervor. He evokes episodes in the life of the Virgin and ruminates on the house as a place of revelation and salvation ("this is where the foundations of human salvation were laid, this is where the secrets of the divine mind were revealed"). The house, Spagnoli reflects, is the most sacred place under heaven, because it is where God became man. In Eden, a woman was made from the rib of a man; here a virgin changed the order of nature and became the mother of God. In the Ark of Noah, a small remnant of humanity was saved; in this place, the salvation of the entire world was effected. Where was God ever more present than in this place? Greater than Bethlehem, greater than Galilee, greater even than the Cross or the Tomb, the house of Loreto was the place where the Word became incarnate. It was a site of graces both mysterious and quotidian, a locus of cosmic mystery, but also Christ's childhood home. His hands had traced every inch of its walls. The pavement had felt the imprint of his little feet. How often had he fasted and prayed beneath its roof? Here the pilgrim might feel closer to Christ than anywhere else in the world.

Another version of the legend was composed in Italian *ottava rima* by Giuliano Dati, author of the metrical *Statione, indulgentie, e reliquie* discussed in the previous chapter. Dati's text was published in Rome around 1492–93. A prolific poet, Dati published numerous topical poems in 1490s Rome: not only his guide to the station churches of Rome, but also saints' lives, accounts of passion plays, newsy ballads on Columbus's discovery of the Americas and the French invasion of 1494, and a ballad of Prester John. His poem on the Holy House overlaps with almost every one of these other productions, providing material for pious devotion, meditation on wonders from the East, and (as his woodcut title page shows) an almost journalistic concern for getting the story straight, geographically and chronologically.

Dati's title-page illustration celebrates the flight of the house in minute detail, tracking its movements from Nazareth to Fiume in Croatia, to Recanati, the grove of "Loreta," and finally to the hilltop of Loreto (figure 6.8).

The poem tracks Tolomei's original narrative closely, adjusting and adding certain elements to fit his poetic genre. Though Dati intends to tell the story of the house and its transfer, he pays equal attention to the *image* of the

Figure 6.8. Giuliano Dati, *Traslazione della sacrata camera di Nostra Donna di Nazareth in Italia appresso la città di Recanati* ([Rome: Andreas Freitag, ca. 1492]). Venice, Fondazione Giorgio Cini, fol. 1r.

Virgin that the house contained. He repeatedly mentions the house being moved along with its image or the image coming to Loreto along with the house.[93] St. Luke painted it from life, he says, and it was *tanto ornato*, presumably with gilding or a precious frame.[94] At one point he even describes what it looked like: the Virgin, like an image of Justice, held scales and a sword.[95]

In Dati's telling, the Virgin also directs the miraculous transfer of her house. In Tolomei's original, angels appear from nowhere and the house seems to make its many moves of its own accord. Dati, by contrast, uses the formula *piaque Maria* or variants to move the action along: the Virgin did not want to stay among infidels, it pleased her that her house should move again, "this great queen did not want the story to remain a secret," and so on.

Dati also expands Tolomei's account of the citizens sent to Nazareth to verify the house's identity into a miniature epic. This includes, quite originally, a journey to Rome and an interview with the pope. After the delegation returns with measurements proving that the house in Loreto would fit on the empty foundation in the Holy Land, the people of Recanati decide to send two of the travelers on to Rome to inform the pope. These "ambassadors" are received kindly at court and invited to share their news. Each then delivers a speech describing the movements of the house around the Adriatic, its arrival in Loreto, the later visions and revelations to local inhabitants, and their recent journey to the Holy Land. This device allows Dati to rehearse his tale a second time; it also widens his narrative frame beyond Loreto to include Rome and endows the legend with papal endorsement far earlier in its history than was probably true. Dati concludes that the pope was delighted with the ambassadors' story and sent them away with his blessing. They returned to Loreto as heroes.

In his closing stanzas, Dati shifts focus to the present day: he mentions that the shrine is now in the care of the Carmelites thanks to an intervention by their cardinal protector in Rome, he refers to the many miracles that have taken place at the shrine in recent memory,[96] and he invites believers and doubters alike to go to Loreto and see more for themselves. There they will find the whole story written up on a placard in the church, alongside a painting of all the miraculous events just as he has described them in verse.[97] He ends by exhorting his readers to remain devoted to the Virgin.

Tolomei

Spagnoli's and Dati's literary treatments contrast starkly with Tolomei's original text. Unlike his later imitators, Tolomei makes few concessions to his readers. He does not address them directly. While he reports how and when the Virgin's house began to move, he ventures no comments on the significance of its travels. He relays the testimony of local witnesses and describes the surveyors' journey to Nazareth, but he makes little reference to later miracles at the shrine or the extent of the devotion that grew up around it. Even the fact that the house was the Virgin's is presented quite rapidly at the start of the narrative and with little comment:

> The Church of Blessed Mary of Loreto was a room in the house of the blessed Virgin Mary, mother of our Lord Jesus Christ, which house was in the territory of Jerusalem in Judaea and in a city of Galilee whose name was Nazareth and in this said room the blessed Virgin Mary was born and there she was raised and later she was hailed by the Angel Gabriel and finally in this said room she raised her son Jesus Christ up to the age of twelve. Finally, after the ascension of our Lord Jesus Christ into heaven, the blessed Virgin Mary remained on earth with the apostles and other disciples of Christ who, seeing that many divine mysteries had occurred in this said room, agreed to make a church out of the said room in honor and memory of the Virgin Mary and so it was done and then the apostles and disciples consecrated that room into a church and there they celebrated divine services. And Blessed Luke the Evangelist made there with his own hands an image in the likeness of the Blessed Virgin Mary which is still there to this day.[98]

Compared with Spagnoli's imaginative meditations on what it meant to stand on the site of the Incarnation, the childhood home of the Christ Child, or Dati's breathless wonder that the Virgin chose Loreto for her home, Tolomei's text does little to engage or excite the reader. It reads more like a legal document as it sets out its claims, buttressed by eyewitness testimony, unadorned by commentary or rhetorical flourishes. It is easy to see how the story might have originated as text on an informational plaque: it offers a public statement of the identity of the house, not a private meditation on its sanctity.

This makes it all the more surprising, then, that Tolomei's text fared so much better on the open market.[99] Dati's text was never reprinted after its first publication. Spagnoli's appeared once in Latin and once in an Italian translation and was reprinted once more around 1514. Many other literary treatments (such as those by Caterina de' Vigri, Giacomo Ricci, or Luigi Lazzarelli) were never printed at all. As the century turned, it was Tolomei's text that proliferated in more than forty editions, Latin, Italian, and German. These are mostly undated but seem to have been produced in Rome just before or after 1500. Julius II issued a plenary indulgence for Loreto in 1507, and many editions may have appeared after that pronouncement.[100] Small in format and inevitably illustrated with a single woodcut on the first page, they were produced by the same printers who produced editions of the *Mirabilia* and *Indulgentiae* in great numbers in these same years: Eucharius Silber and his son Marcellus, Stephan Plannck, Johannes Besicken, and Etienne Guillery.[101]

What made Tolomei's text so popular? It was the first version of the story to be written down, but this would not have been apparent to readers in 1500 or 1510. It reproduced text that was posted on a placard at the site, although none of the printed editions identify it as such. It was short and written in very simple Latin, making it an accessible and easy read for the faithful. Still, it is puzzling to think that Spagnoli's vivid meditations on the house and Dati's lively vernacular verses were somehow less engaging. Was there something particularly appealing about the archaic simplicity of Tolomei's tract that made it more desirable than other, more sophisticated treatments?

Perhaps the text recapitulated something of the actual pilgrim experience. In Rome, printed editions of the *Indulgentiae* featured images of the Veronica as it was seen at St. Peter's, at the moment of exposition, coupled with the liturgical texts that a pilgrim would have heard or recited while standing before it. At Loreto, pilgrims venerated the cult image of the Virgin, left offerings, and were likely shown a sign with Tolomei's text explaining the history of the shrine; it seems that *this* is the text they wanted to carry away with them as a souvenir of their visit. The original placard may have been celebrated as something like a contact relic, since it not only explained the identity of the sacred building but also was posted directly on it, and a reproduction of its text might be thought to participate in the sanctity of the original. The importance that accrued to texts that had been posted in significant places has been noted throughout this study.

Neither Spagnoli's original text nor its Italian translation was illustrated. Dati's poem was printed with a woodcut that summarized the house's travels minutely, but like the text it accompanied, this striking image was never reprinted. By contrast, not only was Tolomei's account reprinted many times; it was also reprinted with many different images. The earliest surviving edition is attributed to Eucharius Silber and assigned to 1486–90.[102] It lacks a title; the first page has a woodcut of two angels carrying the church of Loreto, one on each end of the building, with a seraph supporting the structure from beneath and a gargantuan Virgin and Child rising behind it (figure 6.9). The

Figure 6.9. Translatio miraculosa ([Rome: Eucharius Silber?, ca. 1486–90]).
Munich, BSB, Res/P.lat. 1358#Beibd.3, fol. 1r.

angels seem to be kneeling on the ground and lowering the church into position after its miraculous flight.

Around 1495, Silber reprinted the text with a title on the first page: *Translatio miraculosa ecclesie beate Marie virginis de Loreto*. The accompanying image makes no reference to the flight of the house: a small woodcut depicts just the Virgin and Child in a mandorla, supported from below by a single seraph with extended wings (figure 6.10).[103]

Between 1490 and 1500, Stephan Plannck published three editions of the legend, none of them with an image of the house or its flight.[104] They have, instead, an elaborate cut of the Virgin as the apocalyptic woman clothed with the sun, an iconography also becoming associated with the Virgin Immaculate in these years. Standing on a crescent moon, she holds a scepter in one hand and the Christ Child in the other as solar rays emerge from behind her body and around her head. This cut (figure 6.11) is framed by a thick border of white acanthus leaves and flowers with four figures in the corners: at top left and right, the angel Gabriel and the Virgin Annuciate, and at bottom left and right, two figures carrying banderoles with the prophetic phrases "orietur stella" (Num. 24:17) and "Ecce virgo concipiet" (Isa. 7:14). The figures allude to the mystery of the Incarnation that took place inside the holy house, but this cut was not made specially to illustrate Tolomei's text. It was one that Plannck had used in other Marian contexts before, for example, in several editions of the *Indulgentiae* where it appears as the illustration facing the description of Santa Maria Maggiore.

Around the same time, Johanes Besicken brought out three editions of Tolomei's text using *his* version of this same cut (the Virgin as the woman clothed with the sun, in a thick frame, with the same figures and prophetic banderoles in the corners), a cut that also appears in his editions of the *Indulgentiae*.[105] This cut appears in a contemporaneous edition printed in Eucharius Silber's types, ca. 1500;[106] it reappears in an edition of Etienne Guillery around 1513,[107] and possibly again in another Guillery edition published as late as 1524.[108] Roman printers often traded woodcuts or handed them down to heirs and successors, so the appearance of Besicken's cut in editions by two other printers is not surprising. On the other hand, the recycling of the image suggests that its use was a conscious choice. Silber's son Marcellus would produce another edition of the text with a new cut of the Virgin in the same posture (discussed below). Yet other variants were possible: Eucharius Silber published another edition of the *Translatio* with a small woodcut of the Virgin holding the Child and a scepter, framed in a gothic arch, unaccompanied

Figure 6.10. *Translatio miraculosa* ([Rome: Eucharius Silber, ca. 1495]). Munich, BSB, Inc.s.a 211 m, fol. 1r.

by angels, a house, the moon, solar rays, or any other Marian attributes seen in other Loretan woodcuts (figure 6.12).[109]

Of course, many editions depicted the angelic transfer of the holy house. Besicken himself published an edition after 1500 that included not one but

Figure 6.11. Translatio miraculosa ([Rome: Stephan Plannck, after 1491]). Tübingen UB, Gh 381a, fol. 1r.

Figure 6.12. *Translatio miraculosa* ([Rome: Eucharius Silber, not after 1509]). Munich, BSB, Rar. 1240#Beibd.7, fol. 1r.

two woodcuts of the Virgin atop her house-church: on the title page, a large cut depicts two angels and six seraphs carrying the church over the waves, with the Virgin and Child on top and a pair of angels carrying a banderole over her head.[110] On the next page, a smaller cut reproduces the top half of the scene: the Virgin and Child above the church, with two angels overhead. Besicken would print two more editions of the text with similar (but not identical) versions of this smaller cut, both inset into the start of the text (figure 6.13).[111]

Shortly after 1500, Eucharius Silber acquired a new and more elaborate woodcut of the angelic transfer, which he used in four different editions of Tolomei's text.[112] In it, two angels carry the church with the Virgin and Child behind it, and two seraphs support it from beneath, while two more angels lift a large scroll inscribed "S. Maria de Loreto" over the Virgin's head (figure 6.14).

But the other tradition still persisted. Around 1505, Silber produced an edition with a striking new cut of the Virgin as the woman clothed with the sun,[113] which he reprinted once before his death in 1509.[114] Silber's son Marcellus then took over his father's press; he produced two more editions of the text between 1509 and about 1513 with this same cut (figure 6.15).[115] Marcellus Silber had another cut made of the Virgin as the woman clothed with the sun, which he used in another edition of the *Translatio miraculosa*, dated after 1513 (figure 6.16),[116] and in an edition of a different Loretan text, an Italian-language *Oratione* (published ca. 1512).[117] Not long after, he brought out another edition of the *Translatio miraculosa*, ca. 1515, using his father's cut of the Virgin on the church with the scroll over her head.[118] Also attributed to him is an edition with a small cut of the house carried by tiny seraphs with the Virgin and Child above, a much rougher production than his usual style.[119]

An unidentified Italian printer, likely Venetian, printed Tolomei's text with a woodcut similar to Silber's earliest (figure 6.9 above) a total of three times.[120] Another Venetian printer, Paolo Danza, published a more elaborate cut depicting the landing of the house near Recanati, with additional figures witnessing its arrival and the city labeled by name.[121] Finally, a rather unrefined cut of the flying house appeared in three editions printed by the young Antonio Blado, some of his earliest work in Rome (figure 6.17).[122]

The variation in these editions suggests a basic instability in the iconography of the new cult. While the legend was associated with an image of a house or church in flight, surmounted by the Virgin and Child, it seems printers and readers alike were equally satisfied with an image of the Virgin and Child alone, perhaps with angels nearby, perhaps in the guise of the apocalyptic

Tranflatio miraculofa ecclefie bea te Marie virginis de Loreto.

Ecclefia beate Ma rie de Loreto fuit camera domus beate virgis Marie mauris domini noftri Jefu xpi que domus fuit in par tibus Dierufalez iudee et in ciuitate galilee cui nomen Nazareth:et in dicta camera fuit bea ta virgo maria nata et ibi educata z poftea ab angelo Gabriele faluta »i z demuz in dicta ca / mera nutriuit filiu fuu Jhefum xpm vfcp ad etatez duodecim annoz:Demu poft afcenfionem domini noftri Jhefu xpi in celum: remanfit beata virgo maria in terra cu apoftolis z aliis difcipulis xpi.Qui videntes multa mifteria diuina fuiffe facta in di cta camera:Decreuerut de comuni confenfu omniu de di cta camera facere vnam ecclefiam in honoe z memoiaz beate virginis Marie z ita factuz fuit:Et deinde apoftoli et difcipuli illam camera confecrauerut in ecclefiam: z ibi celebrauerut diuina officia:Et beatus Lucas euangelifta cum fuis manibus fecit ibidem vnam ymaginem ad fimi litudinem beate virginis marie:que ibi eft vfcp hodie:De mum dicta ecclefia fuit habitata z honorata cu magna de

Figure 6.13. Translatio miraculosa ([Rome: Johann Besicken (and Martinus de Amsterdam?), ca. 1504]). Munich, BSB, Inc.c.a 357#Beibd.1, fol. 1r.

Figure 6.14. *Translatio miraculosa* ([Rome: Eucharius Silber, after 1500]). Cambridge University Library Inc.7.B.2.27[3660], fol. 1r.

Figure 6.15. *Translatio miraculosa* ([Rome: Eucharius Silber, after 1500]). Oxford, Bodleian Library, Douce 257, fol. 1r.

Figure 6.16. *Translatio miraculosa* ([Rome: Marcellus Silber, after 1513]). Frankfurt-am-Main, Universitätsbibliothek, Inc. oct. 6 Nr. 8, fol. 1r.

Tráſlatio miraculoſa Eccleſie beate Marie virginis De Lo reto.

Figure 6.17. *Translatio miraculosa* ([Rome, Antonio Blado, ca. 1516]). Tübingen, Universitätsbibliothek, Gb 534b, fol. 1r.

woman clothed with the sun, an iconography that was becoming associated with the Immaculata in the later fifteenth century. However compelling Tolomei's description of the house's peregrinations may have been, readers seem not to have *needed* to see its particulars illustrated on the title page. What they wanted was an image of the Virgin.

Similar variety characterizes the vernacular editions. The Florentine cleric Bartolomeo da Vallombrossa produced an Italian translation of Tolomei's text after visiting the shrine in May 1483. Shortly afterward, he had a broadside of it printed at the Ripoli press in Florence, of which no copy survives.[123] The earliest surviving editions date to around 1500; in all, seven editions are known, printed in Florence, Venice, and elsewhere in Italy. Four of these editions have woodcuts, each different from the next, depicting the Virgin and Child above a church carried by angels.[124] Two others revert to an earlier type: the Virgin and Child beneath a baldachin (figure 6.18).[125]

In Augsburg, meanwhile, a German translation was printed around 1500, and reprinted in 1510, with an entirely different iconography. In both cases, the illustration shows the Annunciation—the event that had made the little house sacred in the first place (figures 6.19 and 6.20).[126]

In short, despite the popularity of Tolomei's text and its frequent republication, the iconography of the printed legend was by no means stable. The forty-three editions I have seen or for which published descriptions are available are remarkably variable: twenty-one contain a woodcut of the Virgin and Child above a church carried by angels; sixteen have the Virgin alone or standing on the moon; two have the Madonna di Loreto of the older type (the Virgin under a baldachin supported by angels);[127] and two, both German, depict the Virgin Annunciate.

The distribution of these images did not change over time, nor did particular printers prefer one image over another. The Virgin on the crescent moon appears in editions by Plannck (ca. 1491–1500), Silber (1500), and Besicken (ca. 1504), while the Virgin on a flying church appears in editions by Silber (1486–90), Besicken (1504), and Marcellus Silber (1510s). In Florence, Francesco di Dino used a cut of the Virgin on a flying church (after 1483), but Bartolomeo di Libri used a cut of the Virgin under a baldachin (around 1497). Eucharius Silber used his striking flying church cut at least four times (after 1500), while his son Marcellus used his equally striking Virgin on the crescent moon at least three times (ca. 1509 and ca. 1513). This might suggest a progression from a focus on the church to a focus on the Virgin, but already, around 1495, Silber had used a tiny cut of the Virgin in a mandorla, and before

Nota che la chiefa di fancta Maria delloreto fu chame-ra della chafa della uergine Maria madre del noftro fignore Gie fu Chrifto:laqual chafa fu nelle parti di hierufalem di giudea nella cipta di galilea chiamata Nazaret. Et in dec-ta chamera nacque lauergine Maria & quiui fu alleuata & dallangel Ga-briel quiui annunciata: & finalmente in decta chamera nutri elfuo figluolo Giefu Chrifto per infino alleta di xii anni. Dipoi dopo lafcenfione del no-ftro fignore giefu chrifto in ciel rima fe labeata uergine Maria con li apo-ftoli & glialtri difcepoli di chrifto e-quali uedendo che molti mifterii di-uini erano facti in decta chamera di liberorono di commune confentimen to di tucti fare di decta chamera una chiefa in honore& memoria della bea ta uergine Maria & cofi fu facto im-peroche dipoi gli apoftoli & difcepo-li decta chamera confacrorono in chie fa & quiui celebrorono ediuini officii Et beato Luca euangelifta con lefue mani fece quella imagine afimilitudi ne della beata uirgine Maria laquale quiui e hoggi infino aldi. Dipoi dec-ta chiefa fu habitata con gran diuoti one & reuerẽtia da quello populo chri ftiano che era in quelle parti nellequa li decta chiefa ftette per infino che de cto populo fu chriftiano; Ma poi ch

Figure 6.18. La dichiaratione della chiesa di sancta Maria del Loreto et come venne tucta intera ([Florence: Bartolomeo dei Libri, ca. 1497]). Genoa, Biblioteca universitaria, Inc. Gaslini 10, fol. 1r.

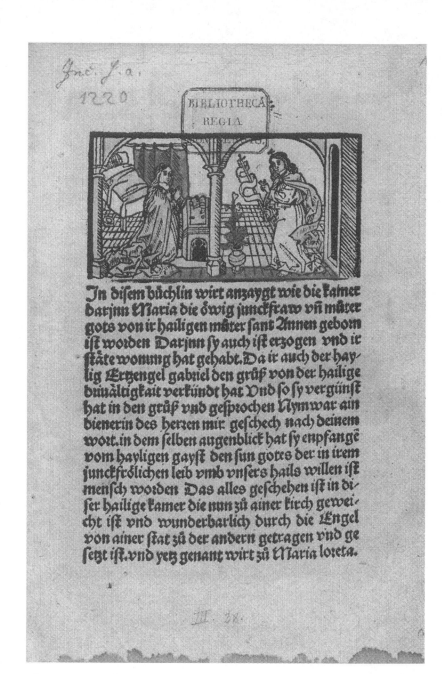

Figure 6.19. Translatio miraculosa (German) ([Augsburg: Johann Schönsperger, ca. 1500]). Munich, BSB, 4 Inc.s.a. 1220, fol. 1r.

Maria zu Loreta
vnſer lieben frawen kamer.

No. 637ª
BIBLIOTHECA
REGIA
MONACENSIS

IN diſem Büchlin wirt angezaigt wie die kamer darinn Maria die ẽwig junckfraw vnd můter gotes von ir hailigen můter ſannt Annen geboren iſt worden/Darinn ſy auch iſt erzogen vnnd ir ſtäte wonung hat gehabt. Da ir auch ð hailig Ertzengel gabriel den grůß von der hailigen drualtikait verkündt hat/Vnd ſo ſy vergünſt hat in den grůß vñ geſprochñ Nÿm̄war ain dienerin des herren mir geſchech nach deinem wort/im ſelben augenplick hat ſy empfangen vom hayligen gaiſt den ſun gottes der in irem junckfrewlichen leib vmb vnnſers hails willen iſt menſch wordñ/Das alles geſchehen iſt in diſer hailige kamer die nun zů ainer kirch geweicht iſt/vñ wunderbarlich durch die engel von ainer ſtat zů der andern getragñ vnd geſetzt iſt/vñ yetz genant wirt zů Maria loreta.

Figure 6.20. *Translatio miraculosa* (German) ([Augsburg, ca. 1510]). Munich, BSB, Res/4 Asc. 637s, fol. 1r.

1509 he used a small, rectangular cut of the Virgin, while his son printed an edition with a small cut of a flying church sometime after 1521, several years after his multiple editions with the Virgin on the moon. Silber (1486–90), Besicken and Amsterdam (ca. 1504), and Blado (1516–24) all reproduced busy, crowded scenes of the Virgin atop her church, carried or sheltered by angels with fluttering banderoles, surrounded by clouds, waves, city walls, solid ground, or stars, but the Silbers, father and son, used a strikingly simple image of the Virgin as the woman clothed with the sun, standing on the crescent moon against a backdrop of solar rays. In the minds of Roman printers—and, presumably, their customers—it did not particularly matter *how* the Madonna di Loreto was portrayed. The legend of her house could be illustrated by Marian imagery of any sort.

Such a mixed bag of evidence from the printed books might seem wholly inconclusive. In fact, the variability reveals something rather important about the cult of Loreto at the end of the fifteenth century. Tolomei, in his legend text, was concerned to establish the authenticity of his shrine: it was, unquestionably, the Virgin's house from Nazareth. Erudite authors like Spagnoli and Dati promoted Loreto as an emblem of Christian resistance against the Turks, as a site of magnificent papal patronage, as proof of Carmelite connections to the Holy Land, or simply as a powerful *locus* in which to imagine the mysteries of the Incarnation. The reading public, however, seems to have been more interested in the Virgin herself and the graces, miracles, and cures she could bestow. They came to Loreto to seek healing, to satisfy vows made in moments of crisis, to ask for the Madonna's help and protection. The story of the house's miraculous flight assured the visitor to Loreto that house and cult image alike were authentic, and therefore sacred and likely efficacious, but the story of its flight was not why visitors came.

So why is the story the thing they took away? I have suggested some possible reasons: Tolomei's text was easy to read, it seemed ancient, and it was associated in a very physical way with the fabric of the church and the experience of a visit to the shrine. As a souvenir, it allowed the pilgrim to take home a (reproduction of a) piece of the place itself. In later centuries, pilgrims to Loreto would buy holy cards with scraps of silk attached, cut from a veil draped on the cult statue overnight, or pottery bowls made from clay mixed with dust from the floor of the house. Tolomei's clunky *Translatio* may have functioned more like a relic than a text: more important for where it had come *from* (the wall of the house) than for what it actually said. (Similarly, around 1525, a new Loreto text appeared in print, reproducing the text of a prayer that had miraculously appeared beneath the cult statue: *Legenda et oratione che fu trovata alli piedi de Santa Maria da Loreto*.)[128] Finally, every edition of Tolomei's text carried an image of the Virgin, whom pilgrims adored, and this may well be a better explanation than any other for the popularity of the text. It traveled with an image of the Madonna di Loreto, variable as she was.

It may be that the appearance of so many, completely different woodcuts illustrating the same text was simply the result of economic or practical choices: if a printer did not have a "flying church" cut, or if it was already in use, he could simply take and use another Marian cut from his inventory. The fact that Besicken and Plannck both illustrated Tolomei's text with what is more familiar as their "Maria Maggiore" cut from the *Indulgentiae* would seem to support such an argument. But this presumes that the default illustration for the *Translatio* was a flying church, and that editions with a different image of the Virgin—more than half of the surviving archive—were somehow improvised or considered second best. A different, more compelling explanation is that printers and readers alike thought of the Loreto legend as a story about the Virgin, not her house; they expected the story to be illustrated with an image of the Virgin, and it didn't particularly matter which Virgin it was.

A well-known argument in the history of art maintains that Renaissance religious pictures should be considered either *immagini* or *storie*.[129] *Storie* narrated episodes in the life of a biblical figure or saint, often several of them in sequence, while *immagini* were visual aids for the veneration of saints and holy figures, who appeared portrait-style with their identifying attributes. According to this argument, there are very few narrative pictures of "The Annunciation," relating the episode in the Gospel of Luke, but a great many images of the Virgin Annunciate, whose attributes could include the Angel Gabriel, a book, or the dove of the Holy Spirit. On this view, woodcut illustrations of the "flying house" were not intended to narrate the story they accompany. Rather, they present elements—a house held aloft by angels, hovering over the sea—that together constituted the attributes of the Madonna di Loreto. Just as a capacious cloak is the attribute of the Madonna della Misericordia, or a child, a devil, and a club are the attributes of the Madonna del Soccorso, or the crescent moon became, eventually, the attribute of the Madonna Immacolata, a house held up by angels was the attribute of the Madonna di Loreto (who could yet be represented in other ways), and it was to her that pilgrims remained devoted.

Iconography

By the early sixteenth century, the image of the Virgin on her house was becoming a common subject for Italian painters (e.g., figure 6.21). But other painters continued to invoke older traditions in painting the Madonna di Loreto. In 1500, Giovanni Battista Rositi painted her under a baldachin held aloft by four angels.[130] So did Luca Signorelli in the Oratory of San Crescenzio in Morra, near Città di Castello (figure 6.22).[131] By contrast, in 1507, a patron of Perugino's in Perugia requested "an image of the glorious Virgin with her Son standing, similar to that of Loreto." Perugino delivered a Virgin and Child about to be crowned by angels, perhaps arranged in the same pose as the cult statue, but with no reference to a flying structure (figure 6.23).[132]

In the same year, Julius II recognized a new confraternity in Rome, organized by the baker's guild and dedicated to the Madonna di Loreto. This group soon began construction of a church next to Trajan's Column, to designs by Bramante and Sangallo.[133] The altarpiece in the church depicts the Virgin and Child atop the holy house, raised aloft by angels, and flanked by the plague saints Rocco and Sebastian, a nod to the thamauturgic character of the original cult.[134] This iconography had already appeared decades earlier, in the cathedral of Atri, and would recur in other sixteenth-century paintings of the Madonna di Loreto as well.[135]

Printers, printmakers, and painters alike explored a variety of representational strategies when portraying the Madonna di Loreto. The Virgin on her house, the Virgin under a baldachin held up by angels, the Virgin with plague saints, the Virgin with no attributes at all except angels floating nearby—all these could be the Madonna di Loreto.[136]

We cannot know exactly what each Renaissance pilgrim went to Loreto for, or how he or she imagined the Virgin of that place. Printed Loreto texts suggest some answers to these questions, but no firm conclusions. What they do provide is evidence for how ambiguously early print interacted with a new and growing Italian devotion, and this has implications for how we understand the larger impact of the technology. In prevailing narratives of the history of the early modern book, print is supposed to have stabilized and standardized visual representations of things—maps, diagrams, and iconographies. But in this case, print received an already variable visual tradition and caused it to diversify even further. Print is also supposed to have encouraged the

Figure 6.21. Saturnino Gatti (attr.), *The Translation of the Holy House of Loreto*, ca. 1510. New York, The Metropolitan Museum of Art, 1973.319.

Figure 6.22. Luca Signorelli, *Madonna di Loreto*, ca. 1507–1510, Oratorio di San Crescenzio, Morra.

Figure 6.23. Pietro Perugino, *Madonna di Loreto*, 1507. London, National Gallery.

development of private devotions and religious interiority, yet in the case of Loreto, mystical and meditative texts went unprinted, while for decades what did proliferate was the *least* interesting or even legible of documentary treatments. Tolomei's *Translatio* functioned more like a talisman than a text, a souvenir of the shrine whose value derived from its origins as a text that had hung on a wall—a contact relic, not a conveyor of content. This, too, seems a surprising phenomenon to emerge from the technology of mass reproduction, although other early printed texts also functioned more like sacramental objects than sources of information.[137]

Finally, early print *could* work in a political way, projecting notions of authority or proprietorship over religious or cultural phenomena. The previous chapter showed how print set the stamp of papal authority onto the pilgrim experience of the Jubilee in Rome. But at Loreto, even as the popes took over the site, sponsored architectural and artistic projects within, and promoted it to pontifical status, printed texts about Loreto (texts mostly printed *in Rome*) seem indifferent to these initiatives. This suggests that printed pilgrim literature was not shaped by political imperatives but was truly responsive to what the market demanded. For Rome, the market demanded papal imagery; for Loreto, it did not.

Loreto was a medieval center of Marian devotion focused on a healing cult statute; in the 1460s, the legend of the church's flight from Palestine grew alongside concerns about the Ottoman threat and plans for a new crusade, and popes sought to appropriate the site to protect and assert their presence in the far northeast of the papal states. The fortified basilica and walls around the town signaled not only readiness against Ottoman attack but also papal authority over the March of Ancona. But pilgrims retained their devotion to the Madonna di Loreto, *effectrix* of miraculous cures, and printmakers and printers followed their lead. By the time Bramante's marble screen around the little house was finished, in 1537, the cult of the Virgin and the cult of the Holy House had begun to fuse with one another. The iconography stabilized (the Virgin and Child atop a flying house), retellings of the legend proliferated, and pilgrims accepted the integration of statue, house, and fortified basilica into a single devotional nexus. But the late medieval and Renaissance elements that shaped Counter-Reformation Loreto looked very different in their earlier stages. Far from stabilizing the cult, early printed editions of the Loreto legend kept its various elements in tension for many decades.

The previous two chapters have explored the invention of religious traditions that were—or were not—enshrined in print in the technology's earliest years. A pilgrim coming to Rome for the 1500 Jubilee could have bought any number of cheap pamphlets celebrating the Veronica, enumerating indulgences for St. Andrew's Head or the Holy Lance, or commemorating the Virgin of Loreto and her flying house. Other texts promoting other new devotions also found their way into print: pamphlets celebrating the canonization of new saints,[138] indulgences for new foundations like the Ospedale Santo Spirito, announcements of new spiritual initiatives like the Jubilee, or the publication of new doctrines like the Marian decrees promulgated by Sixtus IV at the start of his reign. These events, cults, and traditions, some of them mere decades old, were key elements in the papal strategy of drawing pilgrims and pilgrim funds to central Italy. Print codified these new traditions, spread information about them, and helped integrate them into what seemed a timeless landscape of devotion over which the popes had presided since earliest times. But the public had to agree to the fiction: the Veronica and the Virgin of Loreto somehow *caught* right away, while St. Andrew's head, the Holy Lance, and the Virgin's holy house took longer to find their publics. Print could promote new devotions, but it could not, by itself, ensure that they became popular.

CHAPTER SEVEN

Kissing the Papal Foot

In 1484, the tyrant of Bologna, Giovanni Bentivoglio, fell ill. In his sickness he swore that if he got better he would go on pilgrimage to the Madonna di Loreto.[1] On March 21 of the following year, fully recovered, Bentivoglio set out to fulfill his vow. He departed for Loreto with an impressive cavalcade of 160 horses, twelve pack mules, and sixteen gentlemen, all clad in the gray cloaks of pilgrims. Bentivoglio's *brigata* was a noble crew, including young men from the leading families of the city, all sporting heavy gold chains around their necks. The spring excursion to Loreto was an occasion for display as well as devotion, and Bentivoglio had further stops on his itinerary where formal dress would be required. After paying his debt to the Virgin, Bentivoglio set out south: "he went to Rome to kiss the foot of Pope Innocent, who was delighted to see him and from whom he obtained everything he knew how to ask."[2] From Rome, the party carried on to Florence, where they enjoyed Lorenzo de' Medici's hospitality for a few days before returning to Bologna.

Bentivoglio's visit to Loreto reflects the intense devotion that the Virgin's shrine attracted in the late fifteenth century from the sick and the recently healed, as well as the reach of the Loreto cult across Italy.[3] His visit to Rome was more politic. His relations with Sixtus IV had been strained ever since he had sided with Venice against the Church in the War of Ferrara. But Sixtus had died in August 1484, and Bentivoglio now had a chance to form better relations with the new pope, Innocent VIII. The brief that Innocent sent back to the city government of Bologna in April 1485 suggests that he achieved this goal: Innocent had been delighted to see Giovanni follow the example of his ancestors in showing the new pope the reverence he was due. Giovanni was a loyal son of the Church, the pope was well pleased with him, and he wished the city and its "soldier" all the best.[4] In terms of his physical, spiritual, and political health, Bentivoglio's Italian tour had been a success.

The ritual gesture Bentivoglio performed in Rome—kissing the papal foot— had a long history. Chronicles and ceremonial books record instances of foot kissing at the papal court as far back as the fifth century; by the reign of Leo IV (847–55), it was considered a time-honored custom, one often performed after the election or coronation of a new pope.[5] It would continue to play an important part in papal ceremonial for centuries to come. In 1580, Michel de Montaigne was granted an audience with Gregory XIII. Admitted to the papal chamber, Montaigne inched his way slowly across the room, proffering words of admiration and submission at each step, before dropping to his knees and shuffling across a velvet rug to the papal throne. According to his (oddly third-person) diary: "The ambassador who was presenting them knelt on one knee and pulled back the pope's robe from his right foot, on which there is a red slipper with a white cross on it. Those who are on their knees drag themselves in this position up to his foot and lean down to the ground to kiss it. Monsieur de Montaigne said [the pope] had raised the end of his foot a bit."[6] Imbued with biblical symbolism and imperial pretensions, the act of foot kissing was a public acknowledgment of the pope's claim to be vicar of Christ (whose feet Mary Madgalene had bathed with her tears) and divinely ap- pointed sovereign over every earthly prince. The custom was only abolished by Paul VI in the 1960s.

Giovanni Bentivoglio was not the only prominent figure in Rome that spring waiting to place his lips on a papal slipper. Ambassadors from three other European states—Genoa, Brittany, and Montferrat—came to the city in April, and representatives from Milan, Venice, Ferrara, Bavaria, and the Palatinate would arrive in May. Envoys from nine other Italian states had already come to Rome in the winter of 1484–85, and another six would arrive the following year. The papal master of ceremonies, Johann Burchard, recorded most of these visits in his diary, noting the names of the ambassadors, where they lodged in Rome, and when they had their audiences with the pope:

> On Thursday, the 29th of December [1484], a public consistory was held in
> Rome at the apostolic palace of St. Peter, in the first and greater hall. There
> the envoys of King Ferdinand of Sicily . . . offered the customary obedience
> to his holiness. A speech was delivered by the bishop of Gaeta.[7]

On Friday, the 10th of June [1485], the envoys of the duke of Brittany, who
had been received earlier, in a public consistory in the third hall of the palace
aforesaid, presented the proper and customary obedience to his holiness Pope
Innocent VIII, in the name of their duke. A speech was made by the reverend
Lord Robert, bishop of Tréguier, who was the chief dignitary among them.[8]

On Wednesday the 6th of July . . . the envoys—the reverend father, Lord
Johann, bishop of Worms, and the illustrious Lord Bernhardt, count of
Erbestein—presented the customary obedience and reverence to his holiness
in the name of the very illustrious lord, the count palatine of the Rhine, and
the said lord bishop made a speech in a somewhat outlandish accent.[9]

Giovanni Bentivoglio, de facto lord of Bologna, was a private citizen of that
city. On entering Innocent's presence, he merely kissed the papal foot, as
many visiting dignitaries did. The ambassadors of foreign states, by contrast,
not only kissed Innocent's foot but also professed obedience on behalf of
their sovereigns or the republican governments they represented. One member from each embassy would verbally present this obedience in the course
of a formal Latin oration. Burchard calls this a "customary" act, and his accounts imply that both the profession of obedience and the oration in which
it was presented were long-standing elements of the papacy's ceremonial
diplomacy.

In fact, unlike kissing the papal foot, both the act of professing obedience
and the delivery of a formal oration were relatively new innovations in papal
ceremonial. The custom developed after the end of the Great Schism and
only became a regular event in the life of the papal court after Eugene IV's
final return to Rome in the 1440s. The arrival of formal embassies charged
with making this very public act of ritual submission reflects, paradoxically,
the Renaissance papacy's *need* for legitimation by those same powers. With
memories of the Schism still fresh, and the conciliar threat ever lurking, it
was useful—perhaps essential—that the new pope secure from the European
states an acknowledgment that his claim to the throne was legitimate, his
spiritual authority absolute. As with many late medieval or Renaissance innovations, the emergence of the new tradition was accompanied by the claim
that it was something ancient, something that had always been done.[10]

Burchard does not say much about what these ambassadors did in Rome
after their audiences with the pope. Some stayed on to make pilgrim visits
to Roman churches, while others remained at court and participated in more
ceremonies and festivities. Several embassies were in town in December 1485,

and Burchard put their members to work at the pope's Christmas Eve banquet, where they carried washbasins, platters, and carafes of wine to the pope and cardinals at table and presented other diners with trays of gilded pine nuts, candied coriander seeds, and waffles. These minor acts of ceremonial humility further underscored the political submissions they had earlier performed.

There was one further element of the obedience embassy that Burchard never mentions, but for which extensive evidence survives: a majority of the obedience ambassadors who came to Rome during the reigns of Innocent VIII, Alexander VI, Julius II, and Leo X saw their texts conveyed to one or more Roman printshops, where their orations were put into print. At least fifteen of twenty-six ambassadors sent to Innocent, nineteen out of twenty-two who came before Alexander, ten out of nineteen sent to Julius, and six out of seven sent to Leo saw their orations appear in print. A fair number of these texts appeared in multiple editions, not only in Rome but also in the home states of particular ambassadors. All told, nearly a hundred editions of obedience orations were printed in the course of these four pontificates, a substantial portion of all the oratory printed in Rome in these years.[11]

It is not immediately clear why these texts were printed. Were they vanity projects for the orators who had pronounced them? Did the curia have an interest in seeing these professions of submission to the pope broadcast to a wider public? Did anyone actually read them? The archive of surviving editions gives few indications. But, like pilgrim guides for the Jubilee or printed texts honoring the Madonna di Loreto, obedience orations offer another case where the new technology of printing helped to cast an equally new tradition as timeless and unchanging, broadcasting an image of the pope as the supreme authority in the Church, master of Rome and arbiter of Christendom, to whom princes and republics alike hastened to make humble submission.

Across Italy in the fifteenth century, humanists and the states they served multiplied the occasions for ceremonial oratory. In his study of orations delivered during papal liturgies, John O'Malley illuminated the humanist revival of epideictic rhetoric, its structures and its values; John McManamon did the same for funeral oratory delivered in the papal city.[12] The diplomatic oratory of Renaissance Rome has received less attention.[13] Brian Maxson has characterized the delivery of ambassadorial orations (anywhere in Europe) as a form of diplomatic gift, a presentation of intellectual property on a level with the silks and brocades that envoys presented to kings and popes in the fourteenth century.[14] This claim holds especially true for the obedience orations delivered to Renaissance popes. The orator brought, first and foremost, his sovereign's *obedience*, a political gift of supreme importance. He also offered praise of the pope's character, piety, and intelligence; celebration of his ancestors, education, and career; and hopeful predictions for a long and successful reign. At the same time, the orator seized the chance to praise his own prince or the state that had sent him, to demonstrate his rhetorical ingenuity by tracing ancient links between his patrons and the pope, and to press particular points of political need by delivering news from home in a flattering light or importuning aid and support for his state's policies, whether military, fiscal, or political.[15] In this way the obedience orator served his state and raised his own profile even as he laid his gifts of obedience and praise before the pope.

Obedience orations featured high flights of humanist epideictic rhetoric, ceremonial self-deprecation alternating with obsequious praise. In April 1485, the Genoese ambassador to Innocent VIII, Ettore Fieschi, opened by describing his inadequacy for the task: "Terrified, aquiver, completely awestruck, I am so thoroughly atremble that my face has gone pale, my voice fails, my tongue swells and sticks to my teeth. For just as the countenance of immortal God strikes the fragile sight of humans like a lightning bolt so powerfully that, as Scripture attests, no one has ever seen God and lived, so my faltering eyes, like those of the night owl, cannot endure the sublimity of your majesty as it coruscates with divine brilliance like the rays of the sun. But the demands of obedience compel me to press on."[16] Fieschi was one of nearly two dozen ambassadors who came to Rome after Innocent's election to proffer the obedience of their governments, each making a speech along these lines. In his diary, Burchard recorded the size of their retinues and the honors shown their parties; he also noted which ambassadors broke protocol (for example, when some paid a visit to the pope before making their formal obedience) and criticized others for inappropriate dress or actions or rhetorical delivery.[17] On occasion he made ambassadors double up and give multiple speeches at a single consistory, suggesting that their performances were something to be got through rather than enjoyed.[18]

These elaborate displays of diplomatic submission, with their overblown rhetoric and physical prostrations, have struck modern observers as vaguely ridiculous.[19] The texts of some obedience orations run to many thousands of words and must have taken hours to deliver, and in the first year of a new pope's reign, new obedience embassies were arriving at the city gates several times a month.

Both the ceremonial and the rhetoric of the obedience oration suggest an event intended to project papal authority. But the political reality was more complex, as was the genealogy of the ceremony itself. Contemporary sources describe obedience orations as elements of long-standing and unchanging diplomatic tradition. Time and again, Burchard refers to the proffering of obedience as something *consuetam, solitam,* or done *more consueto.* An early practitioner of the form, Aeneas Sylvius Piccolomini (the future Pius II) uses similar language to describe both the obedience orations he delivered in his early career and those that were later delivered to him on his elevation to the papacy.[20] He opened his obedience oration to Pope Calixtus III in 1455 with an almost ironic *praeteritio* reviewing the sorts of themes obedience orators were wont to touch on in their speeches, before moving on to a weightier topic, the crusade against the Turks.[21]

But the custom of proffering obedience to a newly elected pope was not very old at all; Aeneas Sylvius himself was involved in its creation. As a ritual event, the delivery of obedience only really comes into focus with the embassy that Aeneas, then an imperial legate, led to Eugene IV in the early spring of 1447. By the 1480s, at the accession of Innocent VIII, the custom seemed to have been around forever—not least because one of the rhetorical tropes that obedience orators regularly deployed was that they were performing an ancient rite, hallowed by time and repetition.[22]

Origins

Throughout the Middle Ages, Holy Roman emperors and certain kings and princes had sworn oaths of service, fealty, or obedience to the pope at moments of political crisis (when they needed papal support) or, more regularly, as part of their own coronation rites. The rite composed for Henry II in 1014 required the emperor-elect to kiss the pope's foot and swear an oath of fidelity to him and his successors.[23] The counts and, later, kings of Sicily held their domains in fief from the pope, and starting in 1080 each swore an oath of fealty to the pope as the price of his investiture.[24] Otto IV, seeking the support of Innocent III against his rival, Philip of Swabia, had to make various concessions regarding papal lands and privileges and agreed to take a new and more elaborate oath of fidelity to the pope that Innocent himself composed.[25]

In these instances, taking the oath was a necessary element in the making of the prince; it was an act the new sovereign performed to secure the pope's recognition of his claim to his throne. The prince needed the pope's approval, and the oath of loyalty was the price. But there was no corresponding gesture made to a medieval pope on the occasion of *his* elevation. This innovation seems only to have emerged in the aftermath of the Great Schism, after decades in which the legitimacy of the pope himself was in doubt. It was then, too, that the term *oboedientia* took on its particular significance in papal affairs.

When a party of French cardinals deposed Urban VI in 1378 and elected the antipope Clement VII in his place, the College of Cardinals—and eventually all of Latin Christendom—divided into two camps. The "Roman obedience" recognized Urban and his successors, while the "Avignon obedience" supported Clement and his successor, Benedict XIII. (Later, when the Council of Pisa deposed both popes and elevated Alexander V, he and his successor John XXIII commanded the "Pisan obedience.")[26] Throughout the Schism, theorists debated the wisdom of maintaining or withdrawing obedience from one pope or another.[27] *Oboedientia* no longer signified the simple submission of one individual's will to another; it took on a more dynamic sense, the *recognition* by a prince or prelate of a particular candidate's authority, something that could be extended but also withdrawn. This in turn suggested that legitimacy—authority, even—was something conferred on a pope by an outside body, not at the moment of his election but by the consensus of the faithful who ratified the conclave's choice.

In 1417, the Council of Constance deposed the popes of all three obediences and elected Oddone Colonna as Martin V. All of Latin Christendom now belonged to the same obedience, but it was not long before new tensions arose. As the price of his election, Martin promised to observe the conciliar decree *Frequens*, requiring the pope to convene councils at regular intervals. Gabriele Condulmer agreed to the same capitulation when he was elected Eugene IV in 1431, stepping into the office just as the deadline for a new council arrived.

The assembly that convened at Basel in 1431 set itself in opposition to Eugene from the start. By 1439, the council fathers had withdrawn their obedience from the pope, declaring him deposed and setting up a rival, the antipope Felix V, with the tacit support of the emperor-elect Frederick III. Aeneas Sylvius Piccolomini, then an adherent of the council, went to serve Felix V at his miniature papal court in a castle on Lake Geneva. But Felix failed to win European support.[28] Sensing his career coming to a dead end, Aeneas managed a transfer to the imperial civil service of Frederick III, from where he watched the cause of the council and its antipope continue to falter. It took a decade of diplomatic negotiations, many of them led by Aeneas himself, to get Felix to resign and the council and the emperor who backed it to acknowledge Eugene as the true pope. At an imperial diet in Frankfurt in 1447, Aeneas worked out a deal among Frederick, the imperial electors, and the pope: in return for papal concessions over ecclesiastical appointments in the empire, Frederick would repudiate his support for Felix and acknowledge Eugene as pope.[29] In January 1447, the emperor sent Aeneas as part of a legation "to reestablish the [German] nation's obedience to the Holy Apostolic See."[30]

Aeneas sent Frederick a long account of this embassy and retold the episode again in his *Commentaries*.[31] On arriving in Rome, he delivered an oration before the papal court; afterward, the party was supposed to present the emperor's "obedience" to the pope. But Eugene fell ill the same day and took to his bed, and it was not clear that he would recover. Some in the embassy now hesitated, but Aeneas argued that they were charged with presenting Frederick's obedience to this pope and no other. If he died, his successor might not respect the terms of the deal; moreover, they had no orders to treat with a new pope. Another colleague agreed with him: after having come so far, there was no point in giving up at the last minute. If nothing of Eugene remained alive but the little toe of his left foot, he would pledge obedience to that.[32]

The ambassadors entered the sickroom and made their submission to Eugene as he lay in his bed. The ailing pope handed them his bulls of concession in return. "Then," Aeneas says, "the entire city erupted in celebration; the people in high spirits lit bonfires and rang bells and sounded trumpets. The next day public business was suspended and a general thanksgiving was decreed." The cardinals carried relics in procession from San Marco to the Lateran, where mass was said, "during which there was a long and elaborate sermon in praise of Frederick and Eugene."[33]

In both accounts, Aeneas stresses how novel his mission was and how well he had done to bring it off, even as members of his delegation argued over credentials and precedents, where the obedience should be offered, to whom, and under what conditions. Practically everything about the embassy was fraught: the secret negotiations at Frankfurt and in Rome, the internal arguments about the embassy's validity, the hurried submission to Eugene in bed, an event chronologically and spatially separate from the oration Aeneas delivered at court the day before and the sermon delivered at the Lateran the

day after, and the celebrations that broke out across the city after everything
came off as hoped. Aeneas and his party seem to have invented new diplomatic procedures and ritual behaviors on the spot.

Eugene IV never rose from his bed. Sixteen days after Aeneas presented
him with the obedience of the Germans, he died, and Tommaso Parentucelli,
a scholar with long connections to Florence, was elected Nicholas V. Various
Italian states sent embassies to congratulate him. The Florentine embassy
was led by the humanist Giannozzo Manetti, who contributed another key
element to the newly developing tradition.[34] Manetti's biographer, Vespasiano da Bisticci, notes that on arriving in Rome, unlike all previous Florentine ambassadors, Manetti was invited to speak at a public, not a private,
consistory. Feeling tremendous pressure to deliver an oration that would
satisfy the tough rhetorical critics of the curia, Manetti stayed up all night
perfecting his speech. The next day, his performance was acclaimed a triumphant success. His oration was so impressive, in fact, that the Venetian cardinals quickly wrote home to tell *their* ambassadors they had better include
an orator in their party when they came to Rome. Thereafter, says Vespasiano, "all others who have spoken [at the papal court] since have taken the
structure of their own orations from Messer Giannozzo, being a new custom
because the oration that day was so outstanding."[35]

Venice duly sent Nicholas an obedience embassy with an orator, as did
France, Poland, and Burgundy. At the election of Calixtus III in 1455, six Italian states sent orators to present their obedience in a formal speech as Manetti had done: Lucca, Naples, Siena, Bologna, Florence, and Venice.[36] One
of the most accomplished of Manetti's imitators was none other than Aeneas
Sylvius: in 1455, he returned to Rome to present Frederick III's obedience to
Calixtus. He did so quite quickly at the start of a long speech mostly sounding
the alarm about the Turkish threat and praising Calixtus as the figure who
could lead Christendom against it.[37] Aeneas used the occasion to promote his
favorite cause—the crusade—but he also offered the pope the requisite praise
("a brilliant and salfivic star sent from heaven, under whom the bark of Peter,
miserably sinking, will at last sail to safety"),[38] and he stressed Frederick's
authority as he extended his obedience to this latest pope, the third he had
recognized in just over a decade.[39]

Elected Pope Pius II three years later, in 1458, Aeneas received several
obedience embassies while he stopped in Siena on his way to the Congress
of Mantua. The legations came "to pay him reverence according to ancient
custom."[40] The imperial ambassadors offered him Frederick's obedience in a

public ceremony in the Duomo, while the envoy of George Podiebrady, king of Bohemia, delivered his obedience in secret so as not to antagonize his Hussite subjects.[41] Pius does not mention that these ambassadors delivered orations, but a little later the French bishop Jean Jouffroy delivered an oration in Rome on the occasion of Louis XI's renunciation of the Pragmatic Sanction of Bourges, which was framed (by Pius, at least) as the restoration of the French king to his proper obedience.[42] The oration seemed integral to the event, and Aeneas could not resist comparing Jouffroy's lackluster performance to his own. At the conclusion of the French envoy's oration ("long awaited and much desired"),[43] he says, the pope made a pithy reply that all present admired for its brevity and wit.[44]

Obedience in Print

Various orators congratulated Paul II after his election in 1464.[45] Although Sweynheym and Pannartz were by then working in Subiaco and Sixtus Riessinger would soon set up shop in Rome, no obedience oration to Paul II was printed during his reign.[46] By contrast, at least three orators who offered obedience to Sixtus IV saw their orations printed: Bernardo Giustiniani, representing Venice; Laszlo Vetési, speaking for Matthias Corvinus of Hungary; and fifteen-year-old Giovanni d'Aragona, Ferrante's son, who spoke for Naples. (Young scions of noble families often figured in obedience embassies. Lorenzo de' Medici was sent to congratulate Paul II in 1466, at the age of seventeen,[47] and Niccolò Maria d'Este, nephew of Ercole, and Antongaleazzo Bentivoglio, son of Giovanni, both delivered obedience to Alexander VI on behalf of their families.)

   Giovanni d'Aragona's oration was printed in Rome by Giovanni Filippo de Lignamine, in an edition rather different from those of later pontificates and one that, for this reason, is worth examining closely.[48] The book begins with a letter from the printer to the young prince, followed by Giovanni's oration to the pope and another oration, longer than the first, that the boy delivered to his father on his return to Naples. It concludes with a set of elegiac distichs by de Lignamine celebrating Giovanni's achievements. The oration to the pope was fairly straightforward: Giovanni blushed at the thought of his youth and his inadequacy for the task at hand, but as a loyal son of the Church and of his royal father, he was determined to do honor to both. Ferrante's joy at the news of Sixtus's election was difficult to imagine and impossible to describe. The king knew that the cardinals could not have chosen a better candidate; Sixtus was the best man to lead the Church. Ferrante had chosen

Giovanni, from all his sons, to come to Rome to congratulate the pope and offer him the customary (*de more*) obedience, to acknowledge him as the legitimate pope whom all nations should revere.[49] Ferrante put all he had at the pope's disposal: resources, men, naval and land forces, everything—including his very life. Finally, Giovanni was prepared to kiss the pope's foot; he also hoped to discuss some other matters with the pope in private, at a later time.

In this short oration—just over two pages in its quarto edition—Giovanni touched on almost all the topics that later obedience orators would treat at greater length: his modest qualifications for the task, his sovereign's joy at the pope's recent election, the pope's countless virtues and the triumphs that awaited him, an oblique reference to the political transactions surrounding the ceremony, and, finally, the formal offer of his sovereign's obedience, to be sealed by a kiss of the papal foot.

The texts that frame this short speech suggest something about the circumstances of its publication. Giovanni's second oration, delivered to his father and his court, echoed the main points of his oration to the pope but with the panegyric values reversed. Giovanni had blushed to address the papal court but was warmly received and shown every honor. Was this excellent reception a tribute to Giovanni's virtues? By no means! All was done to show the regard the new pope had always had and would continue to have for King Ferrante, noblest of kings, whose praises would never cease to be sung, and so on.

That de Lignamine published both texts, along with his own fulsome dedicatory letter and verses, suggests that it was he who took the initiative to publish the edition and that, in doing so, he was seeking patronage or some other reward from the Neapolitan court. De Lignamine was a Sicilian with long connections to the house of Aragon; in his letter, he reminds Giovanni of these links, as well as his recent publishing activity on behalf of Sixtus (he had just published two tracts the pope had composed before his elevation, one on the blood of Christ and the other on the power of God).[50] He also notes that he had secured a copy of Giovanni's oration by writing to the boy's tutor, the Sicilian humanist Pietro Ranzano, who probably composed the orations in the first place. Dropping as many names as he could into this short dedication, de Lignamine put himself at the center of the enterprise—another example of the editorial opportunism that, as we saw in chapter two, was so typical of Italian humanist editions of the early 1470s.

Bernardo Giustiniani's oration was more substantial and would be much better known in later years. Giustiniani had come to Rome to discuss how

Venice and Rome might cooperate on plans for a military expedition against the Turks after the fall of Negroponte. His oration is mostly concerned with describing the Turkish threat and Venice's long history of defending Christendom against Ottoman advances in the eastern Mediterranean. Technically, his speech was also an obedience oration. At the start, Giustiniani offered the requisite words in praise of the pope and touched on the joy the Venetians had felt on hearing of his election, and he concluded with a formal offer of Venetian obedience. The doge and Senate of the republic recalled that the papacy had been founded "not by any human council" but by Christ himself, they recognized Sixtus as the "legitimate" occupant of the papal throne, and Giustiniani and his fellow ambassadors were pleased to offer their state's "humble and devoted obedience."[51]

Giustiniani's fame as a diplomat and orator was already prodigious and would only grow in later years.[52] In 1483, he composed the Venetian replies to Sixtus IV's interdict, texts that would find their way into print in Venice, Padua, and Westminster (see chapter four), and a collection of his orations and letters was printed in Venice in the 1490s.[53] The printing history of this oration likewise suggests Giustiniani's outsize fame: it would be published ten times in the next two decades. Four editions appeared in Rome, published by three different printers; it was also printed twice in Venice, once in Padua, and once in Mondovì in Piedmont; it was later reprinted in Rome by Johannes Gensberg around 1474 and by Stephan Plannck in the mid-1480s.

The Hungarian bishop Laszlo Vetési delivered his oration to Sixtus on February 2, 1475.[54] His speech, like Giustiniani's, is long, filling twelve leaves in Johannes Schurener's quarto edition. Vetési, too, treats the question of the Turks, describing their raids on Hungarian territory, Matthias Corvinus's countless battles against them, the urgent need for a new crusade, and the opportunity Sixtus now had to win glory by leading Christendom to victory over its ancient foe. Vetési's oration was printed twice in Rome by Schurener in the late 1470s and once more by Stephan Plannck in the mid-1480s.[55]

The Roman editions of both Giustiniani and Vetési's orations foreshadow the large body of orations later printed for Innocent VIII and Alexander VI. Unlike de Lignamine's edition of the young Giovanni d'Aragona's speech, they include no dedicatory letter, no laudatory verses, and no additional texts. Most editions of Giustiniani's speech do not even identify it as an obedience oration (instead titling it *Oratio habita apud Sixtum IV*). Both speeches were probably printed for the extended commentary each author offered on the question of the Turkish threat. These particular legates may have wanted to publicize their calls for a new crusade and may have played a part in getting them published.[56] Moreover, the fact that Plannck reprinted both orators' texts more than a decade after their delivery suggests that orations such as these could circulate and be read as rhetorical models well after topical interest in their contents had subsided. In the mid-1480s, Plannck would print an edition of Aeneas Sylvius's obedience oration to Calixtus III from 1455,[57] and several orations for Innocent and Alexander were reprinted years after their delivery as well, as we will see.

The printing of obedience orations during Sixtus IV's reign was hardly an organized effort. The upwardly mobile de Lignamine printed young Giovanni d'Aragona's oration as a way of currying favor at the court of Naples. Giustiniani's text was an instant hit thanks to its political relevance, its rhetorical polish, and the fame of its author. Laszlo Vetési's text likely circulated in print for similar reasons. Finally, only these three orations—out of several delivered to Sixtus in the early years of his reign—are known to have been printed. At this early date, while the delivery of obedience orations already seemed an established tradition, publication in print was still the exception rather than the rule.

After the election of Innocent VIII (1484–92), the majority of obedience orations delivered to the new pope appeared in print. Embassies from every Italian state and most European countries came before Innocent, the first not long after his election in August 1484 and the last in 1487.[58] By now there was a set routine: ambassadors approached the city via Monte Mario and the Porta Viridaria, or sometimes along the via Flaminia and Porta del Popolo.[59] Depending on the rank of their sovereign, they were met at the Milvian Bridge or at a milestone on the road into the city or at the city gate itself, either by cardinals or by members of their households. They entered the city in a cavalcade and were taken straight to their lodgings, sometimes the palace of a cardinal from their home state, sometimes the palace of some noble family. Legations from smaller states might stay at one of the inns near Campo de' Fiori. By "custom," they remained sequestered until the time came for their presentation to the pope at a public consistory. Depending on the state represented or the rank of the ambassadors, this would meet in either the first or third *aula* of the Vatican Palace (the modern Sala Regia or Sala Ducale), and after kissing the pope's foot, presenting obedience in a Latin oration, and receiving a short speech in reply, the ambassadors were free to stay in Rome or return home. Most of the major European powers had resident ambassadors in Rome by this time, but they were not expected to offer obedience; the task fell, instead, to envoys sent especially for the occasion.

Extensive negotiations took place before and after the delivery of obedience, as the pope made political or fiscal concessions in exchange for public recognition by the foreign state. In Rome to present the obedience of Casimir of Poland to Innocent VIII in May 1486, the Polish ambassador Jan z Targowisk secured a bull granting the king three-quarters of the profits from the sale of a recent indulgence raising funds for a new crusade.[60] After the Genoese dazzled Julius II's court with the splendor of their embassy in 1504, the pope sent them the Golden Rose that year.[61]

Negotiations were sometimes conducted by the resident ambassador in advance of the obedience party's arrival; sometimes the obedience ambassador himself secured concessions, privileges, or other benefits during his stay in Rome. Ambassadors then used the orations to further advance their principals' claims to titles and territories, to put the best possible "spin" on recent controversies or events, and to position their states as best they could in the eyes of the new papal regime.

While the most serious matters were left for private negotiations, even the ceremonial motions of offering obedience could be provocative. When, in May 1504, the English legation presented Julius II with the obedience of Henry VII, "king of England and France," the resident French ambassador lodged a protest, claiming that the obedience of France was not theirs to give. In April of the next year, a French embassy made submission in the name of Charles VIII, "king of France, Jerusalem, and Naples." Now Spain objected: the Spanish ambassador reminded the court that Ferdinand of Aragon had just taken possession of the Kingdom of Sicily, including Naples, and the French had no right to speak for the Neapolitans or offer their obedience. (In both cases, Julius ordered the protesters to keep silence and made no comment as to the validity of the claims the other side had advanced.)[62]

Obedience embassies were staffed from the highest ranks of a state's nobility, clergy, or civil service. The orators might be men of slightly lower status, distinguished scholars or jurists who could deliver a fine rhetorical performance. Some of the most famous humanists of the age participated in these legations: Bartolomeo Scala and Gentile Becchi for Florence, Francesco Patrizi for Siena, Tito Vespasiano Strozzi for Ferrara, Rudolph Agricola for the

Palatinate, and Johannes Reuchlin for Bavaria all delivered orations to newly elected popes. Their speeches followed a set pattern, based on the structures of classical panegyric.[63] The orator might begin by professing his modesty or inadequacy for the task, or with a meditation on the task before him—what topics obedience ambassadors should cover in their speeches, or why sovereigns sent obedience to the pope in the first place. From here, the orator would move quickly to praise of the pope and admiration at his election, using classical and biblical figures of speech. His rhetoric might be further inflected by his professional background or training: theologians invoked biblical verses, jurists might quote Justinian or Ulpian, and at least one philosopher attempted a scholastic disputation (with himself) on the question of "obedience," who owed it to whom, and how it ought to be shown. Humanists would anatomize the virtues of the new pope or recount legations sent to pay homage to the emperors of ancient Rome. All would agree that there was no better man for the office than the one who occupied it now: none more pious, diligent, prudent, or wise, none more able to steer the bark of Peter over stormy seas.

The orator would next attempt to describe the joy that his king—or prince, or duke, or senate—had felt on hearing the news of the new pope's elevation. There had been sadness at the death of the last pope and concern for the state of Christendom (oppressed as it was by internal strife and external foes), but these had given way to relief and delight when news arrived that this particular man had been chosen. Here, a talented orator might identify connections between his home state and the new pope, drawing on ancient or more recent history or individual genealogies. Thus, Ettore Fieschi, speaking for Genoa, claimed Innocent VIII as a native son, latest in a long line of distinguished Genoese popes,[64] and Antongaleazzo Bentivoglio, addressing Alexander VI on behalf of Bologna, recalled how the new pope had spent his formative years at the city's university.[65] The Sienese recalled the favors that Alexander's uncle, Calixtus III, had shown their state in decades past;[66] Pietro Cara, representing Savoy, by contrast, traced his master's royal house back to Alexander the Great, suggesting a mystical connection binding Savoy to Alexander VI.[67] Naturally, this section of the oration might also include praise of the home state, stressing noble ancestors, glorious history, long records of service to the Church, and the like. The French kings were *Christianissimi* and had always shown the popes the greatest honor. The Florentines had long defended the popes against emperors seeking to check papal power; the king of Naples was, not only by law but also by choice, the pope's most loyal vassal; the Knights of Rhodes were his most ready defenders. Portuguese orators stressed that their country had never chosen the wrong side in a schism: no Portuguese orator had ever presented his king's obedience to anyone other than a legitimate pope.

From here, the orator turned to the current concerns of his prince or republic; he might put certain requests or pleas to the pope in the form of further praise. Drawing on the categories of honor and utility, the orator could draw attention to the very *necessity* of this particular pope's election by, for example, rehearsing the problems facing Christendom, the need for concord in Europe, or the urgency of the Turkish threat. Guillaume Caoursin, addressing Innocent VIII on behalf of the Knights of St. John just a few years after the Ottoman siege of Rhodes, gave an impassioned account of his order's battles against the Turks, beseeching the pope for help as they held the line for Christendom in the eastern Mediterranean.[68] Speaking to the same pope, Jan z Targowisk of Poland described King Casimir's recent battles against Tartars and Turks even as he angled for a greater share of the revenues raised in his kingdom from crusade indulgences. Patrizi's oration for Ferrante of Naples stressed the military service the Aragonese kings had always shown the pope, while subtly reminding him of Ferrante's need to defend himself against the claims of the French. The Portuguese ambassadors to Innocent VIII, Alexander VI, and Julius II all included accounts of recent conquests in Africa and the Indian Ocean, both as proof that a golden age was dawning in which the Gospel would reach the ends of the earth and as a way of very publicly reiterating their claims to these lands in the presence of the pope.

Finally, and close to the end of the speech, the orator would make the formal presentation of his state's obedience. There was no set formula for this, but orators usually found ways to describe the pope as *verus, indubitus, legitimus* (or elected *legitime*); they almost always made some reference to the kissing of his foot; and they included some form of the phrase *oboedientiam praestare* or *oboedientiam exhibere*.[69] Frequently, this was done via elaborate periphrases or circumlocutions that might suggest mild embarrassment or even resistance to the act but more likely were meant to display the orator's skill. Emphasizing Innocent's legitimacy, Caoursin delivered the Grand Master Pierre d'Aubusson's obedience to Innocent VIII: "[D'Aubusson] declares that you, Pope Innocent, the eighth of this name, are the *one, true, indubitable* vicar of our lord Jesus Christ, successor of Peter the prince of the apostles, pastor of the Catholic Church, and Roman pontiff. He loves you; he respects you; and to you, as the *true vicar* of Christ, he reverently offers on bended knee, after kisses of your blessed feet, the pure obedience and chaste love of a loyal son."[70] Ettore Fieschi hit many of the same notes: "Therefore we declare that you, Innocent VIII, are the *true and indubitable* pope. To you and to your authority . . . we submit our total and everlasting obedience—the obedience, I say, which faithful Christians are wont to show and indeed ought to show a *legitimate* pope. . . . And whatever measure of strength, power, and honor we have, we lay with willing hearts before your feet already now covered with our kisses."[71] The Portuguese ambassador Vasco Fernandes took the formula to another level altogether in the lengthy conclusion to his oration to Innocent VIII:

[King Joao] acknowledges that Your Beatitude has been sent from on high to mankind as a kind of bright and twinkling star, so that with this oarsman, or better, captain, the secure bark of Peter might not sail without a definite course over the face of the earth. He acknowledges you to be the *true pontiff*. . . . [He] who, in the midst of the many schisms in former times, was never disloyal to the Roman and *true pontiff* (something I am not sure ever happened to any other imperial or royal house), . . . a most formidable king, one who is most compliant with the Roman Church and a propagator of the Christian faith, before whom the kings of the Ethiopians bow down and to whom they offer presents every year, acknowledges that Your Beatitude is *in truth the vicar* on earth of Him concerning whom David in Psalm 71 said, "And he shall rule from sea to sea, and from the river unto the ends of the earth. The Ethiopians shall bow down before him and his foes shall lick the dust. . . . And all kings shall adore him, and all nations shall serve him."[72]

Amplification was one of the highest values of humanist rhetoric, to be sure. But these flowery passages did more than just praise the pope. Stressing the new pope's legitimacy, recognizing him as the *true* vicar of Christ, obedience ambassadors lay a wall of diplomatic protection around the bishop of Rome. The Christian princes recognized *this* man as pontiff, whom no council could challenge. But the very fact that such statements needed to be made underscores the weakness of the papacy in the postconciliar period. The Christian princes could bestow their obedience, but they might also take it away.

Publishing Obedience

A small group of Roman printers, most of whom did other work for the papal chancery and court—printing bulls, chancery regulations, and administrative manuals—also handled the printing of obedience orations. The extraordinary consistency of their format and imposition suggests something like a direct line of production from the papal court to the press. Curial officials may have taken an ambassador straight from his obedience ceremony to the printshop to have copies readied for distribution. Ambassadors may even have had copies printed in advance, to be passed around so the audience could follow along.[73]

Most orations were not only printed but also reprinted, sometimes more than once. We have seen how Giustiniani's 1471 oration for Venice appeared in ten different editions; Bartolomeo Scala's oration for Florence to Innocent VIII was printed six times; and Jason de Mayno's oration for Milan to Alexander VI was printed four times, as was Ettore Fieschi's for Genoa to Innocent VIII. Moreover, for such small and nearly ephemeral publications, large numbers of copies survive: Pierre Cadrat Carré's speech to Innocent on behalf of Charles VIII of France, for example, was reprinted with Ghinucci's for Siena in an edition by Plannck, of which over seventy copies are listed in ISTC.[74]

Almost all were first printed in Rome, and very often they were reprinted there, either by the same or by other printers. Only later might reprints appear in other cities. For the pontificate of Innocent VIII, Stephan Plannck was the constant: fourteen out of the fifteen orations for which printed editions survive include an edition by Plannck (in total, he printed or reprinted twenty-one editions of orations for Innocent VIII), and a study of Plannck's edition of Rudolph Agricola's oration suggests that his was the first and better edition of that text.[75] Plannck was followed by Bartolomaeus Guldinbeck (six editions for Innocent VIII) and Eucharius Silber (five), with Schoemberger responsible for a single commission. During the reign of Alexander VI, Plannck continued to print obedience orations, but his activity was quickly matched by Andreas Freitag, who began operations in Rome shortly after Alexander's coronation in the late summer of 1492. Among Freitag's earliest editions are six of the first ten obedience orations delivered to Alexander between October 1492 and January 1493. Silber also printed orations to Alexander; he would become practically the exclusive printer of obedience orations delivered to Julius II. One major Roman printer—Johannes Besicken, the Basel printer who produced so many of the editions in the Jamometić affair of 1482–83, and who moved to Rome in the early 1490s—seems to have printed very few obedience orations for any pope: one or two for Alexander VI are all that survive.

A small portion of obedience orations were reprinted in the home state of the legation, suggesting an attempt to publicize the event of the embassy and the contents of the oration to the reading public there. Giustiniani's oration to Sixtus IV was reprinted in Venice and Padua, Bartolomeo Scala's oration to Innocent VIII was reprinted in Florence, Giovanni Cataneo's oration to Alexander VI for the Gonzaga of Mantua was reprinted in Parma, Jason de Mayno's oration for the Sforza was reprinted in Pavia and in Milan, and Pietro Cara's oration for Savoy was reprinted in Turin (all were printed in Rome first). Others were reprinted even farther afield, underscoring the interest distant readers might take in these speeches, probably for their rhetorical value rather than their political interest. Giustiniani's oration to Sixtus IV was reprinted in distant Mondovì; Scala's oration for Florence to Innocent VIII was reprinted in Venice, as was Johannes Reuchlin's oration to Alexander VI on behalf of the duke of Bavaria. Sebastiano Badoer's oration for Venice was reprinted in Leipzig, and Angelo da Vallombrosa's oration to Julius II on behalf of Siena was printed in Munich, apparently the only edition of this speech to survive.

A few orations were printed with short paratexts by the orator or by some friend or associate, which help to introduce or frame the text in some way. Fieschi's oration to Innocent VIII was printed in Rome four times; twice it appeared with a poem by Titus Veltrius addressed to the house of Fieschi (*ad Fliscam domum*), in which the poet described the honor Ettore had brought his family and state. The orations to Alexander VI on behalf of Bologna and Ferrara were delivered by junior members of those cities' ruling families (Antongaleazzo Bentivoglio and Niccolò Maria d'Este); in print, each young man's oration was presented by a humanist scholar (his ghostwriter?) in a prefatory letter addressed to a friend and describing the promise of his young protégé. All three editions of Antongaleazzo Bentivoglio's oration to Alexander VI were printed with a letter from Hernando de Salazar to his former professor, Giovanni Gaspare da Sala, in which de Salazar praises the young Bentivoglio's eloquence and bearing. Da Sala, a professor of jurisprudence at Bologna, actually participated in the embassy to Rome in August 1492 and hardly needed a printed copy of the speech; his choice must have made some social sense to de Salazar that we cannot discern.[76]

Niccolò Maria d'Este's oration on behalf of Ferrara was printed twice with a celebratory poem and letter from Giulio Capella to Luca Ripa. In his letter, Capella revealed that he had taken the initiative to have d'Este's oration printed. Both editions of this oration include title pages with shaped text—

unusual for this genre—which may reflect Capella's input into the design of one or both editions. Angelo da Vallombrosa's oration to Julius II for Siena was published by Johannes Koechel in Munich, who in a brief prefatory letter explained how he had come by a copy of the speech and why its eloquence and erudition made it worth printing.[77]

Several obedience orations for Alexander VI were printed with letters from the orators to their sovereigns. Fernando de Almeida included a letter to King Joao II of Portugal describing his mission; Giacomo Spinola of Genoa included with his oration a letter to Ludovico Maria Sforza, duke of Bari and then chief enforcer of the Milanese occupation of Genoa, in which he described his delicate task: presenting the free government of Genoa's obedience to the pope while acknowledging its submission to its new Milanese lords. The obedience ambassador representing the Knights of Rhodes to Alexander VI, Marcus Montanus, included a poignant letter to his predecessor Guillaume Caoursin, the obedience ambassador to Innocent VIII, recounting his mission and his hopes to have performed as well as Caoursin had done a decade before.

The inclusion of prefatory letters like these suggests additional layers of editorial intervention and appropriation. But they were not the norm; most printed editions present the obedience oration on its own without further framing or commentary. The text alone was enough to publicize the sovereign's crucial role in validating the election of the new pope, the political points that the orator had put forward on behalf of his state, and the orator's achievement in speaking for his country with all the eloquence and honor he could muster.

The obedience oration, a seemingly formulaic rhetorical event, which orators presented as time-honored and traditional and papal officials professed to find tedious, was, in fact, only recently devised, an invented tradition that served the political needs of both the pope and the states that came to submit to him. It was not only performed live at the papal court but also captured in print and broadcast across Europe. The ambassadors, literal "talking heads," who spoke for their distant sovereigns, mimicked St. Andrew's head in their performance of exaltation and debasement before the pope. Did Pius II, in staging the reception of St. Andrew's head in the spring of 1462, with its arrival at the Porta del Popolo and its speech inside St. Peter's, actually conceive of it as an obedience embassy, with Andrew presenting the bishop of Rome with the submission of all Greece?

However that may be, the printed orations carried a particular message—

the *legitimacy* of the pope—across Europe at a time when critics were increasingly using the press to contest the pope's right to wield absolute power in the church. Like indulgences, bulls of interdict and excommunications, and announcements of new Jubilees, the obedience oration was another documentary genre that communicated the power of the pope over rival *loci* of European authority. The publication of these speeches in print further reproduced and spread that message across the continent. Even so, it is not clear that the popes had a hand in their publication. Delivering a speech *coram papa* was one of the pinnacles of humanist rhetorical achievement. Individual orators, or their tutors or ghostwriters, stood to benefit from advertising the accomplishment of speaking at the papal court. Just as a prince or republic had much to gain from staging an act of submission to the bishop of Rome, so too did individual speakers from the publication of their words of humble obedience. Consciously or not, the Renaissance papacy exploited humanists' interest in celebrating their own eloquence to create a powerful tool of propaganda—all for a diplomatic gift that the popes needed to have offered to them rather more than the European powers needed to give it.

Packaging Obedience

Important embassies could be commemorated in other printed productions as well. The Knights of St. John were early adopters of the press as a propaganda tool; during the Ottoman siege of Rhodes in 1480, their appeals for aid appeared in print across Europe, as did indulgences sold to support their cause. Guillaume Caoursin, author of the most widely printed account of the siege of Rhodes, was also the ambassador who came to deliver obedience to Innocent VIII in early 1485; his oration was printed in Rome several times, by Bartholomaeus Guldinbeck, Eucharius Silber, and Stephan Plannck.[78] Near the end of the decade, Caoursin compiled various diplomatic transcripts and documents into his *Rhodiorum historia*, which was printed in Ulm a few years later, in 1496.[79] The volume is illustrated with woodcuts of important moments in the history of the order: the siege of 1480, the transfer of Cem Sultan from Turkish custody to Rome, the angelic transfer of the arm of St. John from the Turks to the knights in 1492, and various diplomatic missions that the order undertook across Europe in support of their mission, including Caoursin's embassy of 1485. In the volume, the text of his oration is printed in full and illustrated with a full-page woodcut showing Caoursin as he holds forth to Innocent on his papal throne (figure 7.1).[80] Here, the ambassador's

Figure 7.1. Guillaume Caoursin, *Historia Rhodiorum* (Ulm: Johann Reger, October 24, 1496). Boston Public Library, Rare Books and Manuscript Department, Q.403.97, sig. h2v.

interview with the pope is publicized as evidence for the importance of the order, its close relationship with the pope, and the duty of Christians everywhere to support it.

A different sort of compilation is the volume entitled *The Election and Coronation of Alexander VI* produced by the Roman jurist and auditor of the Rota, Girolamo Porcari, for Ferdinand and Isabella of Spain after the election of the Spanish pope Alexander VI in 1492.[81] The edition, 110 leaves long, was printed by Eucharius Silber in September 1493; unlike most of Silber's and Plannck's spare editions of single obedience orations, the *Commentarius* is laid out as a luxurious product, with margins almost as wide as the text block, blank spaces left for rubricated initials, and even two leaves of printed *corrigenda* at the end and a register of quires. Porcari was an ardent supporter and client of the Borgia pope, who would later write verses in praise of the Borgia bull, a satirical dialogue against Savonarola, and an elegy for the murdered duke of Gandia.[82] His history begins with letters to the Spanish sovereigns and to Alexander VI and a fulsome preface describing Alexander's election, coronation, and *possesso* (into which Porcari weaves some references to and praise of his own family, an old Roman one). He then presents the texts of eight obedience orations delivered to Alexander by ambassadors from the Italian states (Siena, Lucca, Florence, Genoa, Venice, Naples, Mantua, and Milan), with a preface for each describing the ambassadors and the pomp in which they arrived, as well as, at the end of each oration, a transcript of the reply the pope gave to each orator.

Whether the Catholic sovereigns asked him for this work or whether Porcari compiled it of his own accord, the book provides a snapshot of the diplomatic landscape of Italy in 1493. Porcari's collection of texts provided a quick overview of the state of the peninsula on the very eve of the Italian wars, a survey of alignments, political characters, and vulnerabilities. At the same time, Porcari presented himself as an assiduous chronicler of papal glory and a reliable source of information for the Spanish crown.[83] There are many layers of audience and readership at work here: the ambassadors' speeches, first delivered to the pope and court in Rome, found a wider reading public when they were published as individual pamphlets. Porcari then appropriated these for his own purposes, compiling and republishing them as a diplomatic gift of his own to the Spanish court.

Lionello Chiericati, whom we last met in 1482 publishing a letter in Rome on the Jamometić affair, was a frequent orator at the papal court and an assiduous commentator on diplomatic events under Innocent VIII and Alexander VI. He delivered Innocent's funeral oration in 1492, a speech that was printed by both Plannck and Silber. These pamphlets must have traveled; a few years later, a Westphalian editor put the text of Chiericati's speech together with Sebastiano Badoer's obedience oration to Alexander VI in a quarto edition he entitled *Beautiful Orations Delivered in Rome, One in St. Peter's before the Cardinals for the Funeral of Innocent VIII, and a Second before the Present Pope Alexander VI for the Delivery of Obedience*. Not much is known of Johannes de Velmede, but his collection was printed in Leipzig by Martin Landsberg, a printer who would later become one of the main conduits for distributing Julius II's decrees in Germany, as we will see in the next chapter.[84] Courtly and diplomatic orations were packaged here as examples of good style and texts of topical interest to German readers (not unlike what Pietro Carmeliano did with Sixtus IV and Bernardo Giustiniani's epistolary exchange in Westminster in 1484).

Meanwhile, Michele Ferno, a Milanese jurist and humanist resident in Rome and a member of Pomponio Leto's circle, produced his own volume commemorating the accession of Alexander VI, also printed by Silber. Ferno's text, although entitled *Letter on the Arrival of the Italian Legations*, is sixty leaves long. His central essay is prefaced by letters to and from a friend of his in Milan, Jacopo Antiquari, a page of panegyric verse, and a letter from Ferno to Cardinal Federico Sanseverino. Ferno presents the main text as a response to a request from Antiquari "that I should write down when and in what order each of the obedience embassies came to Rome and presented themselves to the pope when they were heard by him."[85] Ferno then recounts the entire first year of Alexander VI's reign, from the death of Innocent VIII to the conclave, his election, coronation, *possesso*, his first consistory, and the arrival of embassies from the Italian states and northern powers to proffer obedience. Ferno, like Porcari, strove to glorify the new pope even as he displayed his erudition in polished prose addressed to his scholarly friends. Unlike Porcari, he did not quote the text of any oration, even in part, but described instead their members, their costumes and retinues, the cavalcades they led through the streets of Rome, their courtesy in approaching the cardinals and pope, and the quality of their orations. All of them were remarkable, brilliant, gleaming with eloquence, and fully consonant with the great dignity and erudition of the new pope. The glittering sequence of embassies indicated not only that a brilliant pontificate had dawned but also that the new pope could count on the support of the Christian princes, whether for promoting concord in Europe or advancing on the Turks.

Porcari's and Ferno's activities raise the question whether the phenomenon of the obedience embassy had become mere spectacle by the end of the fifteenth century. Certainly, under Julius II and Leo X, the legations grew larger, the cavalcades longer, and the rhetoric more overblown. The popes of the High Renaissance, as well as the courtiers surrounding them, touted themselves as inaugurating a new golden age in which new discoveries abroad and artistic achievements at home would herald the triumph of Catholic culture over barbarism and heresy alike. The learned orators who arrived in Rome bearing their sovereign's obedience seemed only too happy to join in this flattering discourse. In 1513, Manuel I of Portugal sent ambassadors to offer obedience to Leo X along with an Indian elephant, Hanno, brought to Europe by Portuguese explorers. Hanno delighted the new pope and became one of his favorite pets, with a special stable built for him in the Belvedere Garden of the Vatican Palace. Hanno's story is well known, but the ritual context of his presentation less so. On reaching Rome, the Portuguese legation led the great beast in procession to an open consistory at Castel Sant'Angelo. At his mahout's command, the pachyderm bent down on one knee, lowered his trunk, and proffered Portugal's obedience to the vicar of Christ.[86]

Given the spectacle of it all, it may be difficult to detect the important work the obedience embassy performed in the political economy of Renaissance Europe. The Renaissance popes had a remarkably precarious hold on their office. From the death of Eugene IV to the coronation of Julius II, the threat of war or a new council, or both, was near constant. Internal conspiracies and external invasions were frequent enough that popes could be glad of—even desperate for—professions of loyalty and support from the secular princes. Nicholas V faced a conspiracy of Roman noblemen who tried to overthrow his regime; Calixtus III and Pius II each faced doctrinal challenges from the Hussites in Bohemia, while Pius quashed a rebellion by Roman barons and struggled to suppress the Pragmatic Sanction of Bourges in France. Paul II faced down a conspiracy of his own, supposedly launched by the Roman Academy, and Sixtus IV, as we have seen, saw Lorenzo de' Medici threaten a council in the course of the Pazzi War (Lorenzo's exact words: he was ready to *levar l'ubbidientia*).[87] Venice threatened Sixtus with a council in the course of the War of Ferrara, and Andrija Jamometić actually attempted to convene a council in Basel in 1482. At the accession of Innocent VIII, Jamometić was still languishing in his prison cell. Savonarola's call for a reform of the church would dominate the early pontificate of Alexander VI. In 1511, a group of French cardinals withdrew obedience from Julius II as they joined a new council, orchestrated by Louis XII of France, at Pisa. Rivals and critics of the papacy could invoke the threat of a new council at any time and, with it, the threat of deposition, the elevation of an antipope, and the renewal of schism.

Securing the obedience of the Christian princes at the outset of a pontificate—and ensuring that news of their profession was widely known—was a sensible priority for the Renaissance popes. Indeed, for all its ritual debasement of the ambassador, the ceremony disguised an essential weakness of the postconciliar papacy: its need for validation by the European states. The ceremonial humiliations—the prostrations, the foot-kissing, playing waiter at a banquet, making an elephant kneel—all masked the fact that diplomacy at the papal court was a matter of cold calculation, with the pope often paying a high price to win these displays of submission.

The obedience ceremony brought benefit to those offering it as well. (After debasing themselves in their public and grandiloquent way, many obedience orators returned home to generous rewards and gifts from their patrons.)[88] Every printed copy of an obedience text advertised the level of access that a particular state had to the pope, the rhetorical brilliance of its orator, and the gift of legitimation that he had bestowed on the pontiff. Eventually, the texts of their speeches became yet more fodder for *other* humanist commentators, ambitious courtiers like Porcari and Ferno, who appropriated and repackaged obedience texts and inserted themselves into new publications as a way to catch the attention of their social superiors.

Like pilgrims' guides to the relics of Rome, the printed speeches of these diplomatic "talking heads" captured and projected a particular image of papal Rome at the end of the fifteenth century: a holy city, filled with sacred relics, a city that drew the world to it, ruled by a pontiff who demanded obeisance and submission, but a city that at the same time depended on its visitors, diplomats and pilgrims alike, to keep the mirage afloat. Only recently emerged from a century of crisis, the papacy drew on brand-new technology to support its traditional claim of spiritual supremacy. But the printed texts that were produced reflect the fact that the popes could not entirely control what pilgrims or diplomats made of their visits to Rome. Miracle cures or personal revelations, diplomatic concessions or personal advancement—these were the souvenirs that a visitor to Rome could come away with once he had made his "submission," whether at the tomb of the apostle Peter or before his successor on the papal throne.

Postscript: Printing by Innocent VIII and Alexander VI

What, in the meantime, did the popes themselves print? Although the production of printed books in Rome increased steadily under both Innocent
VIII and Alexander VI, both popes were surprisingly reticent about using
the press. This was a marked change from the administration of Sixtus IV,
who was, as we have seen, an enthusiastic publisher of bulls, briefs, treaties,
and other documents. Elsewhere in Europe, printers published bulls issued
by Innocent and Alexander, mostly indulgences or texts promoting indulgences and the crusade against the Turks. A smaller number of bulls and
briefs awarded privileges to or advertised indulgences available at particular
churches, and a handful of German-language publications, issued in the name
of the pope, offered dispensations from fasting during Lent—the delightfully
titled *Butterbriefe*—with the proceeds going to benefit local churches or the
restoration of St. Peter's in Rome.[89] Neither Innocent nor Alexander can have
been involved in the production of these publications.[90] In Rome, meanwhile,
publication under both popes was pedestrian. Books printed in Innocent's
name consist almost exclusively of administrative handbooks and collections of bureaucratic regulations—fifty-four editions in total. Under Alexander, twenty-five editions of *Regulae cancellariae* and other administrative
publications appeared in Rome. Toward the end of his reign, printers both in
Rome and abroad produced a stream of announcements concerning the 1500
Jubilee.

Very few publications by Innocent VIII and Alexander VI might count
as "political." Both popes saw their annual bull *In coena Domini* published
in Rome several times.[91] Innocent's condemnation of Pico della Mirandola's
900 theses was printed in Rome, once;[92] his brief condemning the capture of
Maximilian by Flemish troops in 1488 was printed once in Germany, with the
startling title *Maledictio adversus Flamingos*.[93] Alexander VI was the subject
of an ingenious press campaign organized by Savonarola and his followers
in Florence, who published dozens of vernacular tracts calling for reform in
the church and castigating the pope for Italy's woes.[94] It is an extraordinary
and hard-to-explain fact that only one example of a papal riposte to Savonarola survives in print: an Italian translation of Alexander's brief of May 12,
1497, concerning Savonarola's excommunication, and this was printed in Florence, not Rome.[95]

In these years, other figures stepped in to speak for the popes. Lionello
Chiericati was hard at work in the 1490s, publishing not only funeral oratory

but also speeches in praise of peace treaties Innocent and Alexander struck with the Italian and northern powers.[96] These were printed by Silber, Guldinbeck, Besicken, and Plannck and reprinted several times in the north. Bartolommeo Florido, a papal secretary, published an oration on another of Alexander's pacts, struck with Venice and the Sforzas in 1493; this was printed by Silber.[97] Adriano Castellesi, also a secretary of Alexander's, delivered an oration on yet another peace treaty, published in Rome by Andreas Freitag.[98] Alexander may not have cared much for the press, but he cared a great deal about peace, or at least about orchestrating peace treaties to protect Rome and his family's interests in the early years of the Italian Wars. Nearly a dozen editions of orations in praise of papal treaties survive from his reign.

While Sixtus IV grasped the political possibilities of print, and while authors in the service of Innocent and Alexander saw increasing numbers of their sermons, orations, and panegyrics run through the press, papal use of the press for political ends only revived in earnest with the thunderous reign of Julius II. After 1503, papal policy shifted from diplomacy to war, and in place of peace treaties, papal bulls of warning, privation, interdict, and excommunication began to fly off Roman presses once again.

CHAPTER EIGHT

Brand Julius

JULIUS. Enough words, I say. If you don't hurry up and open the gates, I'll
    unleash my thunderbolt of excommunication with which I used to terrify
    great kings on earth and their kingdoms too. You see, I've already got a bull
    prepared for the occasion.
PETER. Just tell me, please, what you mean by all this bombast about bulls,
    bolts of thunder, and maledictions. I never heard from Christ a single one
    of these words.
JULIUS. You'll feel their full force, if you don't watch out.[1]

The satirical dialogue *Julius Exclusus*, first published in 1517, portrays Julius
II (1503–13) as a Jupiter on Olympus, hurling the "thunderbolts" of his bulls
at anyone who displeased him. As the newly deceased pope approaches
heaven, St. Peter interrogates him before the pearly gates, refusing to let the
pontiff through unless he can prove his virtue. Every turn in the conversation
only confirms that Julius was too violent, temperamental, and power-mad to
deserve any semblance of salvation.

Thunderbolts of excommunication, punishment, and vituperation were
not the only documents the late pope issued in pursuit of his worldly goals.
Among other examples of his sacrilege, Peter recalls how the souls of Julius's
dead troops have been preceding him into the afterlife, each carrying a papal
dispensation written out on parchment hung with a heavy seal. The mercenaries, approaching heaven with their certificates of cheap grace, were "always trying to break in by force, using leaden bulls to force their way."[2]

Throughout the sixteenth century, the papal chancery would continue to
issue manuscript bulls of indulgence and dispensation to all manner of individuals, not just soldiers of fortune. But increasing numbers of papal bulls

were also published in print. Dozens of Julius's thunderbolts, his most menacing and punitive ecclesiastical decrees, were printed in the decade in which he reigned. This marked a shift from how the previous two popes had operated. Few of Innocent VIII or Alexander VI's bulls appeared in print, as we have seen; very few were political in nature, and not many of any sort were printed in Rome. Julius's practice, by contrast, harked back to the early printed pamphlet wars that his uncle Sixtus IV had waged in the late 1470s and early 1480s.

Giuliano della Rovere had been one of Sixtus's closest political advisors.[3] When he was elected Julius II in November 1503, he resumed his uncle's aggressive foreign policies, as well as his tactic of deploying ecclesiastical penalties as weapons of state. Julius and his chancery also revived—and then dramatically expanded on—Sixtus's practice of publicizing his bulls against political opponents and rivals in print. A key contributor to this initiative was likely his secretary, Sigismondo dei Conti, who had served Sixtus as a curialist and ambassador and represented the pope in diplomatic missions— and pamphlet wars—with the Venetians during the War of Ferrara. As Julius's private secretary, Sigismondo drafted and signed many of his most solemn political bulls. And while most of the papal bulls printed in Rome during Julius's reign are not signed by their printers, a few that do bear colophons mention Sigismondo by name as the agent who ordered their publication.[4] This high-placed curial official, who had seen Sixtus outwitted by Venetian use of the press two decades earlier, likely took a leading role in the effort to get Julian bulls into print and may well have influenced how they were designed and executed.

Nearly forty different bulls, briefs, and other ecclesiastical documents appeared in print in Julius's reign, in over a hundred editions published both in Rome and abroad.[5] There was, for example, his bull excommunicating Giovanni Bentivoglio of Bologna, issued during his campaign to reassert control over the northern papal states and printed three times in and after 1506;[6] a formidable *monitorium* threatening the people and clergy of Venice with excommunication, printed at least fourteen times in the months after Julius joined the League of Cambrai against Venice in the spring of 1509;[7] and four different printed bulls of warning, excommunication, and interdict against Alfonso d'Este, duke of Ferrara, and the French armies then occupying Milan, printed in sixteen different editions in 1510.[8] In 1511, when a group of cardinals sympathetic to France withdrew from Rome and attempted to call Julius before a new council of the Church at Pisa, Julius issued yet more bulls of excommunication and privation, stripping the cardinals of their offices and incomes, a half dozen texts printed and reprinted in some seventeen editions.[9] Julius's ultimate response to the Pisan *conciliabulum* was to convene a council of his own, Lateran V, which opened in Rome in 1512 and whose deliberations and decrees were published in even greater numbers. Including the bull announcing the council in July 1511, eight different bulls were published in the seven months before Julius died in February 1513, in twenty-seven editions, alongside letters, orations, and tracts written in support of the council by subordinates and allies.[10]

The scale of Julian publication was greater than Sixtus's or any other previous pope's. All told, some twenty-three Julian bulls regarding political or ecclesiastical controversies appeared in over eighty editions, published in Latin as well as Italian, French, and German, printed not only within the papal states in Rome and Bologna but also in Venice, Ferrara, Nuremberg, Augsburg, Leipzig, Burgos, Paris, and Lyon. In addition, the Julian chancery continued to issue administrative documents: collections of bureaucratic rules and regulations, legal decrees governing the conduct of ecclesiastical business, ordinances for the municipal government of Rome, and bulls stipulating procedures for the election of the next pope. Increasing numbers of these found their way into print as well: more than twenty distinct editions, mostly printed in Rome and in Latin, with a few published elsewhere in Italy and some in Italian translation.

These tallies of Julian publications confirm the image of a pope who did not hesitate to draw the spiritual sword against his enemies, to hurl thunderbolts from his Vatican Olympus. Printed for the whole of Europe to read, they performed a dual function, delivering both private, legal sentences and public and embarrassing rebukes. Julius's successor, Leo X, would deploy many of the same print strategies. The Fifth Lateran Council continued its work for another four years under Leo, with decrees and other documents regularly published in print; Leo also pursued his own political and military initiatives across Italy and published bulls in support of them. When Martin Luther launched his critiques of the Roman Church toward the end of Leo's reign, he faced a papacy well versed in both the potential and the perils of print. Leo's thunderous bulls against Luther, printed in multiple editions that circulated throughout Europe, were but the latest in a long string of publications that Renaissance popes had aimed against their rivals and critics.

The corpus of Julian printed decrees has received little critical attention, far less than the extraordinary cultural patronage for which Julius is perhaps most famous.[11] Massimo Rospocher has studied the political image of Julius as it appeared in printed panegyrics, satirical verses, prophesies, laments, and other vernacular texts published in Italy and north of the Alps.[12] Rospocher sheds important light on popular political discourse in early sixteenth-century Europe, but he concentrates less on Latin-language publications, including those produced by the pope himself. In a study first published in 2005, Nelson Minnich called attention to the competing images of the pope articulated in the decrees of Lateran V and the rival Pisan Council organized by Louis XII; Minnich examines both the bulls and briefs issued from the Lateran and the proceedings of the Pisan assembly, but he does not consider how they circulated in print or participated in a larger contest of printed polemics.[13] Here, I focus not on popular texts about Julius but on official decrees issued by him, and I am concerned not only with their arguments but also with how they were put into print: how they looked, what visual and rhetorical strategies they deployed, and what these features can tell us about papal communication strategies on the eve of the Reformation.

Julian bulls, briefs, decrees, circular letters, orations, and commentaries conveyed the political declarations that Rospocher's popular authors and readers responded to; their printing history reveals the presentational strategies that Minnich's officials deployed in order to reach them. Nevertheless, it would be a mistake to imagine any real dialogue with popular readers: these documents did not promote or participate in exchanges with the reading public at large. Rather, they were one-sided declarations (thunderbolts!) issued by the pope to promote his initiatives and punish his opponents, or they responded to provocations published *by* those opponents, who were almost always major political powers themselves. In this sense, Julius's use of the press was again like that of his uncle Sixtus: papal print broadcast the contents of interstate political communications, making it easier for the reading public to see what princes and prelates were saying to one another. As yet, there was still little sense of the public as a body that should be directly addressed when it came to matters of state.

Even so, the corpus of printed publications issued by the Julian papacy was substantial and innovative, and the collection provides a fitting conclusion to this study of papal print. Julian printing was different from the production of the previous century in several ways. There was the scale of the operation, as we have seen, as well as its stylistic innovations; the proliferation of graphic statements of papal authority on title pages and elsewhere; the expanding use of paratexts, both legalistic and rhetorical; the appearance of official decrees in vernacular translations; and the sheer variety of texts that were put into print. At the same time, the *substance* of papal communications remained remarkably constant: the same authoritarian message, the same legal arguments against the presumptions of councils and secular princes, the same insistence on Rome as the seat of the papal monarchy and on the sacred authority of the city's bishop and his documents, no matter what medium they appeared in. The Julian curia found more that it could do with print—graphically, textually, and rhetorically—even as it doubled down on a model of papal authority whose roots extended back deep into the manuscript age. Here, more than in any other chapter of this book, we see the papacy using the revolutionary medium of print in ways that were innovative but reactionary, experimental but doctrinaire.

For the first time in the history of papal print, visual elements played an important role in articulating papal authority. The printed bulls of previous popes had been visually undistinguished. Either large bifolia or slightly longer, quarto pamphlets, they were usually printed in a single fount of type (usually Gothic, and usually quite small), with little in the way of titles or other paratexts to announce their contents, and no visual elements at all. They made no attempt to mimic the format of a handwritten bull but reproduced only the text of the document, omitting most of the diplomatic formulas, signatures, and graphical signs that would have lent authenticity and authority to the original document. Although a few bulls printed north of the Alps in the fifteenth century included visual elements, Roman printed bulls of the period were almost always pure text.[14]

Under Julius, Roman printers began to develop and deploy new graphical and textual signs to identify and authenticate their printed bulls. Some of these replicated established manuscript practices, but some had no precedent at all in earlier tradition. For example, Roman printers headed their editions of Julian bulls with new, descriptive Latin titles. The titles, which grew longer over the course of Julius's pontificate and often editorialized on the contents of the subsequent document in subtle or not-so-subtle ways, appeared above or before the start of the bull and were often set in different type, larger type, shaped type, or some combination of all three (figure 8.1). Printers also commissioned and printed woodcuts of the pope's arms, placing these on the

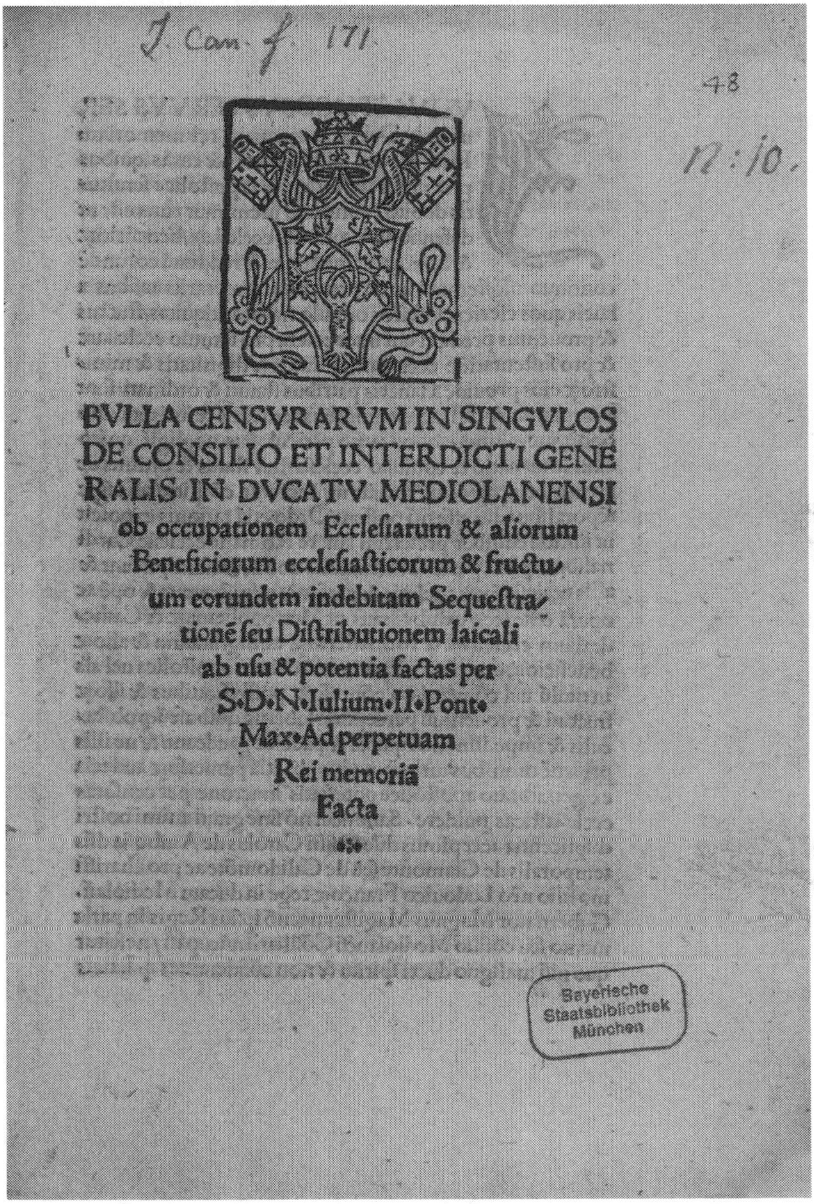

J. Can. f. 171

48

12:10

BVLLA CENSVRARVM IN SINGVLOS
DE CONSILIO ET INTERDICTI GENE
RALIS IN DVCATV MEDIOLANENSI
ob occupationem Ecclesiarum & aliorum
Beneficiorum ecclesiasticorum & fructu-
um eorundem indebitam Sequestra-
tioné seu Distributionem laicali
ab usu & potentia factas per
S·D·N·Iulium·II·Pont·
Max·Ad perpetuam
Rei memoriã
Facta

Bayerische
Staatsbibliothek
München

Figure 8.1. Julius II, *Bulla censurarum* ([Rome, ca. 1510]). Munich, BSB, 4 J.can.f. 171, fol. 1r.

first page of the bull or, later on, on a dedicated title page, another innovation,[15] which was often further decorated with a woodcut border. Printing
the papal *stemma* was a practice that first emerged ahead of the 1500 Jubilee,
with the mass production of editions of the *Indulgentiae*. Under Julius, the
practice migrated from the production of pilgrim souvenirs to the publication of official decrees.

Roman printers in this period borrowed and exchanged both types and
woodcuts very freely; they only rarely signed their editions. This means that
it can be very difficult to sort out dates or attributions for many of these
publications. But it also means that Julian bulls, no matter who printed them,
share a particular textual and visual vocabulary—a style—that was new and,
apparently, intentional.

Julian printed bulls also display intense concern for establishing their authenticity. Perhaps paradoxically, given the visual innovations just described,
Julian publications tend to reproduce in type, in ever greater detail, the manuscript notes inscribed on the original documents, recording how a bull had
been signed, registered, authenticated, and published on the doors of Roman
basilicas and other *loca consueta* in Rome. This attention to ancient protocol
might seem a conservative step, rather in tension with the visual and typographical advances taking place at the same time. Later in this chapter I consider the implications of this practice in more detail. Here, it is enough to
say that all of these elements—visual as well as textual—represent departures
from the way fifteenth-century printers had put papal bulls into print.

Finally, the Julian chancery also had new types of document printed—not
only bulls of excommunication and interdict, as Sixtus issued, but also bulls
of warning and privation, briefs (open letters) to the Christian princes, reissues of documents published by earlier popes and excavated from the papal
archives, and speeches and letters by diplomats, theologians, and court orators in support of Lateran V. These publications extended the documentary
field on which ecclesiastical and diplomatic disputes might be pursued, increasing the reach of papal communications and expanding their character.

The Julian secretariat and chancery seem to have directed much of this
activity as part of a coordinated print strategy. They published texts that
advanced the pope's aggressive ecclesiastical and foreign policy while also
branding papal publications as particularly Julian—emblazoned with his
name, his titles, and the distinctive della Rovere oak. Inside, however, these
visually striking publications became ever less legible, as drifts of legal formulas, certifications, and other statements of authority piled up around the

central text. Why these two innovations occurred together is a puzzle. Replete with references to the ancient traditions and material practices of the papal chancery, branded with papal insignia, and self-consciously invoking the sacrality of the ancient Roman spaces where they had first been displayed, Julian printed bulls were at the same time strikingly modern and thunderously old-fashioned.

An Eventful Pontificate

A survey of the five political episodes in Julius's reign that generated most of his political print reveals both the challenges the pope faced in his dealings with foreign powers and the proliferation of printed texts that accompanied them. These episodes were Julius's campaigns to reassert papal authority over Romagna in 1504–6, the War of Cambrai against Venice in 1509, his 1510 disputes with Alfonso d'Este and the French forces occupying Milan, the calling of a new church council at Pisa by Louis XII and a group of dissident cardinals in May 1511, and, finally, Julius's own council at the Lateran, which was announced in July 1511 and opened in the spring of 1512.

Julius came to the throne in the autumn of 1503, at a moment when Italy had been roiled by invasions, wars, and revolutions for nearly a decade. After the French invasion of 1494, the entire peninsula had been destabilized, with dynastic regimes in Milan, Florence, and Naples overthrown. Venice remained the sole, truly independent Italian power aside from the papacy itself. The Venetians had exploited the chaos of the early Italian Wars to expand their territorial holdings, seizing towns in Lombardy as well as Romagna, which was nominally subject to the pope. The other agent of chaos had been the papacy itself: encouraged by his father, Alexander VI, Cesare Borgia had his own designs on Romagna and began carving a small state for himself out of papal domains in the region. But after his father's sudden death in the summer of 1503, Cesare's tiny domain dissolved. The Venetians were quick to annex much of it for themselves.

Determined to reclaim papal territory lost to Cesare and then to Venice, Julius deployed a mix of incentives and threats. In 1504, he extended political and economic concessions to persuade Imola, Cesena, and Forlì to recognize him as their overlord once again.[16] The concessions he extended to Forlì were put into print by that city's municipal government (in an edition discussed below), but at this early date Julius himself had not yet turned to the press.

That changed in 1506 when the pope launched a new campaign to regain control of Bologna. Here the problem was not a foreign state annexing papal possessions but an unruly client, the *condottiere* Giovanni Bentivoglio, who ruled the city like an independent lord, paid the pope little heed, and threatened to make his own alliances with other states.[17] In the summer of 1506, Julius declared that he himself would lead a military expedition against Giovanni and his sons in order to liberate the people of Bologna from their tyrannical oppressors. He set out from Rome in late August at the head of a large mercenary force, with the College of Cardinals and much of the papal court in tow. The cavalcade proceeded north over the Apennines and arrived in Forlì in October. Calling on his alliance with France, Julius summoned French troops from Milan to advance on Bologna. On October 10, he also issued a bull announcing the excommunication of Giovanni Bentivoglio and demanding his expulsion from the city; if the people of Bologna did not send him away within thirty days, the entire population would be placed under interdict.[18] As the French army approached, support for Bentivoglio evaporated, and on November 2 Giovanni and his sons fled the city. Julius made a triumphant entry on November 11. The speed of his victory was unexpected. Some weeks earlier, his mobile chancery had sent a copy of the bull against Giovanni to Rome, where Johannes Besicken printed a quarto edition of it, dated November 12.[19] An Italian translation of the bull was also put into print.[20]

Besicken would have had no way of knowing that the Bolognese had already met the conditions stipulated in the bull. Even so, the publication was still useful: advertising Giovanni's status as an excommunicate was politic, since the exiled lord had taken refuge in Milan, not all that far from Bologna, and was plotting his return.[21] Moreover, it seems that Julius or one of his secretaries had the text of this 1506 bull reissued in print some years later, in 1510,[22] when the pope was threatening another northern Italian lord, Alfonso d'Este, with excommunication. Julius could use print not just to punish his enemies but to advertise his ability to do so, using examples from the past as well as the present.

Once Bologna was restored to papal control, Julius returned to the question of Venice and the Romagnol towns it still occupied, especially Rimini and Faenza. He was not alone in wishing to push back against the Venetians; most of the European powers had lost territory of some sort to Venice in recent decades. The republic had taken Lombard towns from the French; parts of Dalmatia from Hungary and Friuli from the empire; Puglia from Aragon; and Rimini, Faenza, and other Romagnol towns from the Church. In December 1508, France, the empire, Hungary, and Aragon formed the League of

Cambrai against Venice, determined to secure return of their lands.[23] Julius, although supportive of the league, was not a signatory to its first pact, but after further negotiations, he signed a bull on March 23, 1509, declaring his adherence.[24] According to its terms, Maximilian was to supply troops for an attack on Venice, but not before Julius had formally requested them. On April 10, Julius duly sent a letter to Maximilian asking for his military support (this letter is discussed in more detail below); the text of the brief was printed as a large broadside in Rome, probably by Eucharius Silber, and again in German translation in Munich, presumably to advertise the fact that the pope had the right to summon the emperor to his aid and the news that he had just done so. Such an announcement would bolster support for the invasion in northern Italy and encourage the Venetians to bow to the league's demands before it was too late.[25]

Venice remained defiant, and the pope increased his pressure on the republic. A few weeks after his letter to Maximilian, Julius issued a *monitorium*, or bull of warning, threatening Venice with the interdict.[26] The text, read out at a consistory in Rome on April 26, gave the Venetians twenty-four days to return the cities and lands they had seized from the pope. If they did not, Julius would withdraw the sacraments from the city and excommunicate all its inhabitants. The bull, signed by the papal secretary Sigismondo dei Conti, was registered on April 27 and posted on the doors of St. Peter's and in Campo de' Fiori the same day. By May 3 it had been printed by the Roman printer Jacopo Mazzocchi.[27] News of these events reached Venice within a few days: Venetian diplomats in Rome sent letters on May 3 and 4, which Marino Sanudo recorded in his diary on the eighth of the month. The ambassadors reported, among other signs of Julius's anger with the republic, that the bull had been printed in six hundred copies "which the pope wants to show to Venice and to the whole world." At least thirteen more editions of this text, in Latin and in German, French, and Italian translations, would be published in Rome, as well as in major cities of the league: Nuremberg, Augsburg, Leipzig, Paris, Lyon, and possibly Ferrara.[28]

Like Julius's bull against the Bentivoglio of 1506, Mazzocchi's printed edition of the *monitorium* was both a legal and a propagandistic text. A clause at the end of the bull—an innovation—declares that, since it would be difficult to circulate the original document as widely as it ought, any copy of the document, "whether written out by hand by a public notary *or printed in Rome by our beloved son Giacomo Mazzocchi*, and marked with the seal of any prelate of the Church, should be taken to have as much authority as the original letter and should be trusted just as if it was the original that was being shown."[29] The idea was not just to send a threat to the Venetians but to let the rest of Italy know it. Despite the bull's status as a bull of *warning* (clearly signaled in most printed editions by the title *Monitorium contra Venetos*), which gave the Venetians more than three weeks to bow to Julius's demands before the interdict came into effect, most observers considered it, and referred to it, as a bull of excommunication. Either the Venetians had no intention of giving in to the pope's demands, or they believed that the pope had no intention of backing down from them.

A team of Venetian canonists set to work on an appeal of the interdict; they drew up a text that protested the penalties leveled on the republic and asserted that since their dispute was with the pope, the pope could not adjudicate it. The question must be decided by a church council. As soon as one could be convened, they would submit their complaint to it. The Venetian government sent this appeal to Cardinal Tamás Bakóc, who, as titular patriarch of Constantinople, possessed an ancient prerogative to call a council of the Church, or so the Venetians claimed.[30] In early May, they also had two copies of their text taken to Rome in secret and posted on the doors of St. Peter's and on a column at the entrance to Castel Sant'Angelo. The pope, in a fury, ordered both torn down.[31] Not long after, on July 1, Julius reissued Pius II's bull *Execrabilis*, which asserted that only the pope could convene a council and banned anyone else from making an appeal to a future assembly. This was not printed in 1509, but it did appear in print the next year, in the course of a different conciliar dispute, as we will see.

The exchange of documents soon gave way to war. On May 14, the armies of the league attacked and overcame the Venetians at Agnadello, in Lombardy, and quickly pushed their forces back almost as far as the Adriatic.[32] It was a catastrophic defeat. Machiavelli later commented that in one day the Venetians lost what it had taken them eight hundred years to conquer. Their possessions in the *terrafirma* were parceled out among the victors.[33] In panic, the Venetians informed the pope that they were ready to return his cities and do everything else he had demanded, but Julius now rejected their overtures and thundered that his interdict was still in force. The republic was forced to grovel: in June, six Venetian envoys came to Rome to beg for a release from excommunication, delivering a letter from Doge Leonardo Loredan, who pleaded with the pope to show mercy to his loyal Venetian flock. Eucharius Silber and Etienne Guillery both printed editions of the doge's letter;[34] it seems likely that the legation commissioned its publication in Rome as a way of further publicizing their willingness to come to terms with the pope.

But the pope held out for further concessions, and even as the Venetians' diplomatic stance remained submissive, more critical voices began to be heard. A letter circulated in Venice, purportedly written by Christ himself and addressed to his unseemly vicar, Julius, castigating him for his violence and greed and the danger he posed to the souls of his faithful.[35] The text, though written in Italian, uses the same Latin formulas for salutation and subscription as a real papal brief: it begins "Jesus Christ, son of the Virgin Mary, to Julius II, our unworthy vicar," and is subscribed at the conclusion: "Issued from our celestial heaven on the 26th day of December in the 1509th year since our birth on earth." After this, Christ's "secretary" adds the customary signature: "I, John the Baptist, wrote this out as ordered."[36]

The charges in the letter draw from a common stock of antipapal critique, the same sort as were penned by Andrija Jamometić or Gentile Becchi about Sixtus IV, and as would be articulated at greater length in *Julius Exclusus*: the pope was a greedy warmonger, obsessed with worldly glory, a sorry successor to Peter, leading his Christian flock down the road to perdition. What is striking here is the parody of chancery practice—the idea that heaven included a chancery where John the Baptist supervised the processing of the Lord's correspondence. It suggests that the proliferation of papal documents in the early sixteenth century, while broadcasting the authority of the pope to punish his opponents to a larger audience than ever before, had also made the trappings of his communications a target ripe for satire.

In February 1510, Julius lifted the sentence of excommunication and made his peace with Venice. The republic had been humbled; now, it was the growing power of the French that alarmed the pope. Julius turned his attention to Louis XII, who controlled much of northwestern Italy through direct occupation or alliances with states like Ferrara and Florence. The project of ejecting the French from Italian soil would be the great preoccupation of the last three years of his pontificate. Julius claimed that his ultimate goal was not just to reduce the power of France but to restore the balance of power in the peninsula, unite the European powers, and clear the way for a new crusade against the Turks.[37]

As the pope's new anti-French policy took shape, the French cardinals in Rome grew increasingly uncomfortable. In June, Cardinal François de Castelnau tried to flee to France and was imprisoned in Castel Sant'Angelo.

Other French prelates, including Cardinals Guillaume Briçonnet, Louis d'Amboise, and René de Prie, lodged formal protests with the pope.[38] Meanwhile, Louis attempted to invoke certain provisions of the old Pragmatic Sanction of Bourges, the concession Charles VII had won from Eugene IV during the Council of Basel, which granted the French king rights over appointments and income from the "Gallican" Church in France. Louis XI had renounced his claim to these privileges during the reign of Pius II, but Louis XII now tried to revive the arrangement as a way of limiting both Julius's ability to coerce the French cardinals and his access to church revenues from France. By July, Julius had lost patience with Louis and, by extension, Alfonso d'Este, duke of Ferrara, who remained a close ally of the French occupiers in neighboring Lombardy. Alfonso's actions were particularly provoking: a feudatory vassal of the Church (he had actually served as commander of the papal troops at Agnadello), the duke continued to press his claims to cities and lands the Venetians had occupied and that he had won from them, even as Julius urged him to renounce them for the sake of his new alliance with Venice.[39] As in his dispute with Venice the previous year, Julius issued bulls as a prelude to military operations. In a bull of August 9, 1510, he excommunicated Alfonso and stripped him of his titles and fiefs. This *Bulla privationis* was printed in Rome, Bologna, and Venice, four times in Latin and once in an Italian translation, *Summario dela scomunicha de Ferrara.*[40]

Moved to protect his ally but even more concerned about the military threat to Milan and Genoa, Louis XII now threatened to renounce his obedience to Julius and call a general council.[41] Already on July 30, 1510, the king had invited the French bishops to a synod in Orleans, "to consult on the liberties and privileges of the Gallican Church."[42] And while Louis was thinking of reviving the Pragmatic Sanction, Maximilian was thinking of copying it: in September 1510, he sent a copy of the original compact to the humanist scholar and historian Jakob Wimpheling, from whom he sought advice on how to institute the same measures in Germany.[43] These murmurings seem to have prompted the publication of some older papal texts in Rome: Julius's reissue of *Execrabilis*, brought out the previous year after the Venetian call for a council, was now printed in pamphlet form (in two unsigned editions, printed after December 17, 1510) as a response to Louis's machinations toward the same end;[44] more such texts would follow in 1511 and 1512, including the text of Louis XI's renunciation of the Pragmatic Sanction and a decree of the Council of Constance that endowed the pope alone with the right to call a new council.

Julius was also preparing for war. A week after excommunicating Alfonso d'Este in August 1510, the pope decided to join his armies in the attack on Ferrara. He left Rome on August 17 and traveled north to the Adriatic, stopping in Loreto on September 8 and arriving in Bologna on September 22.[45] From here, he issued three more bulls (October 9 and 14 and November 4) that were quickly put into print. One imposed an interdict on Milan;[46] one renewed the excommunication of Alfonso and imposed the same sentence against Charles d'Amboise, commander of the French forces in Milan, and his captains.[47] The last threatened an interdict on any city that gave quarter to "rebels" who had been exiled from their cities. On November 15, Sanudo saw printed copies of the second of these three bulls, the one dated October 14, for sale on the Rialto. Versions in both Latin and Italian were available, Sanudo noted, for the low price of a soldo apiece.[48]

Lines were being drawn: Julius had ordered the cardinals who supported France to leave Rome and join him in Bologna, but news arrived that five of them—Carvajal, Borgia, Briçonnet, de Prie, and Sanseverino—had instead gone to Milan to seek Louis's protection. By December, French troops had advanced on Bologna, where Julius had fallen dangerously ill. He was able, nonetheless, to draw up a letter to the princes of Europe decrying Louis's actions and appealing for their help.[49] Unlike his thunderous bulls, this letter, which revealed more of Julius's weakness than he would have wanted publicly known, was not printed. The pope resorted to older forms of communication with "the Christian princes" when policy required it.

The early months of 1511 saw Julius recovered from his illness and ready for battle. In January he left Bologna for Mirandola, where papal troops were besieging the city in drifts of snow. After the city's surrender, he returned to Bologna, then carried on to Ravenna to renew the attack on Ferrara, and finally returned to Bologna in April.[50] On April 16, Maundy Thursday, he issued the traditional bull of excommunications, *In coena Domini*, adding a specific condemnation of every "adherent" of Louis, including the entire city government of Milan; it was printed soon afterward, possibly in Bologna.[51] Undeterred, the French troops renewed their attacks, and on May 15 Julius was forced to withdraw from the city. The commander of the French troops, Gian Giacomo Trivulzio, entered Bologna and restored Giovanni Bentivoglio to power, while Julius retreated to Rimini.

There, on May 28, the pope awoke to find a citation to a new council of the Church, affixed to the door of the nearby church of San Francesco.[52] The document, dated May 16, declared that delegates of the emperor and the king of France had decided to summon a council, to be held in the autumn in Pisa, on the grounds that the old decree *Frequens* required such a meeting and the pope had failed to convene one. The document called on every cardinal to participate and summoned the pope to appear as well. The council would be presided over by a group of cardinals whom the document named: the original five defectors (Carvajal, Briçonnet, Borgia, de Prie, and Sanseverino), as well as Philip of Luxembourg, Adriano da Corneto, Carlo da Caretto, and Ippolito d'Este. Because the pope had given no safe-conducts to couriers who might publish the summons in Rome, publication on the cathedral doors in Modena, Parma, and Reggio would have to suffice. The text was also put into print in multiple editions.

Sanudo noted all these provisions in his diary, in a curious order. On June 30, he recorded that he had heard that the summons to a council had been printed, and he believed that printed copies would arrive in Venice before a handwritten one could; next, he said that the text had also been posted on the doors of churches in Modena, Bologna, and other places, including Rimini, which had infuriated the pope and led him to utter maledictions against the cardinals named in the text.[53] For Sanudo, the important thing was that the text had been printed. His remarks suggest that manuscript copies of documents were approaching irrelevance: fewer of them circulated, and they were slower to produce. The posting of the text on various church doors seemed almost an afterthought, of interest mostly because it had so infuriated the pope.

The publication of this summons to a council, orchestrated by the cardinals but really the work of Louis XII, took place against a backdrop of continuing war in Italy. Throughout the summer of 1511, the French kept pushing Venetian and papal forces back toward the Adriatic Sea; in October, the pope announced a new Holy League, consisting of the Church, England, Spain, the empire, and Venice, arrayed against the French. These events prompted, in turn, a period of intense diplomatic negotiation and propagandizing across Europe. In France, Jean Lemaire de Belges composed an aggressive screed against Julius and in support of Louis's proposed new council, which he dedicated to the French king and which was printed multiple times.[54] The dramatist Pierre Gringore published a parodic work, *Chasse du cerf des cerfs*, which mocked the false humility of the pope's title *servus servorum Dei* (with which every one of his bulls began).[55] In Italy, meanwhile, prophecies, ballads, and short vernacular tracts composed both for and against the pope and the league proliferated in print.[56] There is not space to rehearse the various arguments about Julius's character and policies this literature advanced; suffice it to say that the Latin pamphlets that flew between pope and king and their respective councils were produced amidst a near-constant flow of popular vernacular commentary.

Julius returned to Rome in early summer, stopping at Loreto again; he arrived in the city on June 26, 1511. This was the lowest point of his pontificate, with political and military threats from all sides, opposition from his own hierarchy, dwindling support in Rome, and a new council poised to convene and depose him. But some of the cardinals named in the citation now claimed that their signatures had been added without their knowledge and began to denounce the Pisan Council.[57] Julius moved forward with his own plans: on July 18, a bull was affixed to the doors of St. Peter's summoning the fathers of the Church to a general council that would convene in Rome the following April. The bull denounced the cardinals adhering to Louis's council and declared their citation of May 16 null and void.[58] Within two weeks, the bull had been printed by Mazzocchi, and it would be reprinted another thirteen times over the next year, in Latin as well as Italian and German translations, in Rome, Bologna, and Venice; in Leipzig (where the printer Martin Landsberg regularly reprinted papal decrees); in Nuremberg (where Johann Weissenburger did the same); and even in distant Burgos in Spain.

Between the announcement of the council in July and its opening in May the following year, Julius continued to fulminate against Louis's rival council and to hurl furious bulls against the cardinals supporting it. Six more bulls and briefs were published and printed in Rome in these months: one warned the "schismatic" cardinals to return to obedience; another two stripped them of their offices and titles when they did not; an open letter to the Christian princes explained the pope's rationale for his actions; another bull warned the French bishops that similar penalties awaited them if they participated in the "conciliabulum" at Pisa; and a final one, literally the *Last Bull* (*Bulla ultima convocationis et invitationis*), extended a final invitation to the French prelates to abandon their assembly and join Julius in Rome. This was issued on April 13, 1512. The council had been scheduled to open on April 19, but the Battle of Ravenna, fought on April 11 and a major loss for the league, led to a brief postponement (also announced in print). Lateran V finally opened on May 3.

The council convened in five separate sessions before Julius's death in February 1513. After each, Julius issued more bulls conveying decrees or summarizing deliberations of the council. Many delivered further condemnations of

Louis, the French cardinals, the Pisan Council, and the Pragmatic Sanction, and these were put into print alongside orations and other writings by participants in the council and other supporters. The publication of Lateran V bulls, nearly all of them printed by Marcellus Silber, continued into the reign of Leo X.

It would be a lengthy, perhaps tedious project to trace every contour of these five episodes, every claim and counterclaim put into print during so many political, military, and ecclesiastical disputes. And yet, while the volume of papal publications increased dramatically under Julius II, in some ways little had changed in papal policy, and in the rhetoric of papal communications, since the reign of Sixtus IV. The pope was still trying to win control of the cities of Romagna, for example, places nominally part of the papal state but perennially vulnerable to alienation or conquest. The Pazzi War had been provoked, in part, by Sixtus's wish to procure Imola for his nephew and Lorenzo de' Medici's reluctance to see him have it. Julius's earliest campaigns, in 1504–6, were to reclaim Imola along with Cesena and Forlì, cities that had passed from the Riario to Cesare Borgia and then to Venice in the intervening years.

When Julius joined the League of Cambrai against Venice in 1509, his stated reasons were much like those that Sixtus leveled against Florence in 1478: the civil government of the republic had violated ecclesiastical immunities, claimed for itself the right to make church appointments, and seized territory and property that belonged to the Church. The thunderous charges of Julius's *Monitorium* echo the complaints of *Ineffabilis*, Sixtus's bull against Lorenzo, remarkably closely.

In the early 1480s, Sixtus joined Venice in an alliance against Naples and Florence but then abandoned his Venetian allies, forming a league of Italian states arrayed against the republic and insisting they relinquish their claims on Ferrara. In 1510, Julius played practically the same trick on Alfonso d'Este, who had joined the League of Cambrai against Venice in the hope of reclaiming territories he had lost to the republic. Julius's *Bulla privationis* against Alfonso echoes his uncle's disingenuous briefs to Venice of 1483: in both, the pope claimed the right not only to wage war on his enemies but also to declare to his allies who their enemies must be. In 1483, Sixtus threatened the Venetians with dire penalties if they did not give up their designs on Ferrara and join their former foe in his new Holy League; Julius made almost the same arguments to Alfonso, threatening—and eventually leveling—excommunication on the duke if he did not turn on his former allies, the French. Along the way, both popes issued bulls threatening excommunication of, or actually excommunicating, the military captains who dared sell their services to the enemy of the moment: Roberto Malatesta, in 1479, and Charles d'Amboise, in 1510, were both targets of printed bulls that denied them access to the sacraments on the eve of battle.

Finally, Louis XII's attempts to call Julius before a new council echo those of Becchi in Florence and Jamometić in Basel three decades before. The charges were remarkably consistent. A warlike bishop of Rome—obsessed with territorial conquest, worldly glory, the fortunes of his nephews, and the pleasures of his splendid court, to the neglect of his flock and the health of his church and the crusade against the Turks—must be called to account before a body of faithful bent on reform. Arguments like this could be leveled against *any* Renaissance prelate, and Louis, like Lorenzo (and Frederick III, and the Venetians at times), had fairly worldly reasons for wanting to see the power of the pope reduced. Still, there was the problematic issue of *Frequens* and the fact that the pope *was* bound to call a council: the decrees of Constance and Basel, to which earlier popes had assented, not only allowed for consultation in the government of the Church; they required it. The autocratic claims of both della Rovere popes to supremacy in the Church (and temporal supremacy over much of Italy) grated against other contemporary ecclesiologies that had good foundation in tradition and church law; this opened a space for pamphleteering both politically expedient and sincere. Finally, in both pontificates, the papal response was thunderous: total condemnation of the very concept of a council unless called by the pope, and excommunication and interdict on all who supported the notion. The response to the council summons was to articulate, longer and more loudly, the same vision of papal supremacy that had prompted the summons in the first place.

Print allowed all these arguments to circulate more widely and at greater length than in previous ecclesiastical disputes. From the pope's point of view, not only could the press deliver sentences of excommunication, privation, and interdict; it could also advertise them. Ecclesiastical penalties had real-world consequences: targets did not just lose their access to the sacraments. Debts could be canceled, goods seized, contracts voided, and territories lost. The more people knew about the circumstances of a ban, the more damage it caused. Spreading the news of a papal sentence brought shame and peril on the victim in equal measure. Some thirty years apart, both Gentile Becchi and Marin Sanudo remarked with apprehension that papal bulls were no longer just posted on the doors of St. Peter's but also sold for pennies in Campo de' Fiori and on the Rialto. Print made the shame of papal sanction more public—and thus more powerful—than ever before.

Printers, Forms, and Formats

From the sequence of events that roiled Julius's papacy, we can now turn to the printing of his thunderbolt decrees. Julius had no official printer; it was only in the middle decades of the sixteenth century that the office of papal printer was firmly established.[59] Under Julius, papal offices entrusted documents to several different Roman printers. Johannes Besicken, who printed in Basel for Jamometić and the city council and then produced numerous *Indulgentiae* editions in Rome for the 1500 Jubilee, and Eucharius Silber, a cleric from the diocese of Würzburg who had been printing in Rome since about 1480, were responsible for much of the early work.[60] Around 1510, they yielded to a new generation of printers, their heirs or former partners.[61] Giacomo Mazzocchi began editing texts for Besicken around 1505; Etienne Guillery, from the French town of Lunéville, had performed similar work for Besicken in 1506. They each set up their own shops around 1509, while between 1510 and 1512 Johannes Beplin, of Strasbourg, gradually took over Besicken's operation. Eucharius Silber died sometime in late 1509 or early 1510; his son Marcellus began production of his own editions later in 1510. All of them operated printshops around Campo de' Fiori, close to the chancery in its old premises in the via de' Banchi.[62]

They all did job printing for the curia at various times. Silber, father and son, and Mazzocchi were probably the most frequent publishers of papal material. Silber produced papal bulls, as well as administrative handbooks, schedules of taxes and fees, and other legal and bureaucratic material; Mazzocchi, who, like Guillery, had connections to the Studium Urbis, dealt in more literary material: it was he who started publishing annual editions of Pasquino verse, and he also published a number of learned Latin orations and poetry during Julius's reign. Mazzocchi sometimes claimed (or was granted) a papal mandate or privilege in his editions,[63] but almost every papal document printed in Rome was printed in multiple editions, often simultaneously.

Together, this group of printers developed an entirely new style of printed papal decree: a quarto pamphlet, anywhere from four to thirty-two pages long, featuring title pages with descriptive titles, woodcut borders, and woodcut papal arms (figures 8.2 and 8.3). These editions also included, as never before, extended paratexts derived from the original manuscript—notes on the processes of correction, publication, and ratification in the curia and posting in the streets of the city. The design choices these printers made—and some choices they *could* have made but did not—reveal the creative tensions at work in the Julian publicity machine.

A papal bull was a complex diplomatic instrument of which the core text was only a part. Various paratexts—signatures, graphical signs, and statements describing the processes of copying, registration, and publication that the document had undergone—all surrounded the core text and assured readers or recipients of its authenticity. In the fifteenth century, few if any of these subsidiary elements were reproduced in printed papal bulls. But under Julius, printers began to include certain of these elements in their editions, including textual statements about the *publicatio* of the original document in Rome and authenticating statements describing the material appearance of the original manuscript bull (e.g., remarks on the state of the original parchment, its lead seal, and silk cords).

Other elements of the traditional manuscript bull were more difficult, but not impossible, to reproduce in print. Graphic signs used to certify the document, such as the exaggeratedly tall letters used in the opening sentence; the *rotula*, or hand-drawn circle containing the names of Peter and Paul and the current pope; and the manuscript signatures of the pope, his secretaries, and attending cardinals, presented a challenge to the printer, as did the lead *bulla* itself, and the images of Peter and Paul and the papal name that were stamped into its surface.

In the early sixteenth century, a few Italian printers tried to reproduce some of these manuscript elements. In July 1504, Julius issued two bulls to the citizens of Forlì absolving them of their oaths of allegiance to Cesare Borgia and agreeing that if they renewed their obedience to Rome, he would appoint no new papal vicar to rule them.[64] Three years later, the two bulls were printed together in Forlì in a folio edition by the printer Paolo Guarini (figure 8.4).[65] The edition includes several graphic innovations: a tall woodcut initial *I*, for "Iulius," filling some twenty-four lines of space on the first page; a reproduction of the pope's signature in type, preceded by a woodcut cross; a woodcut facsimile of the papal *rotula* with the textual elements supplied in lines of curved type; reproductions of the signatures of the cardinals that followed the document, set in type but each with a small woodcut cross; notes on three further signatures or ciphers provided by Julius's secretary,

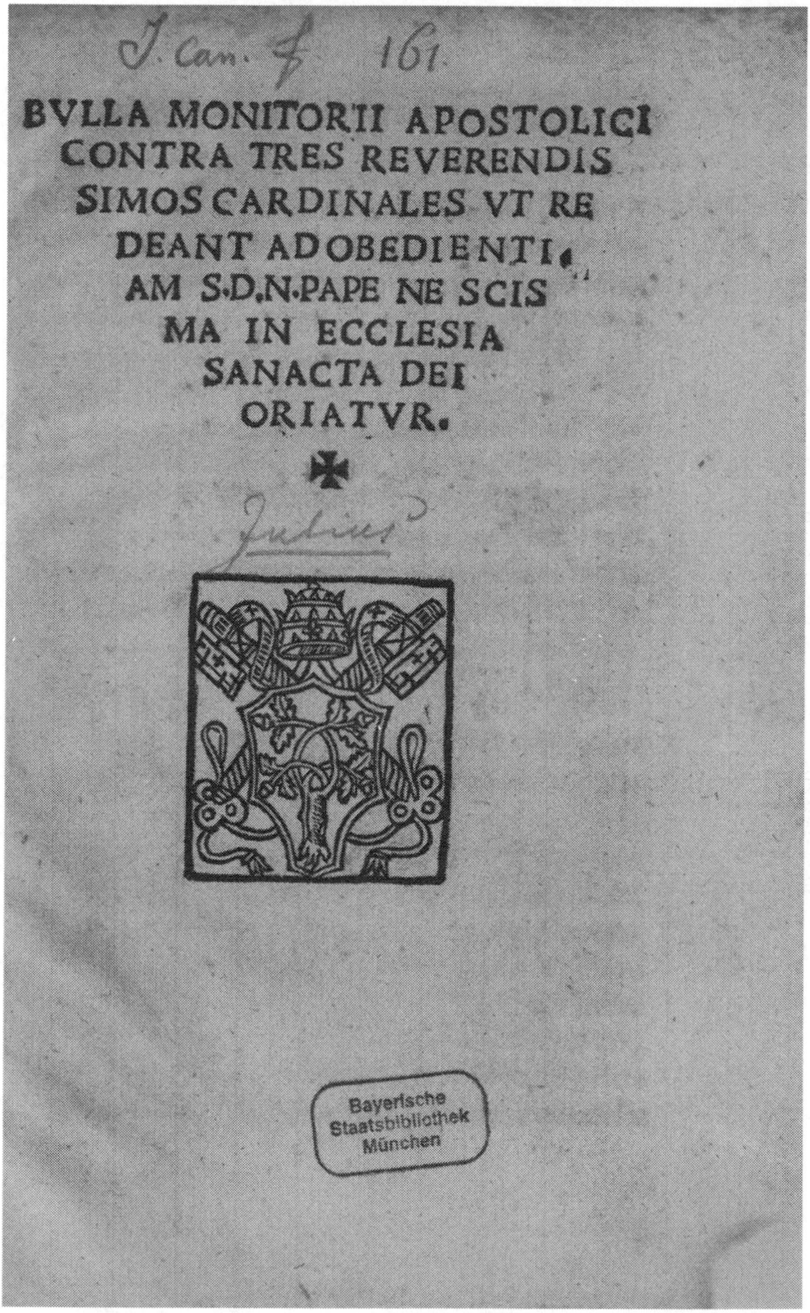

BVLLA MONITORII APOSTOLICI
CONTRA TRES REVERENDIS
SIMOS CARDINALES VT RE
DEANT AD OBEDIENTI
AM S·D·N·PAPE NE SCIS
MA IN ECCLESIA
SANACTA DEI
ORIATVR.

Bayerische
Staatsbibliothek
München

Figure 8.2. Julius II, *Bulla monitorii apostolici contra tres reverendissimos cardinales* ([Rome: Giacomo Mazzocchi or Etienne Guillery, 1511]). Munich, BSB, 4 J. can.f. 161, fol. 1r.

J. Can: f. 166

IVLIVS PAPA.II.

BVLLA INNOVANS ET CONFIR-
MANS CONSTITVTIONEM SIVE
EXTRAVAGANTEM PII .II. CON-
TRA APPELLANTES AD FV
TVRVM CONCILIVM
PER S.D.N.IVLIVM
II.PONT. MAX.
EDITA.

Bayerische
Staatsbibliothek
München

Figure 8.3. Julius II, *Bulla innovans et confirmans constitutionem . . . Pii II contra appellantes ad futurum concilium* ([Rome, 1510]). Munich, BSB, 4 J.can.f. 166, fol. 1r.

Figure 8.4. Julius II, *Bulla: Indulta ac privilegia concessa civitati Forliviensi* (Forlì: Paolo Guarini, 1507). Rome, BNC, MS SS. Apostoli 4 (326), fol. 134r.

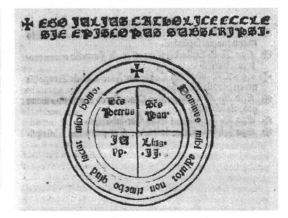

Figure 8.5. Printed *rotule* in editions of Julius II, *Bulla intimationis* ([Rome, 1511]).

Sigismondo dei Conti, at the bottom of the bull and on the back side of it; and, finally, a striking woodcut initial representing the calligraphic *R*, for "Registrata," which would have appeared on the back of each original bull, indicating that its text had been copied into the papal registers in Rome. This elaborate *R* appeared twice, at the end of each bull.[66]

Guarini's graphic experiments created a nearly diplomatic transcription, in print, of his original manuscript bulls. It could be done. But few printers in Rome attempted anything similar. In 1511, Mazzocchi introduced a hybrid, woodcut-and-type *rotula* into his editions of the *Bulla intimationis* announcing the convening of Lateran V (figure 8.5). A bull of Leo X published a few years later includes a clumsy attempt to reproduce the *rotula* in a square constructed entirely of type.[67] But the vast majority of Julian bulls make no effort to reproduce the *rotula* at all.

Under Clement VII (1523–34), printed bulls would regularly include a woodcut *rotula*, as well as new woodcut images of Peter and Paul, a nod to the traditional portrait heads of the saints stamped onto the lead *bulla* of the original; by the middle of the century, the device was commonplace.[68] And in the late 1520s, Roman printers began to print certain bulls as broadsides for posting on doors and walls, just as manuscript bulls had long been displayed. But all this lay in the future. During Julius's reign, most of the visual and graphical elements that printers introduced in their editions of papal bulls had no relation to manuscript precedents. They were entirely new, unique to the printed book, the result of conscious choices to depart from both manuscript and earlier printing conventions.

Printers chose to use Julius's personal *stemma* (an oak tree with intertwining branches hung with acorns), not the *rotula* or the figures of Peter and Paul, as the authenticating image for each printed bull, and to give the papal arms far greater prominence on the cover of the pamphlet than earlier authenticating signs had held on the original document. The arms visually branded the front of the pamphlet—the most visible part of the book when displayed on booksellers' stalls or passed from reader to reader—with the most recognizable emblem of the reigning pope.

The descriptive titles that printers added to their editions further set them apart from their manuscript originals. These titles, which do not vary from edition to edition or printer to printer, seem to have originated in the chancery, not the printshop. Starting as small headlines above the start of the text, by the end of Julius's reign these were lengthy and appeared on their own title page, along with the pope's *stemma* and, sometimes, a woodcut border. The titles rapidly communicated the most important information about the bull to anyone who held it.

Still, neither secretaries, nor chancery officials, nor printers tried to abbreviate the dense text of the bull itself. To the contrary, the quarto format allowed the text of a bull to run to several pages, leaving room for more rhetorical amplification than a single sheet of parchment might. Some pamphlet bulls run to two or more quires. The composers of bulls were able to expand dramatically on their theme; from a compact legal proclamation, political bulls grew into expansive rhetorical productions where the pope could elaborate on his position and address discursive appeals to his readers. Julius's *Monitorium* against the Venetians, published in multiple editions in the spring of 1509, runs to sixteen pages in the Gothic-type edition of Eucharius Silber, and twenty-four pages in Mazzocchi's edition in a larger roman font.[69] One German translation of the document is thirty-two pages long.[70] Mazzocchi's edition of the 1510 bull stripping Alfonso d'Este of his dukedom is also thirty-two pages long. The pope's 1511 bull convening the Fifth Lateran Council, published in eleven different Latin editions, fills anywhere from twelve to twenty pages; an Italian translation is sixteen pages long, while two editions of a German translation published in Leipzig run to twenty pages each.[71] Chroniclers and diarists refer to manuscript copies of these documents being posted on church doors; hence, these very long texts must have been copied out onto parchment at least once, but it seems that the secretaries and scriptors who drafted bulls for Julius were beginning to think of them as texts for publication in print—pamphlets, not parchments.

We can trace the emergence of the new style of Julian bull over the early years of his reign. Johannes Besicken printed Julius's bull against Giovanni Bentivoglio of Bologna in Rome on November 12, 1506, a month after Julius had issued the text in Forlì. Besicken's bull is not all that different from those printed for Sixtus IV a quarter century before. The 1506 pamphlet bears a title, and that title includes a bit of editorializing ("Bull of Pope Julius II issued against Giovanni Bentivoglio, who is oppressing the liberty of the Church in the city of Bologna").[72] Where the original manuscript bull would have displayed the name of the pope in exaggeratedly tall letters at the start of the document, the printed pamphlet has "JULIUS" in capital letters centered over the start of the text, and a woodcut initial *A* (for the formulaic opening phrase, *Ad futuram rei memoriam*). On the whole, it is a pretty plain piece of printing (figure 8.6). At the conclusion, the final formulaic phrases of the bull text are omitted (the usual sentence that begins *Si quis autem* is replaced with a simple "etc."); omitted also are the date of the document and any information about its publication except that it was issued in Forlì. Besicken does include his own name in the colophon, which also gives the date of printing (not the date of the document): November 12, 1506. In short, this is a printed pamphlet that reveals almost nothing of its origins in manuscript, but neither does it exploit the medium of print in any creative way: no variation in type sizes, almost no variation in typesetting, and no graphic elements beyond a single woodcut initial.[73]

As Julius's reign progressed, printers moved beyond Besicken's simple typography, experimenting with new forms of authenticating imagery and text. In 1510 or 1511, Julius's 1506 bull against the Bentivoglio was republished in Rome by Johannes Beplin, who seems to have used Besicken's 1506 edition as his base text (figure 8.7).[74] The Beplin edition is an altogether more sophisticated piece of printing: the title appears centered over the top of the first page in a mix of large Gothic type and the smaller roman type used for the text. Below it, a large, thickly bordered, criblé woodcut of Julius's *stemma* of the della Rovere oak, surmounted by a papal tiara, backed by crossed keys, and festooned with tasseled cords, swirling ribbons, and the episcopal pallium, takes up most of the rest of the page; underneath, the text is prefaced by "IVLIVS . &c" centered on its own line, and then starts with an elaborate, seven-line, fretty woodcut *A* for *Ad futuram rei memoriam*.

The transition from Besicken's edition to Beplin's, accomplished in just

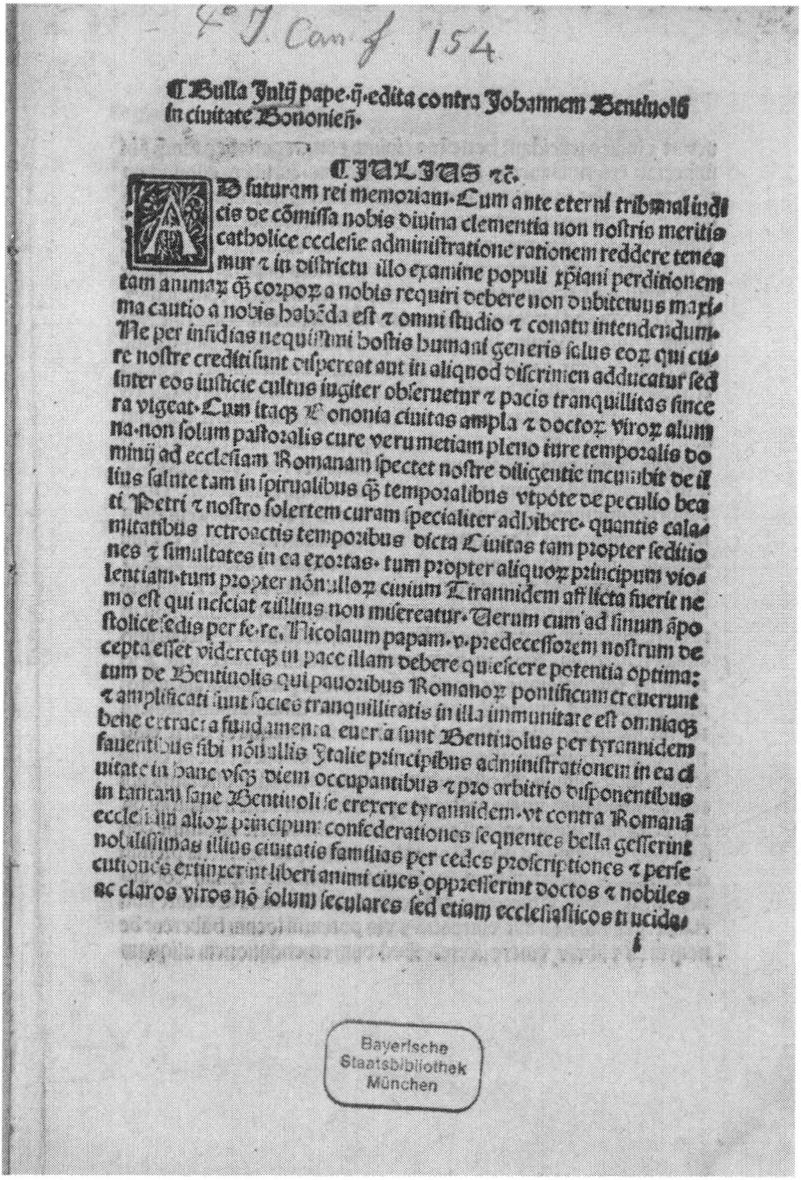

4° J. Can. f. 154

¶ Bulla Julij pape·ij·edita contra Johannem Bentiuoli in ciuitate Bononieñ·

CIVLIVS ꝛc.

B futuram rei memoriam·Cum ante eternl tribunal iudi cis de cõmissa nobis diuina clementia non nostris meritis catholice ecclesie administratione rationem reddere tenea mur ꝛ in districtu illo examine populi xpiani perditionem tam animaꝝ ꝗ corporꝝ a nobis requiri debere non dubitemus maxi ma cautio a nobis habẽda est ꝛ omni studio ꝛ conatu intendendum· Ne per insidias nequissimi hostis humani generis salus eoꝝ qui cu re nostre crediti sunt esperear aut in aliquod discrimen adducatur sed inter eos iusticie cultus iugiter obseruetur ꝛ pacis tranquillitas since ra vigeat·Cum itaꝗ ꝝ ononia ciuitas ampla ꝛ doctoꝝ viroꝝ alum na·non solum pastoralis cure verumetiam pleno iure temporalis do minij ad ecclesiam Romanam spectet nostre diligentie incumbit de il lius salute tam in spiritualibus ꝗ temporalibus vtpote de peculio bea ti Petri ꝛ nostro solertem curam specialiter adhibere· quantis cala mitatibus retroactis temporibus dicta Ciuitas tam propter seditio nes ꝛ simultates in ea exortas· tum propter aliquoꝝ principum vio lentiam·tum propter nõnulloꝝ ciuium Tirannidem afflicta fuerit ne mo est qui nesciat ꝛillius non misereatur·Uerum cum ad sinum apo stolice sedis per se·re, Nicolaum papam·v·predecessorem nostrum de cepta esset videretꝗ in pace illam debere quiescere potentia optima: tum de Bentiuolis qui pauoribus Romanoꝝ pontificum creuerunt ꝛ amplificati sunt sacies tranquillitatis in illa immunitate est omniaꝗ bene extracta fundamenta euersa sunt Bentiuolus per tyrannidem fauentibus sibi nõnullis Jtalie principibus administrationem in ea ci uitate in hanc vsꝗ diem occupantibus ꝛ pro arbitrio disponentibus in tantam sane Bentiuoli se erexere tyrannidem·vt contra Romanã ecclesiam alioꝝ principum confederationes sequentes bella gesserint nobilissimas illius ciuitatis familias per cedes proscriptiones ꝛ perse cutiones extinxerint liberi animi ciues oppresserint doctos ꝛ nobiles ac claros viros nõ solum seculares sed etiam ecclesiasticos ti ucide·

Bayerische
Staatsbibliothek
München

Figure 8.6. Julius II, bull against Giovanni Bentivoglio (Rome: Johannes Besicken, 1506). Munich, BSB, 4 J.can.f. 154, fol. 1r.

Bayerische
Staatsbibliothek
München

Figure 8.7. Julius II, bull against Giovanni Bentivoglio ([Rome: Johannes Beplin, ca 1510?]). Munich, BSB, 4 J.can.f. 155, fol. 1r.

four years, occurred during one of the most tumultuous periods in the history of the Renaissance papacy: a sequence of wars, diplomatic crises, alliances made and broken, interdicts leveled, princes and cardinals excommunicated in quick succession, and the prospect of two rival councils that threatened to throw the Roman Church into schism for the first time in a century. Julius used his bulls to manage (and at times provoke and prolong) these crises, even as he also continued to administer the more quotidian affairs of the Church.

The first signs of a stylistic change date to 1508 and raise some doubts about the dating of certain editions of Julian bulls and other texts connected to his papacy. In 1508, Eucharius Silber published an edition of the *Indulgentiae ecclesiarum principalium urbis Romae*, the popular pilgrim guide (discussed in chapter five), with a colophon indicating the year of printing, but not the month or day.[75] After the glut of *Indulgentiae* editions produced for the Jubilee of 1500, Roman printers held off producing new editions for some time. This seems to be the first Silber had produced since 1500, and, oddly, it features the *stemma* of Alexander VI in the usual position, despite the fact that the Borgia pope had been dead for five years. Had Silber still not got around to commissioning a cut for the new pope's arms? In fact, he had, but the *Indulgentiae* edition gives us a likely date *terminus post quem* for just when he had the new cut made. Silber included a woodcut of Julius's arms in several editions of Julian documents from around this time. In 1507, Julius issued a bull establishing a new class of apostolic scriptors, venal offices he proposed to sell to raise money for his military campaigns. Silber printed this text with a title in larger Gothic type and a small, quite simple new woodcut of the pope's *stemma* on the first page (figure 8.8).[76] The document was dated 1507, but the edition itself is not dated. It is unlikely that Silber printed it before the edition of the *Indulgentiae* dated 1508. Had he done so, it would be hard to explain why he then produced an *Indulgentiae* with the *stemma* of the previous pope. More likely, the 1507 bull was printed a year or more after its issue, and after the *Indulgentiae* of 1508.

Very similar was Silber's edition of *In coena Domini* for 1508. The annual publication of excommunications had yet to be printed in Julius's reign. Silber brought out an edition of the bull for 1508 with a headline title and the same small woodcut *stemma* as appears in the bull on scriptors.[77] The woodcut also appeared in two Silber editions of texts delivered to or before the pope in 1508.[78] Silber then issued another three editions of the *Indulgentiae* with the new Julian cut: one dated 1509 without indication of the day or month; one dated October 7, 1509; and one dated 1511.[79]

The woodcut also appeared in Silber's 1509 edition of Julius's *Monitorium* against Venice. Silber would reuse it in several more Julian publications over the next few years: in the bull stripping Alfonso d'Este of his dukedom in 1510;[80] in his bull of interdict against Milan issued later that same year;[81] in

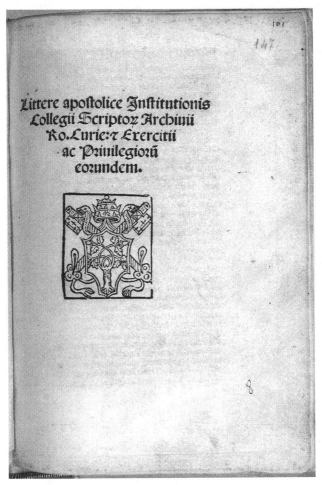

Figure 8.8. Julius II, *Littere apostolice institutionis Collegii scriptorum Archivii Romane Curie* ([Rome: Eucharius Silber, 1508?]). Rome, Biblioteca Nazionale Centrale, 1875.31(8), fol. 1r.

the bull levying further penalties against Alfonso, Charles d'Amboise, and other military captains serving France;[82] in the 1511 bull announcing the start of the Lateran Council;[83] in his bull of warning against the three dissident cardinals, calling them back from the Pisan Council to attend the Lateran Council instead;[84] in an oration delivered in Rome by Angelo da Vallombrosa on September 8 of the same year condemning the Pisan Council and praising the Lateran assembly;[85] and in the brief Julius sent in October to the Christian princes explaining why he had demoted the cardinals adhering to Pisa.[86]

The woodcut would continue to appear in Silber editions published in 1512 and even in one volume published after the pope's death in 1513.[87]

Meanwhile, Silber's innovation attracted an imitator: Etienne Guillery, who had arrived in Rome a few years earlier and worked for Besicken before striking out on his own.[88] Guillery would frequently print the same texts as Silber. His first editions date to 1508–9 and include an edition of the same bull *In coena Domini* as Silber published, with a headline title in roman type and his own version of a small woodcut papal *stemma*. He likely copied Silber's design. But the influence could go both ways: in 1509, Guillery published an edition of Leonardo Loredan's letter begging Julius to show mercy to the defeated Venetians, complete with his own address to the reader; Silber then republished both texts—including the preface by Guillery—in an edition of his own.[89] This was not piracy but rather the result of the near-constant collaboration that characterized Roman printing in these decades. All five major printers active in Rome in Julius's pontificate regularly reprinted documents the others had printed first and often exchanged types and woodcuts. It is nearly impossible to declare the authorship of any edition unless it is signed (and even then, misattributions are possible). At any rate, Guillery's little woodcut of the papal *stemma* would reappear in his editions of the Julian reissue of Pius II's bull *Execrabilis* (after December 17, 1510), of the bull announcing Lateran V (*Bulla intimationis*, July 18, 1511), and of the *Bulla monitorii* against the three dissident cardinals (July 28, 1511), as well as in an undated edition of Julius's bull of January 14, 1505, on simoniacal elections, which was reissued and reprinted several times in Julius's pontificate in increasingly complex editions.

Another innovation was the title page itself. As the fifteenth century turned into the sixteenth, printers everywhere were experimenting with a more complex style of *mis-en-page*: title pages, frontispieces, decorative borders, multiple fonts, and shaped typography.[90] All these elements were adopted in Rome too, perhaps a little later than in Venice, Paris, or Nuremberg. But Rome's embrace of the possibilities, when it happened, was dramatic; title pages were used not just for literary, scholarly, or legal and administrative publications but also for the pope's political communiqués.

Of the five editions of papal documents Guillery printed with his little Julian *stemma*, four start the text of the bull on the first page, underneath a title in shaped text and the *stemma*. Only the *Bulla monitorii* of July 28, 1511, has a true title page, on which title and *stemma* appear alone. Hereafter,

Guillery's editions increasingly featured true title pages. About half of Silber's editions with his little *stemma* bear true title pages as well.

In about 1510, either Johannes Beplin or Marcellus Silber, now operating his father's shop, introduced a new woodcut, the more complicated criblé woodcut that featured in Beplin's edition of the bull against the Bentivoglio.[91] This would appear in sixteen Silber editions between 1510 and Julius's death in February 1512: bulls assailing the Pisan Council, bulls publishing the proceedings of Lateran V, texts by others celebrating and supporting Lateran V, and decrees concerning political topics other than the council.[92] Following this, Guillery commissioned his own criblé cut of the Julian *stemma*, the bottom of the shield rounded rather than pointed, which appeared in his *Bulla super sententia privationis* against the three dissident cardinals (figure 8.9).[93] Guillery's rounded shield also appears in his *Bulla privationis* against Cardinal René de Prie. Finally, in these same years Silber introduced two large woodcut borders for his title pages, both full-page cuts in the form of architectural frames, with which he sometimes surrounded the title of either a papal bull or other text, sometimes with the woodcut *stemma* of the pope included as well (figure 8.10).[94]

Between 1510 and 1511, then, with the emergence of a new generation of printers in Rome, the standard way of presenting a Julian bull was fixed: a large woodcut of the Julian arms, dominating a title page with a descriptive title in large or shaped type, which always included the pope's name, and sometimes a woodcut border as well. This model traveled: the Nuremberg printer Johann Weissenburger had a cut of the Julian *stemma* modeled so closely on that of Marcellus Silber that his editions have sometimes been mistaken for Silber's (see figure 8.15, below).[95] Weissenburger also had a second cut of the Julian *stemma* made, which he used in other editions of Julian bulls.[96] The Venetian printer Franciscus Lucensis printed a small della Rovere oak tree, on a plain shield, on the title pages of his Julian bulls (printed after Julius had made his peace with the Venetians in 1510), alongside the winged lion of St. Mark—a politic move for a printer publishing in the independent republic?[97]

The use of the papal *stemma* would so come to define printed texts as papal—and thus as important—that it even spread into other texts that were not, strictly speaking, papal decrees. A German pamphlet announcing indulgences, entitled *Ablassbüchlein* and published sometime after the accession of Leo X in 1513, bears the *stemma* of the new Medici pope, six balls sur-

Bulla super sententia priuationis in publico confistorio facte p S.D.N. Contra.D.Ber.Caruaialé Guillermú Brizonettú et Fráciscú de Borgia olim S.R.E. Cardinales ad futuram rei memoriam.

Figure 8.9. Julius II, *Bulla super sententia privationis* ([Rome: Etienne Guillery?, 1511]). Munich, BSB, Res/4 J.can.f. 26#Beibd.28, fol. 1r.

mounted by a circle of lilies, but the volume is credited to Julius II.[98] This was an odd enough mix-up (we might expect a printer to be recycling an older cut under the name of a new pope, not vice versa), but stranger still, the pamphlet does not present a papal bull or brief at all, but a German translation of the *Stationes Romae*—one of the many pilgrim texts for which woodcut papal arms had been devised in the first place. Here, the traditional pilgrim guide

Figure 8.10. Julius II, *Bulla super privatione Alfonsi ducis Ferrariae* ([Rome, 1510]). Munich, BSB, 4 J.can.f. 157, fol. 1r.

was recast to conform to the new design, only recently devised, that usually announced the arrival of a papal bull.

Titles were another new element in these early sixteenth-century bulls. Traditional manuscript bulls did not bear titles at all. The ones printed under Julius have titles that not only describe their contents but also editorialize on the topic. The bull of excommunication against Giovanni Bentivoglio further qualified him as one "usurping the authority of the church in Bologna."[99] The *Bulla monitorii* of 1511 targeted "the three reverend cardinals, that they should return to the obedience of our most holy lord the pope, lest schism arise in the holy church of God."[100] The title of the bull of interdict against Milan in 1510 charged that authorities in the duchy had "usurped churches and other ecclesiastical offices and seized and spent the income of the same, without cause, through the exercise of lay power and lay abuse."[101] The very long title given to the 1510 bull excommunicating Charles d'Amboise, commander of the French troops in Milan, made reference to a previous bull of warning, as well as Charles's current offenses against the Church: "A bull declaring the incursion of those censures and penalties set out in the bull of privation of Alfonso d'Este, at that time duke of Ferrara, against Lord Charles d'Amboise, grand master, and other captains and officers of the army of the most Christian king of France, and generally against all who in the defense or aid of the said Alfonso d'Este take up arms against our most holy lord and the holy Roman Church and attack its territories and towns and despoil them."[102] These titles, while hardly snappy, provided readers with a simple and legible précis of the contents of each bull; emblazoned on the title page in large or shaped type, they made clear both the target of the bull and the pope's rationale for leveling penalties against him.

After 1511, Julius's contempt for the goings-on in Pisa was vividly expressed in the titles of the bulls he issued against that assembly. His *Bulla ultima* before Lateran V issued an invitation to the absent cardinals and prelates of France to attend the Lateran Council, "together with a declaration of nullity regarding the things done at the Pisan *conciliabulum*."[103] The bull issued after the second session of Lateran V likewise "approves and renews the condemnation and repudiation of the Pisan *conciliabulum*, and annuls each and every thing that has been done or will be done there."[104] A flash of irony was even possible: "A letter to the Christian princes presenting the most compelling reasons (*although there are many others*) for stripping the heretical and schismatic cardinals of their offices."[105]

New Texts

The Julian chancery also had new types of documents printed to support the pope in his various conflicts. The letter Julius sent to Maximilian in the spring of 1509, mentioned above, is unusual in being the only single-sheet papal document printed in Rome that survives from Julius's reign (figure 8.11). It was also unusual for being a brief, not a bull. In it, Julius invites Maximilian to join him in attacking the Venetian Republic in order to reclaim the papal cities of Rimini and Faenza.[106] "At the start of our reign, the Venetians,

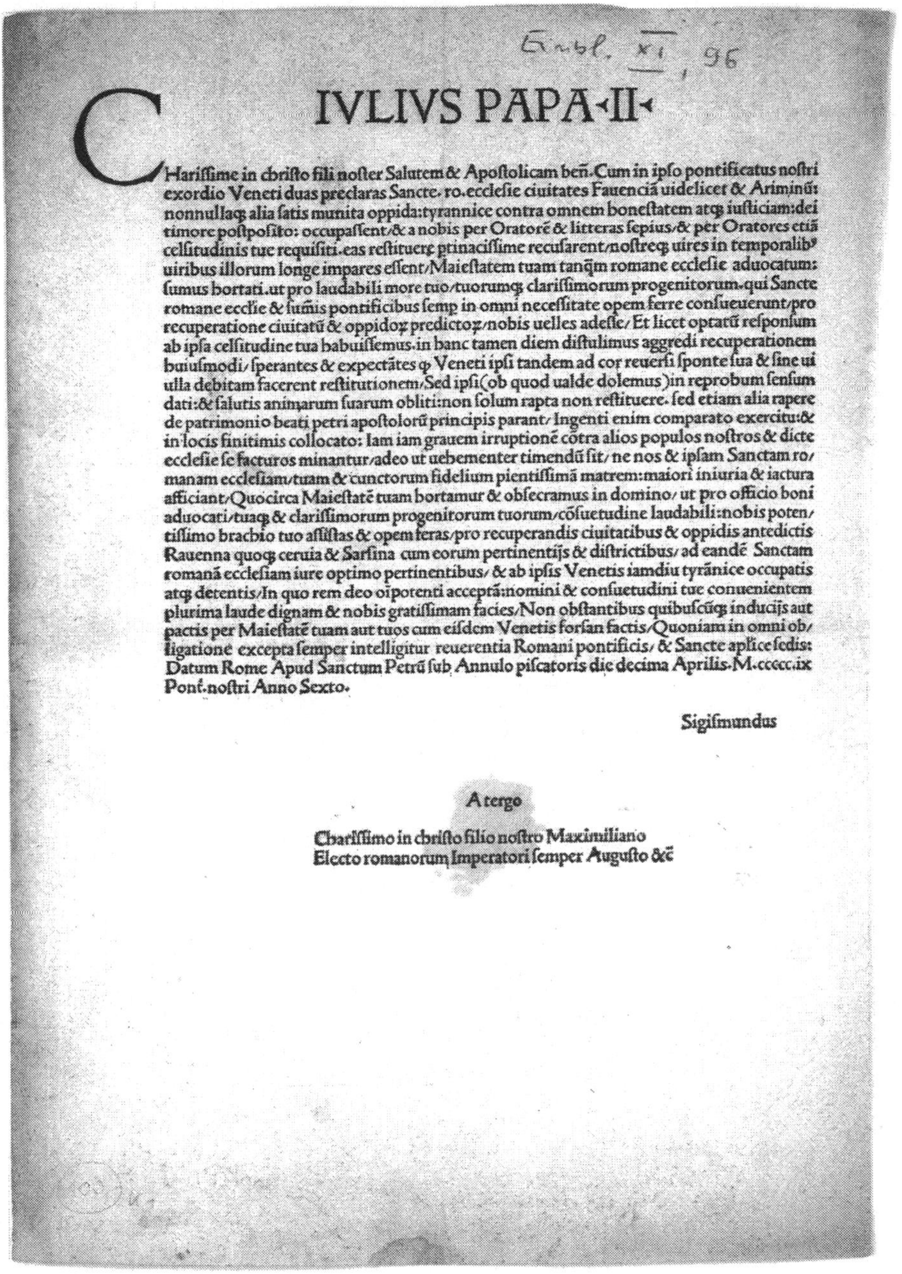

Einbl. XI, 96

IVLIVS PAPA·II·

CHariſſime in chriſto fili noſter Salutem & Apoſtolicam ben̄.Cum in ipſo pontificatus noſtri
exordio Veneti duas preclaras Sancte, ro.eccleſie ciuitates Fauenciā uideſicet & Arimin̄:
nonnullaꝗ alia ſatis munita oppida:tyrannice contra omnem honeſtatem atꝗ iuſticiam:dei
timore poſtpoſito: occupaſſent,& a nobis per Oratorē & litteras ſepius,& per Oratores etiā
celſitudinis tue requiſiti.eas reſtituere̱ ptinaciſſime recuſarent,noſtreꝗ uires in temporalib⁹
uiribus illorum longe impares eſſent,Maieſtatem tuam tanꝗm romane eccleſie aduocatum:
ſumus hortati.ut pro laudabili more tuo,tuorumꝗ clariſſimorum progenitorum.qui Sancte
romane eccſie & ſumis pontificibus ſemp in omni neceſſitate opem ferre conſueuerunt,pro
recuperatione ciuitat̄ū & oppidoꝗ predictoꝗ,nobis uelles adeſſe, Et licet optat̄ū reſponſum
ab ipſa celſitudine tua habuiſſemus.in hanc tamen diem diſtulimus aggredi recuperationem
huiuſmodi, ſperantes & expectātes ꝗ Veneti ipſi tandem ad cor reuerſi ſponte ſua & ſine ui
ulla debitam facerent reſtitutionem,Sed ipſi(ob quod ualde dolemus)in reprobum ſenſum
dati:& ſalutis animarum ſuarum obliti:non ſolum rapta non reſtituere, ſed etiam alia rapere
de patrimonio beati petri apoſtoloꝛū principis parant, Ingenti enim comparato exercitu:&
in locis finitimis collocato: Iam iam grauem irruptiōne cōtra alios populos noſtros & dicte
eccleſie ſe facturos minantur,adeo ut uehementer timend̄ū ſit, ne nos & ipſam Sanctam ro,
manam eccleſiam,tuam & cunctorum fidelium pientiſſimā matrem:maiori iniuria & iactura
afficiant,Quocirca Maieſtatē tuam hortamur & obſecramus in domino, ut pro officio boni
aduocati,tuaꝗ & clariſſimorum progenitorum tuorum,cōſuetudine laudabili:nobis poten,
tiſſimo brachio tuo aſſiſtas & opem feras,pro recuperandis ciuitatibus & oppidis antedictis
Rauenna quoꝗ ceruia & Sarſina cum eorum pertinentiſs & diſtrictibus, ad eandē Sanctam
roman̄ā eccleſiam iure optimo pertinentibus, & ab ipſis Venetis iamdiu tyrānice occupatis
atꝗ detentis,In quo rem deo o̅ipotenti acceptā:nomini & conſuetudini tue conuenientem
plurima laude dignam & nobis gratiſſimam facies,Non obſtantibus quibuſcūꝗ inducijs aut
pactis per Maieſtatē tuam aut tuos cum eiſdem Venetis forſan factis,Quoniam in omni ob,
ligatione excepta ſemper intelligitur reuerentia Romani pontificis, & Sancte apl̄ice ſedis:
Datum Rome Apud Sanctum Petr̄ū ſub Annulo piſcatoris die decima Aprilis.M.cccc.ix
Pont̄.noſtri Anno Sexto.

Sigiſmundus

A tergo

Chariſſimo in chriſto filio noſtro Maximiliano
Electo romanoruṃ Imperatori ſemper Auguſto &c̄

Figure 8.11. Julius II to Maximilian, April 10, 1509 ([Rome: Eucharius Silber, 1509]).
Munich, BSB, Einblatt. XI,96.

against all sense of honor or justice, and forgetful of the fear of God, tyrannically occupied these two cities belonging to the Holy Roman Church," Julius explains. He then reminds the emperor of his duty to serve and protect the bishop of Rome and asks him to come now in the pope's hour of need.

The brief was a documentary form that developed in the fifteenth-century papal chancery as a shorter (hence "brief") and less formal communication than a papal bull. The original would have been sealed not with a lead bull but with wax imprinted with the "fisherman's ring" worn by the pope, just as the subscription on the printed broadside states (*sub anulo piscatoris*), and it would have been signed by the pope's secretary and registrar of correspondence, Sigismondo dei Conti, whose name appears on the document as well. The printed sheet even reproduces notes on how the original brief was folded and addressed (*a tergo . . .*). In other words, this was another printed publication taking pains to invoke the appearance of its manuscript original.

The letter to Maximilian may have been issued in print in order to appeal to popular opinion. But whose? Who ought to know that the pope was inviting the emperor to attack Venice? The people of Romagna might take comfort knowing that the pope's campaign was likely to be successful, given his powerful ally. The Venetians might be demoralized by the same news. There was even more to it than this. When the European powers formed the League of Cambrai in December 1508, Julius was not among the original signatories. He held out for better terms, including the stipulation that Maximilian would respect the pope's sovereignty over the northern papal states by not invading until the pope invited him in. This precondition induced Julius to join the league on March 23, 1509; he then issued his "invitation" to Maximilian on April 10. The brief inviting the emperor to attack Venice presents the pope as the master of Italy, without whose permission no foreign prince could set foot on Italian soil. This broader message was one Julius was most eager to publicize. Copies of the letter circulated not only in Rome and Bavaria (where a German translation was printed) but also in Romagna, informing residents of the captive cities that help was on the way, that loyalty to Venice was pointless and even dangerous, that the pope had powerful friends, and that Venice was isolated and in peril.

Other actors in the drama used print in similar ways. After the victory at Agnadello, Maximilian had printed letters circulated among the Venetians encouraging them to overthrow their government and embrace imperial rule. And poets in Rome and elsewhere produced panegyrics praising the pope for his conquests. After 1510, when Julius reconciled with Venice, joining forces to combat the northern powers, the pope's sometime allies did not hesitate to target him with print. Gringore and Lemaire de Belges published mocking critiques of the pope in Paris, as we have seen. By 1518, Maximilian himself had broken with the papacy and issued an open letter to the Italian cities—in print—excoriating Leo X for ignoring Maximilian's rights on the peninsula. The emperor invited the city-states of Italy to throw off the papal yoke in favor of imperial protection.

Finally, the Julian chancery extended its branding in print to the more prosaic category of administrative bulls. Here, too, the variety and sheer number of publications grew far beyond that of any previous pontificate. Julius published rules for the chancery, lists of officers and magistrates in the city and papal states, schedules of fees for services rendered by different papal departments, bulls condemning simony in future papal elections, bulls regulating fees paid by episcopal appointees, and bulls setting out privileges and immunities for different ranks of curial officials. He published regulations for the colleges of scriptors and abbreviators, a reform of curial offices, an attempt to reorganize the prosecution of legal cases in Rome, regulations for drawing up wills and bequests of property, and bulls condemning bandits, thieves, and anyone violating the immunities of ecclesiastical personnel. Such publications had been commonplace since the popes had returned to Rome from Avignon and resumed the government of the papal states. What was new now was their regular appearance in print and the innovations in format they displayed. They were published in quarto, with title pages bearing the Julian coat of arms and a title in larger type indicating the contents of the pamphlet, precisely the same format as his more topical political bulls.

These publications also projected images of papal authority. A bull penalizing those who violate ecclesiastical immunities, published by Silber around 1510, presents an extraordinary image of the pope, in full regalia, extending his hand over an assemblage of stars and a dove (the Holy Spirit?), while a papal tiara floats, oddly, over a royal crown behind him (figure 8.12). The image derives from the cycle of papal portraits associated with the *Vaticinia de summis pontificibus*, a set of prophecies dating to the late thirteenth century, attributed at times to Joachim of Fiore, which circulated much in manuscript. The prophecy associated with this image refers to the pope illuminating the celestial bodies until one flashes more brightly than all the others. In some traditions of the *Vaticinia*, each prophecy is assigned to a particular pope, and this image and prophecy were sometimes associated with Benedict XII (1334–42).[107] There is no connection at all between the reign of this

Figure 8.12. Julius II, *Bulla interdicti ecclesiastici contra receptatores rebellium* ([Rome, 1510]). BAV R.I.IV 1811(19), fol. 1v.

Avignon pope and Julius II. The cut seems to have been chosen simply to assert that the document was a papal one. (The presence of the woodcut in this edition also suggests that a printed edition of the *Vaticinia* may have been published, or in preparation, in Rome, but no evidence of such an edition survives.)[108]

Lateran V

Julius's various rhetorical, legal, and presentational strategies reached their apex in the suite of fourteen bulls, printed in forty-three different editions, issued in connection with Lateran V. The Julian chancery published these over the last seventeen months of the pope's life, from his *Bulla intimationis* announcing the council on July 18, 1511, to the *Cedula quintae sessionis* published on February 16, 1513, just five days before Julius's death.[109]

Julian bulls proclaiming the council, attacking and excommunicating opponents of the council, and publicizing the work of the council—from positive reform measures to yet more sanctions against political opponents—appeared in print alongside dozens of other editions of orations, sermons, and speeches by clerics, legates, and other dignitaries at the council. Several Roman printers produced series of these bulls, although Marcellus Silber was the most frequent and possibly the original publisher; each used his own consistent visual and textual apparatus. These collections of printed texts carried the deliberations and the very *fact* of the council far beyond its meeting rooms at the Lateran, underscoring its legitimacy and superiority over any rival assembly, and branding it as a particularly papal enterprise as well.

The effort was also a response to a rival publication campaign that Louis XII was supporting in Paris. Propagandists like Lemaire de Belges and Gringore were not the only authors creating antipapal content in 1511–12. Louis was also sponsoring a group of theologians at Paris and elsewhere in a vast academic project of locating, compiling, providing commentary for, and then printing the *acta* and other documents from previous councils, including Pisa I, Constance, and Basel.[110] It seems clear that the motivation to issue and then print a papal bull after every session of Lateran V—and to print each text *as* a bull, a solemn decree issued by Julius himself, not a set of conciliar proceedings issued by the assembly as a whole—came as a response, at least in part, to this ambitious French publishing campaign.

Another innovation was the recycling or reissue of older bulls and documents. Louis XII's council aimed at securing him more control over revenues and appointments in the Church in France, moving toward the establishment of something like a Gallican Church.[111] In response, the Julian chancery dug into the archives, reissuing and then printing documents that had been issued against earlier challengers. These included Pius II's 1460 bull *Execrabilis*, outlawing appeals to a general council by anyone other than the pope;[112] Louis XI's 1462 renunciation of the Pragmatic Sanction of Bourges;[113] a decree of the Council of Constance, in which the right to call a future council was vested exclusively in the pope;[114] and even the *Donation of Constantine*, the supposed foundation of all papal claims to temporal power over the Roman Empire in the west.[115] These reissues of earlier ecclesiastical legislation (real or confected) appeared in the same types and formats as the new bulls. There

Figure 8.13. Julius II, bulls and a brief for the Fifth Lateran Council, all printed in Rome by Marcellus Silber, 1511–12. Munich, BSB, 4 Conc. 97a; 4 Conc. 98; 4 J.can.f. 151, all fol. 1r.

was no legal reason at all to reissue these; they appear in print solely to remind the public that the question had been settled and needed no reopening.

When the council opened in May 1512, there was an established format for printed papal bulls. Julius's *Bulla intimationis* of 1511 set the basic pattern: quarto format, papal coat of arms, a title in large type, and even, at the end, a primitive woodcut approximation of the papal *rotula*, as well as a record of all the cardinals who had signed the original bull, with a woodcut cross next to each name. A similar bull (but without the signatures) was issued after every session of the council once it opened: *secunda, tertia, quarta,* and *quinta.* The title page of the bull for the first and second sessions included a description of what was discussed. By the third, fourth, and fifth sessions, the descriptive titles had fallen away; the title pages for these bulls simply indicate the number of the session, along with the papal coat of arms (figure 8.13).

At the same time, other publications related more or less explicitly to the council were also appearing in the same format: briefs from Julius to the emperor and other Christian princes, reissues of historical documents, and sermons and orations delivered by other clerics at the council itself. Silber published the oration delivered by Egidio da Viterbo at the third session of the council in exactly the same style as Julius's own conciliar bulls, with the same large Gothic type on the title page and criblé woodcut of Julius's arms.

Silber published another edition of Egidio's oration, along with others delivered at the council by Bernardo Zane, Cristoforo Marcello, and Tommaso da Vio (*alias* Cajetan), as a series of identical pamphlets, all using the same, large architectural border on the title page that Silber had previously used for Julius's bull against Alfonso d'Este, his *Bulla intimationis* announcing the start of Lateran V, his bull warning the cardinals and clergy of France against participating in the Pisan Council, and an administrative bull reforming the offices in the curia.[116] An oration delivered by Simone da Begno at the sixth session of the council, under Leo X, was printed by Silber with a different architectural border, the same one he used for one of Julius's last bulls, issued in the fifth session of the council on the question of future papal elections.[117]

Angelo Fondi da Valombrossa, the Sienese hermit and visionary who wrote and preached extensively against the Pisan assembly on Julius's behalf, had a collection of letters, some addressed to Julius in support of the council and some to Cardinal Carvajal and Louis XII condemning Pisa, printed by Silber in Rome (figure 8.14). Fondi's edition includes a preface to Julius in which he presents himself as the sole driving force behind the publication. He modestly justifies putting his letters in the hands of a "venal bookseller"; he aims not for profit but to spread his words of support for the council and against Pisa as widely as possible; and he hopes that the letters, once printed, will find their way more easily to Julius himself.[118] Even so, Silber published the book with the Julian criblé *stemma* on the cover, in precisely the same style as the pope's official decrees. Even a work that openly declared itself unknown to the pope could be branded by the printer to look like an official publication.

Fondi also published his *Oratio pro concilio Laterano*, delivered in September 1511, with Silber, who included the Julian *stemma* on its title page as well. This edition traveled: the *Oratio* was printed again in Nuremberg, where Johann Weissenberger had made an almost identical copy of the criblé *stemma* cut that he used to adorn editions of both conciliar decrees and texts by conciliar supporters (figure 8.15).[119]

Weissenberger used this woodcut, and a second that he also commissioned, for a series of pamphlet publications about the council, including the *Bulla intimationis* of 1511 and *Bulla ultima* of 1512 (figure 8.16). Interestingly enough, he also printed a copy of the Pisan *Convocatio*, or citation, the document Louis XII put his cardinals up to issuing as a prelude to the Pisan Council. A standard typographic and decorative vocabulary was emerging—a form of branding—that allowed any document connected with the pope and his council, or any council for that matter, to be recognized as such.

Epistole Angeli Anachorite Uallisumbrose.

℮ Iulio Secundo Pont.Max.

℮ Francorum Regi.

℮ Bernardino : tunc Cardinali sanctæ Crucis.

Pro Christiana vnitate seruanda.

Figure 8.14. Angelo Fondi da Vallombrosa, *Epistole* ([Rome: Marcellus Silber, after October 25, 1511]). Hesburgh Libraries, University of Notre Dame, Special Collections, Rare Books Medium BX 4705. A5927 A4 1511, fol. 1r.

Figure 8.15. Angelo Fondi da Vallombrossa, *Oratio pro concilio Lateranensi* ([Nuremberg: Johann Weissenburger, 1511]). Regensburg, Staatliche Bibliothek, 999/Patr.544f, fol. 1r.

Julius had need of these auxiliary shows of support, for Louis XII had his own publicists and theorists working on antipapal orations, dialogues, and tracts, which were put into print almost as fast as the supporters of Lateran V could publish theirs in praise of the assembly. Throughout 1511 and 1512, conciliar theorists published tracts and *consilia* arguing for the supremacy of the

Figure 8.16. Julius II, *Bulla ultima invitationis* ([Nuremberg: Johann Weissenburger, 1512]). Munich, BSB, 4 Conc. 102, fol. 1r.

council over the pope. Zaccaria Ferrerio, whom Louis had tasked with editing the *acta* of previous councils (published by Jean Petit in Paris), composed his own *Apologia sacri Pisani concilii* in September 1511.[120] Filippo Decio, the canonist who provided Louis XII with much of the legal reasoning for convening the Pisan Council, issued a *Consilium . . . pro reverendissimis cardinalibus*, encouraging the entire College of Cardinals to attend. This prompted a barrage of pamphlets from Angelo Fondi and Cajetan out of Rome. The former composed an *Apologeticum* for Julius in the autumn of 1511, printed by January of the next year; the latter produced a tract *De auctoritate papae et concilii* in October 1511. Fondi, finding certain flaws in Cajetan's reasoning, then published a letter against him (February 1512), as well as another attack on Louis: *Exhortatoria ad regem Franciae ut cesset persequi pontificem et ecclesiam*, in March 1512.[121] These publications, again clad in the livery of Julian ecclesiastical decrees, hammered home the notion that only the pope's council could be considered legitimate. All arguments in its favor were issued under the aegis, and with the apparent blessing, of the pope.

The length of Julian bulls, the innovations in their visual presentation, and the multiplying types of document that were put into print all suggest how innovative Julius's secretaries, chancery, and printers could be in their use of the technology. Printing changed the way bulls read, how they looked, and how they were used, as they increasingly were directed not just against other political actors but also to the court of public opinion. By applying consistent graphic and typographical conventions, print also allowed peripheral documents by theologians, jurists, and other supporters of the pope to be assimilated to papal decrees, expanding the *corpus* of official, or quasi-official, publications. In this sense, we might say that print functioned here in a truly "revolutionary" way, transforming both the content and the form of highly traditional documents. This has implications for our understanding of the coming conflict between the papacy and Luther, where conventional wisdom holds that it was Lutheran reformers, not the papacy, that first grasped the possibilities of the press.

On the other hand, there is the curious fact that Julian printed bulls grew longer, in part, because they included more of the paratextual apparatus documenting production and promulgation of the original manuscript than previous printed bulls had ever done. Even as their title pages grew bolder, the titles longer and more tendentious, the woodcuts larger and more visually compelling, the interiors of the pamphlets grew cluttered with ever-larger blocks of dense legalese. These paratextual elements could include the signatures of the secretaries who had drafted and then approved the original text; the signature of the official who had registered it; notes on who had read it out in the *Audientia* (where that was a requirement); for very formal bulls, the signatures of the pope and cardinals; for all sorts of bulls, the formulaic paragraphs specifying how the bull took precedence over previous decrees (*Non obstantibus . . .* and *Nulli ergo . . .*) and protecting the bull against interference by others (*Si quis autem* . . .); statements of *publicatio* by heralds detailing how they had posted the original around the city; a statement by the master of heralds authenticating that previous statement; and/or a statement certifying the authenticity of the "copy" (*transumptum*) the reader now held. This last element was not new to printing—there was a set formula that notaries and other chancery officials had long used to assure readers of a manuscript copy of a document that it was an accurate reproduction of the original. When the statement appears in a printed text, it means that this statement was already there in the printer's copy (a manuscript), and that typesetters chose (or were instructed) to reproduce it in print. In fact, this is true of all the above-mentioned elements: all would have been written on various, set parts of the original manuscript bull; different corners of the parchment, as well as different parts of its reverse, were reserved for notes from different officials. None of these elements were added specially to the printed book. What is odd is that earlier printers, in Rome and elsewhere, had somehow seen past or through these paratexts, printing the core text of a bull and little else. Sometimes they even abbreviated the end of the bull itself; many abbreviate or omit entirely the concluding formulas (*Non obstantibus, Nulli ergo*, and *Si quis autem*) that stipulated the validity of the bull and penalties for those who impeded its enforcement. They extracted the core of the bull text from this protective web of procedural verbiage and printed only it.

Under Julius, printers or their patrons began to "see" these elements again and to include them in their publications. Their goal seems to have been to reproduce every possible textual element of the original document. Printers allotted man-hours to setting these passages in type and substantial expanses of paper to printing them. This was no idle choice on their part; unlike the medieval scribe who might keep copying, because bored or distracted, beyond the end of his sample text, printers did not set or print these documentary paratexts by accident. Julius's reissue of *Execrabilis*, forbidding future councils called by anyone but the pope, is only eight pages long, but nearly half of them are devoted to paratexts: the title and a woodcut *stemma* at the start and, at the end, two pages of statements describing how the document was read out and posted in Bologna on three separate occasions. Further notes indicate that Sigismondo dei Conti verified the text and entered it into the chancery register. An edition of the 1511 *Bulla intimationis* announcing Lateran V devotes seven of sixteen pages to paratexts: on the first two pages, a title and woodcut *stemma* and an authentication statement by Girolamo

Ghinucci; on the final five pages, internal publication formulas, Julius's signature, Sigismondo dei Conti's signature, a woodcut-and-type rotula, and two long statements about the document's *publicatio* in Rome. Scattered comments in other printed bulls (perhaps added by a chancery official or by the printer) even specify *where* on the original document these peripheral texts had appeared.[122]

The proliferation of these formulas might seem a sign of insecurity or uncertainty about the status of a printed text, something that secretaries or chancery officials either felt or feared readers might feel in the absence of the original manuscript decree with its lead seal, silk cords, and handwritten markers of authentication. These markers themselves had evolved in late medieval documentary manuscripts as a result of anxiety over forgery, a concern that a genuine document should be able to authenticate itself long after the people who had drafted and signed it were gone.[123] But if Roman officials had concerns about the status of print, we should expect to see them manifest much sooner after the advent of the technology. And yet, in the fifteenth century, papal decrees and instructions were printed with few, or even none, of these elements. They really only appear as major elements of printed publications during the reign of Julius II.

The explanation for their appearance lies not in some new Julian anxiety about print, I think, but rather in political insecurity. The purpose of these bulls—and the purpose of printing them—was to protect papal prerogatives against external threats: secular lords alienating the property of the Church, foreign princes seeking to limit papal power in their lands, the ever-present threat of a new council that might hobble or even depose Julius in favor of a more pliable candidate. Against these opponents the Julian chancery issued its thunderbolts in print, to assert the pope's rights and the ancient foundations upon which they lay. There were various ways to do this. Bold woodcuts and rhetorical titles on the cover were one; making the full weight of the curia's bureaucratic procedures visible to the reading public was another. On this view, it is doubtful whether anyone was actually meant to *read* these lengthy paratexts; few can have worried that a printed Julian bull was forged. But invoking the traditional processes by which a papal decree became law, and the various markers of authority it attracted along the way, was a powerful statement in itself. In this sense, the drifts of dense legal Latin functioned rather more as symbolic text than a conveyor of actual meaning. It was armor for the document, a sign of its inviolability, and the inviolability of the institution that had issued it. The pope who donned a silver breastplate to breach the walls of snowy Mirandola, and thought nothing of demolishing old St. Peter's to replace it with something bigger and better, would fortify his legal decrees with all the verbal reinforcement his bureaucracy could offer. Legal prose became a visual weapon. It also conjured up a scene of authoritative efficiency: Rome was a city of sacred basilicas and doors, but also a city of busy officials, drafting, registering, stamping, posting, guarding, reading out, and printing the pope's decrees. The European powers who dared cross that pope should beware the reach of his grasp, the extent of his authority, the power of his word.

Julian publicity campaigns still had at their heart a highly traditional, even retrograde vision of the Church, its leadership, and their relationship to the faithful. Julius flooded Europe with printed publications that insisted on an ecclesiology of previous centuries, based on principles of papal supremacy over cardinals and temporal princes, on the one hand, and salvation by works (specifically, the purchase of indulgences), on the other. Subsidiary concerns included the defeat of the infidel, the pope's absolute authority over his vassals in the papal states, and the subordination of the council to the pope. These were the ideas that papal print broadcast to the European reading public at the start of the sixteenth century. Even Lateran V, ostensibly convened to reform the Church in head and members, was more immediately concerned with quashing the French king's challenge to papal authority and his rival council at Pisa. The Lateran documents that were printed focused on process: on the need for every cardinal to adhere to Julius's council, on the pope's precedence over secular rulers, on his exclusive right to call a council in the first place, and on his right to condemn the rival assembly.

Leo X

The transition from Julius II to Leo X brought dramatic changes to Rome, as well as to the politics of the Holy See. A triumphal arch at Leo's *possesso* famously proclaimed, "Venus has had her time, and Mars his; now Minerva will have her day." The age of Venus, that of the libidinous Alexander VI, had given way to the endless martial activity of Julius II's long decade on the papal throne. Leo X, the Medici pope, cultured Florentine and patron of arts and letters, would inaugurate a Palladian age of wisdom and culture, a golden mean between the sybaritic indulgence of the one and the explosive violence of the other.

How true was this prediction? Certainly there was a change in tone at the papal court, from Julius's hot temper and readiness to take up arms to Leo's milder, more cultured and diplomatic political character. One continuity between the two pontificates, however, was their shared administration of Lateran V. Leo inherited the council and its goals of reforming the Church, establishing peace in Europe, and promoting a crusade against the Turks. Leo did not alter this basic agenda; he kept the council meeting in periodic sessions until its close in 1516, with many more decrees and declarations printed, the pamphlets now bearing the arms of the Medici pope. Church historians usually consider Lateran V as a single event stretching across the two pontificates. Its decrees were collected and published in a single volume in 1521.[124] The major modern historiographical debates around it have focused not on internal changes from one pontificate to the next but on its overall success or failure: how many prelates actually attended, whether its agenda was realistic, whether its measures were effective, whether they were actually implemented anywhere beyond the city of Rome, and, above all, why this council called to reform the Church did so little to forestall the emergence of the more momentous reform movement that erupted just a year after its close.

The idea that the transition from Julius to Leo saw changes in papal policy and style, and the idea that the Lateran Council was, by contrast, an event marked by continuity rather than change, can be tested against the evidence of printed official documents from one pontificate to the next. The decrees of Lateran V, and the presentation of those decrees, were deliberate projections of the political character of each pope. While Julius publicized the work of the council in a way that emphasized his absolutist claims to power, Leo's publications emphasized the consultative aspects of the assembly, with frequent nods to the need for the council fathers to approve its work and to the pope's deference to their statutory right to advise and consent.

In many respects, the publications of the council under Leo followed the patterns established by Julius. Decrees and resolutions continued to be printed as quarto pamphlets, with a title page containing the new pope's *stemma* and a descriptive title in large type; inside, the text of each decree was supplemented with information regarding how the original document had been read out, approved, published, or sent for printing. But there were also significant differences. Under Julius, the business of each council session was printed as a bull, a document issued by the pope. Leo adopted a more consultative, indeed *conciliar* posture: from 1513 to 1516, individual decisions of the council were published in a *Bulla concilii* and information was added to both the title and the interior concerning the clerics who had read the text out and how the assembly had approved it. Leo himself was presented in the text as a relatively passive member of the assembly.

Thus, the proceedings of the tenth session of the council, issued in 1515, are entitled *Bulls of the council, on the question of the monte di pietà, read out in the tenth session by the reverend father Lord Bertrando bishop of Adria, ambassador of the duke of Ferrara to the court of Rome*.[125] Under Julius, prelates had read out various decrees for the council to vote on and also approve too, and this information was sometimes recorded in printed volumes. But here, details of the process were integral to the title of the bull. The name and political affiliation of the *lector* anchored the decree in a wider pool of authority than just the pope. In other Leonine bulls, yet other aspects of conciliar process were recorded internally. Whereas Julian bulls established their authenticity by reference to how papal heralds had posted them at the papal basilicas and elsewhere around the city, the Leonine bulls of the council emphasized the moment of conciliar ratification, when the council fathers verbally gave their approval to the text that had been read out. At the close of the decree on the *monte di pietà*, for example, the printed page records the dialogue of acclamation with which the reading of the decree concluded:

> Placent vobis Patres quae per me lecta sunt?
>
> Fuit responsum per omnes: Placet.
>
> Scribatur in forma.[126]

Within these lines of printed text, Leo again appears as an observer, listening and lending his approval, but essentially allowing the council fathers to conduct the business of the assembly without his direction. The same emphases can be seen in Leo's administrative bulls. Because, when Leo was elected, the Lateran Council was already underway, many of the administrative decrees issued in the first year or two of his pontificate were promulgated through the council and issued as conciliar decrees. Instead of a regal Julius dictating instructions to his subjects within the curia and the papal state, Leonine editions present papal policy as something initiated and approved by the assembly.

Despite their typographical sophistication, papal documents of the early sixteenth century rarely included figural depictions of the pope. The few extant examples seem further to underscore the differences between the political persona Julius presented in print and that adopted by Leo. The woodcut of the pope with dove and tiara printed in Julius's illustrated bull of 1510 (figure 8.12) emphasizes the role of the pope as both pontiff and prince, his fingers extended in a gesture of blessing and command. A northern Italian edition of the 1509 *Monitorium contra Venetos* (in Italian translation) likewise shows an aggressive Julius, seated on a raised throne and extending a finger in warning at a group of Venetians.[127] Another woodcut, printed with an Italian translation of the 1511 bull inaugurating the council, shows the pope seated before his *stemma* and holding forth as he gestures at an assembly of cardinals and scribes (figure 8.17).

Printed portraits of Leo, by contrast, stress the role of the pope *in council*. In a woodcut illustration in Silber's edition of Giovanni Battista Gargha's oration at the eighth session of the council, the pope presides impassively over the assembly from the far rear of the scene, with ranks of cardinals, bishops, scribes, and ambassadors shown at work around and in front of him (figure 8.18). In an illustration for Guillery's edition of a Leonine bull on the striking of treaties, the pope occupies the very position that the title of the bull specifies ("per . . . Leonem X edita et per . . . Cardinalem de Farnesio lecta"): the pope sits in a chair, his hands at rest, and listens as the cleric in the pulpit reads out the text (figure 8.19).[128] The contrast with the Julian way of legislating could not be clearer.

This is not to say that the *conclusions* of the Lateran Council were any less autocratic under Leo. Among many other decrees, one bull of the sixth session tackles the question of printing head-on. Despite the benefits that printing had brought to the world, the bull asserts, certain printed books "not only fail to edify, but promote errors in faith as well as in daily life and behavior." Therefore, the pope, with the approval of the council, ordered that bishops and inquisitors should review and censor all books before their publication; none should be printed without ecclesiastical approval.[129] The circulation of French pamphlets against Julius in 1511–12, among others, was likely on the minds of the council fathers, who especially warned against texts "attacking the good name of individuals, even very eminent ones."

As Leo's pontificate progressed, other issues than the Lateran Council came to preoccupy him. Leo may have felt the need to present a more collegial image in the early years of his reign, but he began to shift from that posture as his troubles increased. In 1516, several cardinals mounted a conspiracy to overthrow him, a gambit Leo angrily suppressed. 1516 also saw the outbreak of the War of Urbino, the first time the pope had recourse to Julius's favorite means of penalizing a military opponent: spiritual threats. Leo issued a *monitorium* against Francesco Maria della Rovere, Julius's nephew, whom he had installed as duke of Urbino and whom Leo's Medici relations were attempting

𝕭olla de la 𝕹otitia 𝕯el Concilio de 𝕽oma a san͜to Joane laterão. Fata per el Santiſſimo Signoꝛ noſtro Julio. Papa Secũdo.

ſ.

Figure 8.17. Julius II, *Bolla de la notitia del concilio de Roma* ([Rome: Marcellus Silber, 1511]). BAV R.I.IV 2107(10), fol. 1r.

Figure 8.18. Giovanni Battista Gargha, *Oratio in octava sessione Lateranensis concilii* ([Rome: Marcellus Silber, after May, 6, 1514]). Santiago de Compostela, UL, Res L 5129(42), fol. 1r.

BVLLA SVPER TREVGIS

& Induciis quinquenalibus inter Principes Christianos per.S.D.N.Do.Leonem.X.Pont.Max.Editat& in Eccl.sia Diuę Marię Virginis de Minerua,per R.D.A. sancti Eustachii Diac. Card. de Farnesio lecta, & publicata. Die XIIII.Mensis Martii. M.D.XVIII.

✠

Figure 8.19. Leo X, *Bulla super treugis et induciis quinquennalibus* ([Rome, 1518]). Munich, BSB, 4 P.lat. 550#Beibd.1, fol. 1r.

to dislodge. He issued the same type of document as Julius used against the Venetians in 1509 and the dissident cardinals in 1511: not a bull of excommunication but a bull of warning, putting the target on notice that unless certain conditions were met, a bull of excommunication would follow. It was published in the same format as a typical Julian thunderbolt: the name of the accused, the name of the pope, and the papal *stemma* on the cover; inside, there were no bishops, no council fathers, no consensus voting recorded in the text (figure 8.20).[130]

Lateran V concluded just seven months before Luther released his Ninety-Five Theses to the public. Leo attempted, unsuccessfully, first to ignore and then to silence the German theologian. In 1520, he published the bull *Exsurge domine* condemning Luther. This was posted on the doors of St. Peter's and the chancery and carried instructions that it should be posted on the doors of churches in Brandenburg, Meissen, and Merseburg. The document leaned heavily on its thirteenth-century forebears: "We decree that the publication of this same letter, on those same doors, should render the aforesaid Martin and all others whom this document concerns, notified as to its contents, from the moment of their publication in that way, just as if the text had been read personally to them. For it is not plausible that anyone could pretend ignorance of such a sentence once it has been made public in such a way." But the text also appeared as a quarto pamphlet, published by Giacomo Mazzocchi, with an explanatory title and the pope's *stemma* surrounded by woodcut border (figure 8.21). This format was copied by other printers who published multiple editions across northern Europe.[131]

In these same years, Silber, Mazzocchi, and Guillery printed other pamphlets against Luther prepared by papal theologians and members of the curia: two texts by Silvestro Prierias in 1518,[132] orations by Tommaso Radini Tedeschi and Giovanni Antonio Modesti in 1520,[133] Charles V's decree against Luther of 1521,[134] texts by the theologians of the University of Paris and Henry VIII's defense of the seven sacraments in the same year,[135] and Cajetan's defense of papal supremacy in 1522.[136] These Latin texts, framed as dialogues, orations, or theological tracts, echoed arguments about the supremacy of the pope, his ability to dispense grace, and his immunity to any kind of judicial action from his church that had dominated papal communications since the close of the Schism and had featured in papal print for nearly half a century. They also appeared with the same design elements as papal bulls: the same style of title page, with woodcut borders and shaped type, as had featured in Julian and Leonine bulls and the decrees of Lateran V for the past decade.

Car. f. 222^d.   238

n:45.

Bayerische
Staatsbibliothek
München

Figure 8.20. Leo X, *Monitorium penale contra Franciscum Mariam ducem Urbini* ([Rome: Marcellus Silber, after March 1, 1516]). Munich, BSB, 4 J.can.f. 222 d, fol. 1r.

Figure 8.21. Leo X, *Bulla contra errores Martini Lutheri et sequacium* (Rome: Giacomo Mazzocchi, [1520]). Munich, BSB, Rar. 1477#Beibd.6, fol. 1r.

The Reformers' use of the press would soon outstrip that of the papacy's.[137] Cardinal Guillaume Briçonnet, one of the cardinals Julius excommunicated for the "heresy" of attending the Pisan *conciliabulum* in 1511–12, went on to become a fierce critic of Lutheran writings. In a letter to the faithful in France of 1523, he bemoaned the fact that "almost the entire world is filled with [Luther's] books and the people, enchanted by novelty and license, are impressed by the liveliness of his style." Lutheran propaganda was more varied and versatile; it drew on vernacular narrative traditions and potent visual vocabularies; above all, it touched on genuine concerns among the European faithful that papal propaganda had failed to address over the previous fifty years.

Julius and Leo flooded Europe with printed publications that insisted on the ecclesiology of previous centuries: papal supremacy, salvation by works, defeat of the infidel, the Donation of Constantine, the council beneath the pope. These were the ideas that papal print broadcast to the European reading public at the start of the sixteenth century. With relentless monotony, they published their bulls declaring supremacy in matters both spiritual and temporal. Their critics would assail these claims using the same medium but with far greater flexibility and to devastating effect. The papacy used the press—frequently, often cleverly, sometimes with tremendous impact—for years and even decades before the start of the Reformation. But the constraints of papal ecclesiology often restricted the pope's use of new technology to broadcasting a remarkably retrograde vision of his office and the Church over which he presided. It would take another council (Trent) to reform the message, and not just the medium by which the pope addressed his faithful.

Conclusion

In 2017, the five hundredth anniversary of Luther's protest against the theology of indulgences, the residents of Rome awoke one winter morning to find the city plastered with posters criticizing the pope. Over an unflattering portrait of Pope Francis, captured in an uncharacteristic scowl, large letters posed a question in Roman dialect: "But where's your mercy, Fran?"[1] There followed some commentary on Francis's recent attempts at reform in the hierarchy, which his critics denounced as high-handed and peremptory. The posters, while embarrassing to the Vatican, were put up beyond its borders in the streets and squares of Rome itself. The Comune di Roma, ostensibly provoked by the posters' failure to pay the usual tax on public advertisements, quickly covered them up with large white stickers that denounced them as *affissioni abusivi*.

Three aspects of this minor municipal tempest recall the media wars of the Roman Renaissance. The first is the perennial potency of *affissioni* (in Latin, *affixiones*; in English, broadsides, posters, or placards) in Roman civic life. While the international fellowship of Francis's detractors can—and does—publish critiques of his pontificate in books and magazines, on websites, and in blogs, there was something especially outrageous, electrifying even, about the physical manifestation of their critiques across the doors and walls of the Roman center. Like those Venetian and French spies who posted overnight calls to a new council on the doors of Roman churches in the fifteenth and sixteenth centuries, the provocateurs of 2017 knew how effective it would be to embed their attacks on the pope in the urban fabric of the Eternal City, under the nose of the pope. In this regard, little had changed from the fifteenth century to the twenty-first.

Second is the fact that the protest, however striking, was only really effective because it was amplified by other media. Italian newspapers and online news sites reported on the event.[2] Some websites included video footage of Roman streets and thoroughfares lined with the posters, and most speculated about who might have been responsible and what recent acts of curial housecleaning might have provoked their attack. A few online commenters grasped the historical echoes of the gesture: "Pasquino's back!" one posted below a newspaper account; "How we've needed him!"[3] Other commenters weighed in with their favorite *pasquinades* from the Roman past.

Media amplification of posted political material also recalls something from the early years of print in Renaissance Rome. Papal bulls and *pasquinades* alike appeared first on Roman doors, walls, bridges, and statues, but their messages were greatly amplified by the new medium of print. Printed papal bulls and collections of satirical verse spread the message—and advertised the *fact*—of Roman postings widely; the new medium broadcast the message further, but the message was potent, ultimately, because it had first been displayed in Rome. The city—this city—had a strange power, to ratify messages, to amplify them, to make them matter, no matter what medium finally transmitted them to the world.

Finally, there was the papal response to the printed critique. In 2017, this came not in a thunderous bull or a polished brief but in a tweet. Fr. Antonio Spadaro, editor of the Jesuit journal *La Civiltà Cattolica* and a friend and ally of the Jesuit pope, denounced the paper posters in a series of digital posts on Twitter. (It's surely no coincidence that the social media platforms of the twenty-first century have adopted the vocabulary of medieval communication: users put up posts on their walls, comment below the posts of others, and see them taken down if they offend too much.) Shortly afterward, Spadaro stitched his tweets together into a longer post that he put on his Facebook wall. Criticism of the pope, Spadaro said, merely showed that his reforms were working. Well-organized and well-funded critics of the pope had orchestrated the campaign. They had tried (but failed) to make it look like popular dissent: "It's a sign that he is doing well, that he's hitting some nerves. These posters are acts of intimidation, written in fake *Romanesco* to suggest that they come from the people. But regular people don't argue about the Order of Malta, or the canonical *dubia* of cardinals, for goodness' sake. There are corrupt people behind this, powerful forces mounting plots to separate the pope from the love of the people, which is his great strength. But they're having the opposite effect."[4]

Spadaro's spin was impressive, but equally remarkable was his choice of medium. Like a Renaissance chancery official, the Jesuit did not hesitate to seize on a new form of communication to defend the pope, his authoritarian prerogatives, his pastoral relationship with his followers, and the ancient institution he served. Critics of the papacy might delight in their clandestine postings; defenders would dismiss these as nefarious agitation and take to even newer media to propound their own lines of attack. Indeed, Spadaro's use of Twitter was only the latest in a long line of advances in papal communications, from the adoption of the printing press in the fifteenth century—the topic of this book—to the establishment of *L'Osservatore Romano*, the Vatican newspaper, in the 1860s and the founding of Vatican Radio in 1931. The Vatican began posting papal encyclicals online in the early 2000s.[5] Under Benedict XVI, the Holy See launched its Twitter feed (@Pontifex).

Papal officials in the Renaissance exploited the new technology of printing with movable type for a variety of ends: to publicize the pope's bulls of excommunication and interdict, to circulate open letters to the Christian princes, to make events like the Jubilee and the Fifth Lateran Council come alive for both those who participated and those who observed from afar, and to project as widely as possible the image of Rome as the papal metropolis, capital of a muscular and imperial Christendom, a city whose sanctity lent luster to the occupant of the papal throne and vice versa. From Paul II's first Jubilee announcement, printed in 1470, to Leo's bull against Luther of 1520, the first half century of papal print featured a steadily increasing stream of announcements, invitations, condemnations, and excommunications, all aimed at preserving the pope's place at the center of Catholic Christendom, translating medieval notions of papal monarchy into the idiom of early modern popular debate. All the while, critics of the papacy exploited the new technology to challenge the pope's authority and to subvert the very means and media he used to communicate with his faithful. They lampooned and parried papal pronouncements, forged documents of their own, and published appeals to popular opinion more nimbly than the papal chancery could. Thus, printed debates over the pope's spiritual powers and political authority were taking place across Europe in the late fifteenth and early sixteenth centuries, long before the start of the Reformation. The arguments were not particularly salutary: the first printed political pamphlets took aim at Turks and Jews, argued for the rights of fairly unsavory princes, and defended local ecclesial rights, not great points of principle. At the same time, the Christian faithful and the Christian princes alike looked to Rome as a source of grace and favor: spiritual graces for the next life, political favors in the here and now. Pilgrims, poets, diplomats, and spies all participated in the world of Roman publishing on their own terms, producing and consuming printed literature that reflected the majesty of the pope and his city but also *deflected* it to other purposes and projects, whether personal or political.

My narrative stops here, just before Luther burst on the European scene, for deliberate reasons. By surveying the half century from 1470 to 1520, it is possible to grasp just how much papal print there was before Luther. (Given the constant stream of pamphlet bulls, briefs, and decrees out of Rome, it seems less likely, then, that it was print alone that caused Luther's message to succeed.) Keeping our eyes on the landscape before 1517 also lets us focus on what sort of print it was that surrounded and amplified the policies of the Renaissance papacy. Here, despite their embrace of a revolutionary medium, the message remained resolutely traditional: papal primacy, salvation by works, the dangers of heretical thinking, the crusade against the Turks. Print certainly did not open up, or loosen, papal thinking on these core issues of ecclesiology, diplomacy, or financial management. The medium could not, by itself, modernize the institution, or even the message. Papal print was an art that functioned like so many others deployed by the Renaissance popes: architecture, painting, sculpture, music, liturgy, poetry, history-writing, the minting of coins and medals, the staging of festival entries, military pageants, ritual processions, and any number of other cultural expressions. The institution had an extraordinarily sophisticated sense of itself and its prerogatives— perhaps more fully developed since and because the popes had returned to Rome under such precarious circumstances and still found themselves vulnerable to attack from sacred and secular quarters alike. The arts, even the typographic arts, were there to express and amplify this vision ever more broadly and loudly. Even legal decrees could borrow from the common stock of visual and ceremonial language and contribute to the general project.

From the advent of print to the Reformation was, *sub specie aeternitatis*, a vanishingly short stretch of time: a scant six decades, or about three generations. Peter Schoeffer of Mainz, who helped Gutenberg print his forty-two-line Bible, went on to print Diether von Isenburg's challenge to Pius II, questioning the pope's authority to appoint bishops in Germany. Two decades later, still in Mainz, Schoeffer printed a broadside edition of Andrija Jamometić's *Retraction*, the document he signed after his arrest in which he withdrew his charges against Sixtus IV and encouraged the faithful to resume their obedience to the pope. Where Diether had seemed triumphant, Jamometić was despondent, having failed to realize his promised revolt against the power of

Rome. In Basel, Johannes Besicken printed Jamometić's fiery attacks against Sixtus, as well as the city council's more measured attempts to disarm the papal legate, Geraldini. In Besicken's broadsides, Jamometić condemned the practice of selling indulgences as "great extortions." And yet, within a decade, Besicken would move to Rome and find extraordinary success printing pilgrim guides for the 1500 Jubilee, texts that set out how many indulgences could be found in the papal city and where the best ones were on offer. By 1509, Besicken had two apprentices or junior partners: Etienne Guillery and Giacomo Mazzocchi. In 1521, Guillery would print the official *Acta* of Lateran V; by then, Mazzocchi had already printed *Exsurge Domine*, Leo's bull condemning Luther. The point is not just that early printers were opportunistic and willing to print for any customer, regardless of his politics; it's also that the changes print wrought on European politics happened at a breathtaking pace. In less than three generations, the baton was passed from Schoeffer to Besicken to Mazzocchi, from Pius to Sixtus to Leo, from Diether to Jamometić to Luther. In those intervals, the popes had come to see how the press could be used to attack them, how they could use it to protect themselves, and how very urgently they needed to do so. The printing press could not, by itself, forestall what was to come. The printing revolution in Rome helped broadcast a brighter image of the pope than ever before, on the very eve of his eclipse in European politics.

AAV	Archivio Apostolico Vaticano
ASR	Archivio di Stato, Rome
Basel SA	Staatsarchiv Basel-Stadt
BAV	Biblioteca Apostolica Vaticana
BL	British Library, London
BMC	*Catalogue of Books Printed in the XVth Century Now in the British Museum*, 13 vols. (London and 't Goy-Houten, 1908–2007)
BNC	Biblioteca Nazionale Centrale
BSB	Bayerische Staatsbibliothek
Burchard	Johannes Burchard, *Liber notarum*, ed. Enrico Celani, *RIS*[2] 32.1–2
DBI	*Dizionario biografico degli Italiani* (Rome, 1960–)
EDIT-16	*Censimento nazionale delle edizioni italiane del XVI secolo*: edit16 .iccu.sbn.it/web_iccu/ihome.htm
GW	*Gesamtkatalog der Wiegendrucke* (Leipzig, 1925–38; Stuttgart, 1968–)
GWM	GW Manuskript: www.gesamtkatalogderwiegendrucke.de
Hain	Ludwig Hain, *Repertorium bibliographicum*, 2 vols. (Stuttgart and Paris, 1826–38)
IGI	*Indice generale degli incunaboli delle biblioteche d'Italia*, ed. T. M. Guarnaschelli, E. Valenziani et al., 6 vols. (Rome, 1943–81)
Infessura	Stefano Infessura, *Diario della città di Roma*, ed. Oreste Tommasini (Rome, 1890)
ISTC	Incunabula Short-Title Catalogue: istc.bl.uk
JWCI	*Journal of the Warburg and Courtauld Institutes*
LL	Floriano Grimaldi, *Il Libro Lauretano, secoli XV–XVIII*, 2nd ed. (Loreto, 1994)
Malipiero	Domenico Malipiero, *Annali Veneti*, ed. Tommaso Gar and Agostino Sagredo, *Archivio storico italiano*, 1st ser., 7 (1843–44)
Navagero	Andrea Navagero, *Storia della repubblica veneziana*, ed. L. A. Muratori (Milan, 1733)
Pastor	Ludwig Pastor, *A History of the Popes from the Close of the Middle Ages*, trans. F. I. Antrobus et al., 40 vols. (London, 1891–1953)

PL	*Patrologiae cursus completus, series latina,* ed. J.-P. Migne, 221 vols. (Paris, 1844–64)
*RIS*²	*Rerum Italicarum scriptores. Racolta degli storici italiani dal cinquecento al millecinquecento,* ed. G. Carducci and V. Fiorini (Città di Castello and Bologna, 1900–)
Sander	Max Sander, *Le livre à figures italien depuis 1467 jusqu'a 1530; essai de sa bibliographie et de son histoire,* 6 vols. (New York, 1941; repr., Nendeln/Liechtenstein, 1969)
Sanudo	Marino Sanudo, *I Diarii,* 58 vols. (Venice, 1879–1902; repr., Bologna, 1969–70)
Schreiber	Wilhelm Ludwig Schreiber, *Manuel de l'amateur de la gravure sur bois et sur métal au XVe siècle.* Tom. 5: *Un catalogue des incunables à figures imprimés en Allemagne, en Suisse, en Autriche-Hongrie et Scandinavie* (Leipzig, 1910–11)
SCJ	*Sixteenth Century Journal*
Setton	Kenneth M. Setton, *The Papacy and the Levant, 1204–1571,* 4 vols. (Philadelphia, 1976–84)
Sheehan	William J. Sheehan, *Bibliothecae Apostolicae Vaticanae incunabula,* 4 vols. (Vatican City, 1997)
Statuti	*Statuti della città di Roma,* ed. Camillo Re (Rome, 1880)
Tinto	Alberto Tinto, *Gli Annali tipografici di Eucario e Marcello Silber* (Florence, 1968)
USTC	Universal Short Title Catalogue: ustc.ac.uk
VD16	*Verzeichnis der im deutschen Sprachbereich erschienenen Drucke des XVI. Jahrhunderts* (Stuttgart, 1983–)
VE15	Falk Eisermann, *Verzeichnis der typographischen Einblattdrucke des 15. Jahrhunderts im Heiligen Römischen Reich Deutscher Nation* (Wiesbaden, 2004)
Walsh	James E. Walsh, *A Catalogue of the Fifteenth-Century Printed Books in the Harvard University Library,* 5 vols. (Binghamton, NY, and Tempe, AZ, 1991–95)

Introduction

1. Piccolomini later saw quires of the Bible at Wiener Neustadt; Janet Ing, *Johann Gutenberg and His Bible: A Historical Study* (New York, 1988), 66–70; Martin Davies, "Juan de Carvajal and Printing: The 42-Line Bible and the Sweynheym and Pannartz Aquinas," *The Library*, 6th ser., 18 (1996): 193–215.

2. Ing, *Gutenberg*, 62–5; and Stephan Füssel, *Gutenberg and the Impact of Printing*, trans. Douglas Martin (Aldershot, 2005), 25–9.

3. Here and throughout I refer to incunabula editions by their ISTC numbers. For the thirty-line indulgence using the type of the forty-two-line Bible: ic00422400; the thirty-one-line indulgence using the type of the thirty-six-line Bible: ic00422600; the so-called *Türkenkalendar*: it00503500; Calixtus's *Bulla Thurcorum*: ic00060000 (Latin) and ic00060100 (German).

4. George Painter, "Gutenberg and the B36 Group: A Reconsideration," in *Essays in Honor of Victor Scholderer* (Mainz, 1970), 292–322; Janet Ing, "The Mainz Indulgences of 1454/5: A Review of Recent Scholarship," *British Library Journal* 9 (1983): 14–31; Blaise Agüera y Arcas, "Temporary Matrices and Elemental Punches in Gutenberg's DK Type," in *Incunabula and Their Readers: Printing, Selling, and Using Books in the Fifteenth Century,* ed. Kristian Jensen (London, 2003), 1–12; Joseph A. Dane, *Out of Sorts: On Typography and Print Culture* (Philadelphia, 2011), 32–56.

5. Norman Housley, "Indulgences for Crusading, 1417–1517," in *Promissory Notes on the Treasury of Merits: Indulgences in Late Medieval Europe*, ed. R. N. Swanson (Leiden, 2006), 277–307; Eckehard Simon, *The Türkenkalender (1454) Attributed to Gutenberg and the Strasbourg Lunation Tracts* (Cambridge, MA, 1988). The latter title is the invention of a modern editor; the printed text is headed *Eyn manung der cristenheit widder die durken* (An Appeal to Christendom against the Turks).

6. Falk Eisermann, "The Indulgence as a Media Event: Developments in Communication through Broadsides in the Fifteenth Century," in *Promissory Notes on the Treasury of Merits*, 309–30.

7. Andrew Pettegree, *Brand Luther* (New York, 2015), with a review of previous scholarship; Robert W. Scribner, *For the Sake of Simple Folk: Popular Propaganda for the German Reformation* (Oxford, 1981), remains fundamental.

8. Jared Wicks, "Roman Reactions to Luther: The First Year (1518)," *Catholic*

Historical Review 69 (1983): 521–62; David V. Bagchi, *Luther's Earliest Opponents: Catholic Controversialists, 1518–1525* (Minneapolis, 1991).

9. *Lux in Arcana: The Vatican Secret Archives Reveals Itself* (Rome, 2012), 111.

10. Thomas Frenz, *I documenti pontifici nel medioevo e nell'eta moderna* (Vatican City, 1989); Gerhard Jaritz, Torstein Jørgensen, and Kirsi Salonen, *The Long Arm of Papal Authority: Late Medieval Christian Peripheries and Their Communication with the Holy See* (Bergen-Budapest-Krems, 2004); Kirsi Salonen and Ludwig Schmugge, *A Sip from the "Well of Grace": Medieval Texts from the Apostolic Penitentiary* (Washington, DC, 2009).

11. Pius II, *Commentarii*, ed. Adrianus van Heck, 2 vols. (Vatican City, 1984), books I–VII, translated as Pius II, *Commentaries*, ed. Margaret Meserve and Marcello Simonetta (Cambridge, MA, 2004–18).

12. Kai-Michael Sprenger, "Die Mainzer Stiftsfehde 1459–1463," in *Lebenswelten Gutenbergs* (Stuttgart, 2000), 107–41; Pastor, 3.164–7, 192–208; Pius II, *Commentaries*, 6.1–3.

13. Frederick III to the imperial estates: if00318100, the document (not the edition) dated August 8, 1461.

14. Printed a total of seven times: Pius II to Adolph of Nassau: ip00654950, ip00655000; Pius II to the Mainz Chapter: ip00655350; Pius II to the clergy and civil officials of Mainz: ip00655400, ip00655410; Pius's bull formally deposing Diether from office: ip00655300, ip00655310. The documents, all Latin, are dated August 21, 1461.

15. Diether von Isenburg to the princes and prelates of the Holy Roman Empire (in German): id00191700; to the imperial estates (in German): id00191710; to Pope Pius II (in Latin): id00191750; documents dated March–April 1462.

16. Adolph von Nassau's "Manifesto" (in German): ia00053200; document undated but written sometime in spring 1462.

17. C. von Heusinger, "Die Einblattdrucke Adolfs von Nassau zur Mainzer Stiftsfehde," *Gutenberg Jahrbuch* (1962), 341–52; Konrad Repgen, "Antimanifest und Kriegsmanifest. Die Benutzung der neuen Drucktechnik bei der Mainzer Stiftsfehde 1461/63 durch die Erzbischöfe Adolf von Nassau und Diether von Isenburg," in *Studien zum 15. Jahrhundert. Festschrift für Erich Meuthen*, ed. Johannes Helmrath et al. (Munich, 1994), 2.781–803.

18. G. P. Carosi, *La stampa da Magonza a Subiaco* (Subiaco, 1965; repr., 1994); Giovanni Andrea Bussi, *Prefazioni alle edizioni di Sweynheym e Pannartz prototipografi romani*, ed. Massimo Miglio (Milan, 1978); Edwin Hall, *Sweynheym and Pannartz and the Origins of Printing in Italy: German Technology and Italian Humanism in Renaissance Rome* (McMinnville, OR, 1991); Johannes Roell, "A Crayfish in Subiaco: A Hint of Nicholas of Cusa's Involvement in Early Printing?," *The Library*, 6th ser., 16 (1994): 135–40. Martin Davies argues that Sweynheym sold printed books in Italy before the general expulsion from Mainz; Martin Davies, "From Mainz to Subiaco: Illumination of the First Italian Printed Books," in *La stampa romana nella città dei papi e in Europa*, ed. Cristina Dondi et al. (Vatican City, 2016), 9–42.

19. Anna Modigliani, *Tipografi a Roma prima della stampa: Due società per fare libri con le forme (1466–1470)* (Rome, 1989), describes an earlier partnership between German and Italian printers, but no products of this partnership survive.

20. *BMC* 4.vii–xvi; Anna Modigliani, "Tipografi a Roma, 1467–1477," in *Gutenberg e*

Roma: Le origini della stampa nella città dei papi (1467–1477), ed. Massimo Miglio and O. Rossini (Naples, 1997), 41–67; Arnold Esch, "La prima generazione dei tipografi tedeschi a Roma (1465–1480): Nuovi dati dai registri di Paolo II e Sisto IV," *Bulletino dell'Istituto storico italiano per il medio evo* 109 (2007): 401–18; Esch, "I prototipografi tedeschi a Roma e a Subiaco," in *Subiaco: La culla della stampa* (Subiaco, 2010), 53–62.

21. Maria Grazia Blasio, *Cum gratia et privilegio: Programmi editoriali e politica pontificia, Roma 1487–1527* (Rome, 1988); Anna Modigliani, "Printing in Rome in the XVth Century: Economics and the Circulation of Books," in *Editori ed Edizioni a Roma nel Rinascimento*, ed. Paola Farenga (Rome, 2005), 65–76.

22. Susan Noakes, "The Development of the Book Market in Late Quattrocento Italy: Printers' Failures and the Role of the Middleman," *Journal of Medieval and Renaissance Studies* 11 (1981): 23–55.

23. Victor Scholderer, "The Petition of Sweynheym and Pannartz to Sixtus IV," in *Fifty Essays in Fifteenth- and Sixteenth-Century Bibliography*, ed. Dennis E. Rhodes (Amsterdam, 1966), 72–3.

24. Concetta Bianca, "Le strade della 'sancta ars': La stampa e la curia a Roma nel XV secolo," in *La stampa romana*, 1–9. Rudolf Hirsch, *Printing, Selling, and Reading, 1450–1550* (Wiesbaden, 1967), 99–101, notes "orations, prayers, bulls, and edicts emanating from the papal chancery" but calls the empire the more interesting producer of official print. William Dana Orcutt, *The Book in Italy during the Fifteenth and Sixteenth Centuries* (London, 1928), 35–48, surveys the history of Roman print via these landmarks: Sweynheym and Pannartz and classics; Ulrich Han and theological works; Stephan Plannck and the *Mirabilia* tradition; Mazzocchi, *pasquinades* and antiquarian scholarship; and Blado, who published the *Index*.

25. Histories of early printing often focus on literary or scholarly publication, tracing the printing of important or difficult texts and emphasizing the role of the editor in their production; e.g., Brian Richardson, *Print Culture in Renaissance Italy: The Editor and the Vernacular Text, 1460–1600* (Cambridge, 2002); Farenga, *Editori ed edizioni*; M. D. Feld, *Printing and Humanism in Renaissance Italy: Essays on the Revival of the Pagan Gods*, ed. Cynthia M. Pyle (Rome, 2015).

26. Orcutt, *Book in Italy*, follows the history of printing in Italy from Rome to Venice, Naples, Milan, and finally Florence. More recently, see Andrew Pettegree, *The Book in the Renaissance* (New Haven, CT, 2010), 49–62.

27. Martin Lowry, *Nicholas Jenson and the Rise of Venetian Publishing in Renaissance Europe* (Oxford, 1991); Lowry, *The World of Aldus Manutius: Business and Scholarship in Renaissance Venice* (Ithaca, NY, 1979).

28. Roberto Ridolfi, *La stampa in Firenze nel secolo XV* (Florence, 1958), remains fundamental. See also Mary A. Rouse and Richard H. Rouse, *Cartolai, Illuminators, and Printers in Fifteenth-Century Italy: The Evidence of the Ripoli Press* (Los Angeles, 1988); Sean E. Roberts, *Printing a Mediterranean World: Florence, Constantinople, and the Renaissance of Geography* (Cambridge, MA, 2013); William A. Pettas, *The Giunti of Florence: A Renaissance Printing and Publishing Family* (New Castle, DE, 2013); Richardson, *Print Culture*, chap. 3.

29. On this period, see Paolo Sachet, *Publishing for the Popes: The Roman Curia and the Use of Printing (1527–1555)* (Leiden, 2020).

30. Italian scholarship on Roman printing has never taken quite such a dim view,

and I am deeply indebted to scholars of the Roman Renaissance such as Massimo Miglio, Paola Farenga, Concetta Bianca, Paolo Veneziani, and Anna Modigliani. In addition to the studies in notes 18–20 above, see *Scrittura, biblioteche e stampa a Roma nel Quattrocento: Aspetti e problemi*, ed. Concetta Bianca et al. (Vatican City, 1980); Massimo Miglio, *Saggi di stampa: Tipografi e cultura a Roma nel Quattrocento*, ed. Anna Modigliani (Rome, 2002); Concetta Bianca, "La stampa a Roma nel XV secolo," *Lezioni bellinzonesi* 4 (2011): 98–105.

31. Hirsch, *Printing, Selling, and Reading*, 52.

32. Peter Stallybrass, "Little Jobs: Broadsides and the Printing Revolution," in *Agent of Change: Print Culture Studies after Eisenstein*, ed. Sabrina Alcorn Baron et al. (Amherst, MA, 2007), 315–41; Stallybrass, "Printing and the Manuscript Revolution," in *Explorations in Communication and History*, ed. Barbie Zelizer (London, 2008), 111–18.

33. Historians have explored how other early modern monarchs and states exploited the power of the press to promote their political interests; see, e.g., Larry Silver, *Marketing Maximilian: The Visual Ideology of a Holy Roman Emperor* (Princeton, NJ, 2008); Cynthia J. Brown, *The Shaping of History and Poetry in Late Medieval France: Propaganda and Artistic Expression in the Works of the Rhetoriqueurs* (Birmingham, 1985); Filippo De Vivo, *Information and Communication in Venice: Rethinking Early Modern Politics* (Oxford, 2007).

34. *Archival Transformations in Early Modern Europe*, ed. Filippo de Vivo et al., special issue of *European History Quarterly* 46 (2016); Randolph C. Head, *Making Archives in Early Modern Europe: Proof, Information, and Political Record-Keeping, 1400–1700* (Cambridge, 2019), 63.

35. From a vast bibliography, Pastor remains fundamental, as is Charles A. Stinger, *The Renaissance in Rome* (Bloomington, IN, 1985, 2nd ed. 1998). On the theme of papal image making, see, more recently, Jan L. de Jong, *The Power and the Glorification: Papal Pretensions and the Art of Propaganda in the Fifteenth and Sixteenth Centuries* (University Park, PA, 2013). On individual popes, see Elizabeth McCahill, *Reviving the Eternal City: Rome and the Papal Court, 1420–1447* (Cambridge, MA, 2013); Caroll William Westfall, *In This Most Perfect Paradise: Alberti, Nicholas V and the Invention of Conscious Urban Planning in Rome, 1447–55* (University Park, PA, 1974); Emily O'Brien, *The Commentaries of Pope Pius II (1458–1464) and the Crisis of the Fifteenth-Century Papacy* (Toronto, 2015); Sixtus IV, *Un pontificato ed una città: Sisto IV (1471–1484)* (Vatican City, 1986); F. Cantatore et al., eds., *Metafore di un pontificato: Giulio II (1503–1513)* (Rome, 2010); Nicholas Temple, *Renovatio Urbis: Architecture, Urbanism and Ceremony in the Rome of Julius II* (New York, 2011).

36. Paolo Prodi, *Il sovrano pontefice: Un corpo e due anime: La monarchia papale nella prima età moderna* (Bologna, 1982), translated as *The Papal Prince: One Body and Two Souls: The Papal Monarchy in Early Modern Europe*, trans. Susan Haskins (Cambridge, 1988), 176–7; Norman Housley, *Crusading and the Ottoman Threat, 1453–1505* (Oxford, 2012).

37. Lucien Febvre and Henri-Jean Martin, *L'Apparition du livre* (Paris, 1958); Febvre and Martin, *The Coming of the Book* (London, 1976); Elizabeth Eisenstein, *The Printing Press as an Agent of Change*, 2 vols. (Cambridge, 1979), abridged as *The Printing Revolution in Early Modern Europe* (Cambridge, 1983, 2nd ed. 2005); Anthony Grafton, Adrian Johns, and Elizabeth Eisenstein, "AHR Forum: How Revolutionary Was the

Printing Revolution?," *American Historical Review* 107 (2002): 84–128; Baron et al., *Agent of Change* (where more than eighty reviews of the book are listed); Adrian Johns, *The Nature of the Book: Print and Knowledge in the Making* (Chicago, 1998); David McKitterick, *Print, Manuscript, and the Search for Order, 1450–1850* (Cambridge, 2003); Joseph Dane, *The Myth of Print Culture* (Toronto, 2003); Asa Briggs and Peter Burke, *A Social History of the Media: From Gutenberg to the Internet*, 3rd ed. (Cambridge, 2009), 18–24; and Elizabeth Eisenstein, *Divine Art, Infernal Machine: The Reception of Printing in the West from First Impressions to the Sense of an Ending* (Philadelphia, 2011).

38. For printed bulls restricting the ability of printers to print without ecclesiastical approval, see Rudolf Hirsch, *"Bulla super impressores librorum, 1515,"* in *The Printed Word: Its Diffusion and Impact* (London, 1978), study XIV.

39. Prodi, *Il sovrano pontefice*; see also Peter Partner, *The Pope's Men: The Papal Civil Service in the Renaissance* (Oxford, 1990); Gianvittorio Signorotto and Maria Antonietta Visceglia, eds., *Court and Politics in Papal Rome* (Cambridge, 2002); Maria Antonietta Visceglia, "The Pope's Household and Court in the Early Modern Age," in *Royal Courts in Dynastic States and Empires: A Global Perspective*, ed. J. Duindam et al. (Leiden, 2011), 239–64; Simon Ditchfield, "Papal Prince or Papal Pastor: Beyond the Prodi Paradigm," *Archivium historiae pontificiae* 51 (for 2013, published 2018): 117–32.

40. Margaret Meserve, "Patronage and Propaganda at the First Paris Press: Guillaume Fichet and the First Edition of Bessarion's *Orations against the Turks*," *Papers of the Bibliographical Society of America* 97 (2003): 521–88; Meserve, "News from Negroponte: Politics, Popular Opinion, and Information Exchange in the First Decade of the Italian Press," *Renaissance Quarterly* 59 (2006): 440–80.

41. Martin Lowry, "Diplomacy and the Spread of Printing," in *Bibliography and the Study of 15th-Century Civilisation*, ed. Lotte Hellinga and John Goldfinch (London, 1987), 124–46, has been a key influence for me on this point.

42. For recent studies on the interlocking worlds of politics and print in early modern Italy, see Brendan Dooley, *The Social History of Skepticism: Experience and Doubt in Early Modern Culture* (Baltimore, 1999); Filippo de Vivo, *Information and Communication in Venice: Rethinking Early Modern Politics* (Oxford, 2007); Massimo Rospocher, ed., *Beyond the Public Sphere: Opinions, Publics, Spaces in Early Modern Europe* (Bologna, 2012); Rouse and Rouse, *Cartolai, Illuminators, and Printers*; Bronwen Wilson, *The World in Venice: Print, the City, and Early Modern Identity* (Toronto, 2005); Rosa Salzberg, *Ephemeral City: Cheap Print and Urban Culture in Renaissance Venice* (Manchester, 2014); Ottavia Niccoli, *Prophecy and People in Renaissance Italy*, trans. Lydia G. Cochrane (Princeton, NJ, 1990); Massimo Rospocher, *Il papa guerriero: Giulio II nello spazio pubblico europeo* (Bologna, 2015); John Gagné, "Counting the Dead: Traditions of Enumeration and the Italian Wars," *Renaissance Quarterly* 67 (2014): 791–840. The world of Roman book printing is less studied, but for printed maps and visual guides, see Rose Marie San Juan, *Rome: A City out of Print* (Minneapolis, 2001); Rebecca Zorach, ed., *The Virtual Tourist in Renaissance Rome: Printing and Collecting the Speculum Romanae Magnificentiae* (Chicago, 2008); and Jessica Maier, *Rome Measured and Imagined: Early Modern Maps of the Eternal City* (Chicago, 2015).

43. Paul Needham, "Prints in the Early Printing Shops," in *The Woodcut in Fifteenth-Century Europe*, ed. Peter Parshall (New Haven, CT, 2009), 39–91.

44. On the complexity of the problem, see Paolo Veneziani, "Besicken e il metodo degli incunabolisti," *Gutenberg Jahrbuch* 80 (2005): 71–99.

45. Dooley, *Social History of Skepticism*; Filippo de Vivo and Brian Richardson, eds., *Scribal Culture in Italy, 1450–1700*, special issue of *Italian Studies*, 66 (2011); Luca degl'Innocenti et al., *The* Cantastorie *in Renaissance Italy: Street Singers between Oral and Literate Cultures*, special issue of *Italian Studies* 71 (2016); Niall Atkinson, *The Noisy Renaissance: Sound, Architecture, and Florentine Urban Life* (University Park, PA, 2016). Conversely, medievalists have argued that some aspects of print culture (standardization of forms, for example, and the strengthening of authorial identity) were developing in European culture well before Gutenberg; see David d'Avray, *Medieval Marriage Sermons: Mass Communication in a Culture without Print* (Oxford, 2001); Daniel Hobbins, *Authorship and Publicity before Print: Jean Gerson and the Transformation of Late Medieval Learning* (Philadelphia, 2009). See also Julia C. Crick and Alexandra Walsham, eds., *The Uses of Script and Print, 1300–1700* (Cambridge, 2004); Julia Boffey, *Manuscript and Print in London c. 1475–1530* (London, 2012).

46. Adrian Armstrong, "Books on the Bridge: Writing, Printing, and Viral Authority," in *Authority in European Book Culture 1400–1600*, ed. Pollie Bromilow (Farnham, UK, 2013), 31–42, at 33.

Chapter 1 • Urbi et orbi

1. Riccardo Burigana, "Una cartolina da Roma: Il viaggio di Martin Lutero nella Roma di Giulio II," in *Giulio II: Papa, Politico, Mecenate*, ed. Giovanna Rotondi Terminiello and Giulio Nepi (Genoa, 2005), 68–78; Michael Matheus, Arnold Nesselrath, and Martin Wallraff, eds., *Martin Luther in Rom: Die Ewige Stadt als kosmopolitisches Zentrum und ihre Wahrnehmung* (Berlin, 2017).

2. An insoluble question; see Volker Leppin and Timothy J. Wengert, "Sources for and against the Posting of the Ninety-Five Theses," *Lutheran Quarterly* 29 (2015): 373–98; Peter Marshall, *1517: Martin Luther and the Invention of the Reformation* (New York, 2017).

3. Francis Haskell and Nicholas Penny, *Taste and the Antique* (New Haven, CT, 1981), 291–6; Anne Reynolds, "Cardinal Oliviero Carafa and the Early Cinquecento Tradition of the Feast of Pasquino," *Humanistica Lovaniensia* 34 (1985): 178–208; V. Marucci, ed., *Pasquinate del Cinque e Seicento* (Rome, 1988); Leonard Barkan, *Unearthing the Past: Archaeology and Aesthetics in the Making of Renaissance Culture* (New Haven, CT, 1999), 210–31.

4. Daniel Jütte, *The Strait Gate: Thresholds and Power in Western History* (New Haven, CT, 2015), 175–208, with emphasis on Luther. Also north of the Alps: Pierre Fournier, "Affiches d'indulgence manuscrites et imprimées des XIVe, XVe, et XVIe siècles," *Bibliothèque de l'École des chartes* 84 (1923): 116–60; Wendy Scase, " 'Strange and Wonderful Bills': Bill-Casting and Political Discourse in Late Medieval England," in *New Medieval Literatures*, ed. Rita Copeland et al. (Oxford, 1997), 225–47; Élodie Lecuppre-Desjardin, "Des Portes qui parlent: Placards, feuilles volantes, et communication politique dans les villes des Pays-Bas à la fin du Moyen Âge," *Bibliothèque de l'École des chartes* 168 (2010): 151–72.

5. Robert Brentano, *Rome before Avignon: A Social History of Thirteenth Century Rome*, 2nd ed. (Berkeley, 1990); Andre Vauches, ed., *Roma medievale* (Bari, 2001); Chris

Wickham, *Sleepwalking into a New World: The Emergence of Italian City Communes in the Twelfth Century* (Princeton, NJ, 2015).

6. Ronald Musto, *Apocalypse in Rome: Cola di Rienzo and the Politics of the New Age* (Berkeley, 2003).

7. *Statuta SPQR*, first printed by Ulrich Han in 1470 (is00722300) and reprinted in revised form by Etienne Guillery in 1519–23.

8. *Statuti*, 1.3 and 1.4, 3–5.

9. *Statuti*, 1.4 and 2.120–1, 5 and 150–1.

10. Different words are used: civil *citationes* were *affigatur in hostio domus*, but criminal notices were delivered by *fossura*. The exact meaning of *fossor* and its cognates is unclear. In classical Latin a *fossor* was a gravedigger, but the 1363 statutes make clear that the *fossor* does something to a wall. Medieval lexical texts use *fossura* as a gloss on the Greek word *skaphetos*, meaning a pilot hole drilled into a wall to receive a bolt or screw: Charles du Cange, *Glossarium ad scriptores mediae et infimae latinitatis* (Niort, 1883–87), 3.581c.

11. Court officials could impose *fossura* when other remedies failed (*Statuti*, 3.38, 3.92, 223, 248).

12. *Statuti*, 2.120, 150–1. For Roman wall paintings see Richard Ingersoll, "The Ritual Use of Public Space in Renaissance Rome" (PhD diss., University of California, Berkeley, 1978), 284–5.

13. *Statuti*, 2.120, 150.

14. Thomas Cohen and Elizabeth Cohen, *Words and Deeds in Renaissance Rome* (Toronto, 1993), 23–7, 99–100.

15. *Statuti*, 2.120, 151. Some historians have misread this statute as indicating a punishment due to debtors, but the text makes clear that this is a penalty meant for the communal official himself.

16. Emanuele Rodocanachi, *The Roman Capitol in Ancient and Modern Times*, trans. Frederick Lawton (London, 1906), 57–119; Musto, *Apocalypse*, 98–9.

17. Carved in the late fourth century BCE and restored by Ruggero Bascapé in 1594, the lion is now in the Musei Capitolini. A drawing by Martin van Heemskerck ca. 1532–37 shows it installed at the top of the stairs to the Senator's Palace, but in engravings from the 1540s and 1550s it is no longer there; see Kathleen Wren Christian, *Empire without End: Antiquities Collections in Renaissance Rome* (New Haven, CT, 2010), 22, 112–13, fig. 84; Rodocanachi, *Roman Capitol*, 80–2.

18. Carrie E. Beneš, "Whose SPQR? Sovereignty and Semiotics in Medieval Rome," *Speculum* 84 (2009): 874–904, at 877–81.

19. Massimo Miglio, "Il Leone e il Lupa: Dal simbolo al pasticcio alla francese," *Studi Romani* 30 (1982): 177–86.

20. *Statuti*, xciv; Anonimo Romano, *The Life of Cola di Rienzo*, trans. Alberto Maria Ghisalberti (Toronto, 1975), 56, 76, 81, 92, 143.

21. Anonimo Romano, *Life of Cola*, 49 (execution of Martino de Porto), 143 (Fra Morreale): the bell was rung, the people assembled, the sentence was read out at the lion, and the prisoner was executed in the piazza nearby.

22. Anonimo Romano, *Life of Cola*, 151–2.

23. *Statuti*, 2.35, 105; Laurie Nussdorfer, *Brokers of Public Trust: Notaries in Early Modern Rome* (Baltimore, 2009), 26.

24. *Statuti*, 2.53, 111; Musto, *Apocalypse*, 158–9.

25. *Statuti*, 3.118, 261: "Sufficiat citatio per mandatarium in scalis curie capitolii et talis citatio valeat et teneat et pro citatione legitima habeatur."

26. Didier Lett and Nicolaus Offenstedt, eds., *Haro! Noël! Oye! Pratiques du cri du Moyen Âge* (Paris, 2003), 11–39, 141–55; Michel Hébert, *"Voce preconia*: Note sur les criées publiques en Provence à la fin du moyen âge," in *Milieux naturels, espaces sociaux*, ed. Elizabeth Mornet and Franco Morenzoni (Paris, 1997), 689–701; Stephen J. Milner, "'Fanno bandire, notificare, et expressamente comandare': Town Criers and the Information Economy of Renaissance Florence," *I Tatti Studies in the Italian Renaissance* 16 (2013): 107–51.

27. Antonio di Pietro dello Schiavo, *Diarium Romanum, RIS²* 24.5.76; *Statuti*, 1.6, 7, and 3.334, 217; Infessura, 162, 184, 190; Burchard, 2.218, 2.300–1; *Regesti di bandi, editti, notificazioni, e provvedimenti diversi relativi alla città di Roma ed allo stato pontificio*, 7 vols. (Rome, 1920–58), 1.1234–1605. Emilio Re, "Bandi Romani del secolo XV," *Archivio della R. Società di storia patria* 51 (1929): 79–101, publishes seven Roman *bandi* issued in the vernacular but headed with Latin text.

28. *Statuti*, 3.80, 275.

29. *Statuti*, 2.132, 158, and 2.168, 174.

30. Dello Schiavo, *Diarium Romanum, RIS²* 24.5.75–6; Paolo dello Mastro, *Il Diario e memorie delle cose accadute a Roma, RIS²* 24.2.86; Paolo di Lello Petrone, *La Mesticanza, RIS²* 24.2.14, 37–8, 49; Infessura, 58; Odorico Rinaldi, *Annales ecclesiastici* (Lucca, 1743), 10.16–17; Burchard, 1.180; Giuseppe Pardi, ed., *Diario Ferrarese, RIS²* 24.7.501, 540.

31. Infessura, 306.

32. *Statuti*, 1.6, 7.

33. *Statuti*, 3.334, 217.

34. Along with the stone one, the commune kept a live one in an enclosure on the slopes leading up to the Campidoglio well into the fifteenth century; Ingersoll, "Ritual," 253.

35. *Statuti*, 3.80, 242.

36. The statutes insist on good behavior from *mandatarii* and *banditores* alike and urge them to think of themselves as employees of the commune, not independent contractors; *Statuti*, 2.164, 176.

37. Filippo Bonanni, *La gerarchia ecclesiastica* (Rome, 1720), 499–500; G. Moroni, "Cursori apostolici o pontifici," in *Dizionario di erudizione storico-ecclesiastica*, 103 vols. (Venice, 1840–61), 19.49–62; E. Rodocanachi, "Les couriers pontificaux du quatorzième au dix-septième siècle," *Revue d'Histoire Diplomatique* 26 (1912): 392–428; Philippe Boutry, "Curseur apostolique," in *Dictionnaire historique de la papauté*, ed. Philippe Levillain (Paris, 2003), 532–3.

38. AAV, Cam. Ap., Div. Cam. tom. 30, 14, "Creatio cursoris ad deferendam litteras" (1464); AAV, Cam. Ap., Div. Cam. tom. 38, 263, "Cursores apostolici cur [habeant] facultatem gestandi in pectore arma pontificis" (1476).

39. Boutry, "Curseur apostolique," 532–3. *Cursori* should not be confused with the similar office of *servientes armorum* or *mazzieri*, on whom see Sergio Pagano, *Gli statuti dei mazzieri pontifici del 1437* (Rome, 2007).

40. Thomas Frenz, *Die Kanzlei der Päpste der Hochrenaissance (1471–1527)* (Tübingen, 1986), 230, records prices for the office of cursor between 1506 and 1526, ranging

from 700 to 900 ducats. The office was cheaper than positions like secretary or scriptor (or even bullator, which cost over 5,000 ducats in these same years; 217), and closer to posts like usher (400–600 ducats; 230) or scutifer (800 ducats or more; 233).

41. Twelve *cursores* accompanied Innocent VIII to the Lateran in his *possesso* of 1484; in 1513, Leo X was accompanied by heralds carrying the *vexilla XII cursorum*. Presided over by the *magister cursorum*, they were part of the papal household.

42. AAV, Cam. Ap., Div. Cam. tom. 40, 257, "Cursoris pontificis officium quid sit."

43. Thomas Frenz, *I documenti pontifici nel medioevo e nell'eta moderna* (Vatican City, 1989).

44. Armando Petrucci, "The Illusion of Authentic History: Documentary Evidence," in *Writers and Readers in Medieval Italy: Studies in the History of Written Evidence*, trans. Charles M. Radding (New Haven, CT, 1995), 236–50, at 239.

45. John F. D'Amico, *Renaissance Humanism in Papal Rome: Humanists and Churchmen on the Eve of the Reformation* (Baltimore, 1983), 25–28. By the fifteenth century, other branches of the curia were also issuing their own bulls, but the chancery handled the most serious bulls, ones most likely to be "published" by the procedures outlined below. See Peter Partner, *The Pope's Men: The Papal Civil Service in the Renaissance* (Oxford, 1990), 20–47; and Frenz, *Die Kanzlei*.

46. Jules Simier, "La promulgation des lois pontificales," *Revue augustinienne* 15 (1909): 154–69; John T. Creagh, "The Promulgation of Pontifical Law," *Catholic University Bulletin* 15 (1909): 23–41, esp. 27–32; René Metz, *What Is Canon Law?*, trans. Michael Derrick (New York, 1960), 41–3; more recently, Brett Edward Whalen, *The Two Powers: The Papacy, the Empire, and the Struggle for Sovereignty in the Thirteenth Century* (Philadelphia, 2019).

47. Simier, "La promulgation," 159.

48. Augustus Potthast, ed., *Regesta pontificum romanorum inde ab anno post Christum natum MCXCVIII ad annum MCCCIV*, 2 vols. (Berlin, 1874–75), 2.1665, no. 20682.

49. Quoted by Simier, "La promulgation," 159: "Idem Palaeologus et alii contra quos processus ipse contigit nullam possint postmodum excusationem praetendere, quod ad eos talis processus non pervenerit vel quod tam patenter omnibus publicatur." The bull, dated May 7, 1281, is in *Bullarum, diplomatum et privilegiorum sanctorum romanorum pontificum taurinensis editio locupletior*, ed. Luigi Bilio et al., 25 vols. (Turin, 1857–72), 4.52–3. See Setton, 1.138; George Ostrogorsky, *History of the Byzantine State*, trans. Joan Hussey, rev. ed. (New Brunswick, 1969), 463–4; Giovanni Villani, *Nuova Cronica*, ed. G. Porta (Parma, 1991), 414. Another bull of Martin's, issued on March 21, 1283, after the Sicilian Vespers, was posted on the doors of Orvieto Cathedral (Bilio et al., *Bullarum editio* 4.54–66) and reissued and reposted on the same doors on April 15: *Regesta pontificum*, 2.1778, no. 22013.

50. Simier, "La promulgation," 164–5; Creagh, "Promulgation of Pontifical Law," 23.

51. Bilio et al., *Bullarum editio*, 4.170–4: Boniface VIII, September 8, 1303, excommunicating Philip of France, posted on the doors of Anagni Cathedral.

52. Bilio et al., *Bullarum editio*, 4.152–5: Boniface VIII, April 16, 1299, against trading with Saracens, posted on the doors of the Lateran; Boniface VIII, April 4, 1303, against anyone attacking pilgrims on their way to or from Rome, also on the Lateran doors. Both bulls were issued on Maundy Thursday, later fixed as the day for issuing the annual bull *In coena Domini*, which covered these and other offenses against the Church; *The*

Papal Bull In coena Domini *Translated into English with a Short Historical Introduction* (London, 1848), 6–9.

53. *Monitoria et declarationes excommunicationis contra Venetos promulgatae a Clemente V, Sixto IIII, et Iulio II Romanis pontificibus* (Rome, 1606), 12; Martin Kaufhold, "Öffentlichkeit im politischen Konflikt: Die Publikation der kurialen Prozesse gegen Ludwig von Bayern in Salzburg," *Zeitschrift für Historische Forschung* 22 (1995): 435–54, at 437–8, dates the origin of the practice to the reign of Clement's predecessor, Boniface VIII (1294–1303), but see note 48 for earlier examples. See also Bilio et al., *Bullarum editio*, 4.522 (Urban V, October 12, 1363: a bull to be posted on the doors of the palace and of the cathedral in Avignon).

54. Bilio et al., *Bullarum editio*, 4.234–6 (John XXII, March 30, 1317).

55. Kaufold, "Öffentlichkeit," 436: "ut idem Ludovicus et alii quos processus ipse contingit nullam possint excusationem pretendere . . . cum sit non verisimile quoad ipsos remanere incognitum vel occultum, quod tam patenter omnibus publicatur."

56. Musto, *Apocalypse*, 69, 87, 122. The church, in the Colonna district of Rome, was protected from interference by imperial troops.

57. Jütte, *Strait Gate*, 186; Anthony K. Cassell, *The Monarchia Controversy: An Historical Study* (Washington, DC, 2004), 35–7.

58. Anonimo Romano, *Life of Cola*, 38; Musto, *Apocalypse*, 130, 134.

59. Musto, *Apocalypse*, 246–7; *Chronicon Estense, RIS²* 15.158.

60. Bilio et al., *Bullarum editio*, 4.605 (Boniface IX, September 18, 1390).

61. Dello Schiavo, *Diarium Romanum*, 24.5.13.

62. Bilio et al., *Bullarum editio*, 4.722 (Martin V, September 1, 1425).

63. Joannes Domenicus Mansi, *Sacrorum conciliorum nova et amplissima collectio* (repr., Paris, 1901), 31B, col. 1410; Joachim W. Stieber, *Pope Eugenius IV, the Council of Basel, and the Secular and Ecclesiastical Authorities in the Empire* (Leiden, 1978), 43–4.

64. Mansi, *Sacrorum conciliorum collectio*, 31B, col. 1420.

65. Bilio et al., *Bullarum editio*, 5.81: Eugene IV, "Cum vectigalia," November 26, 1444.

66. AAV, Bandi Sciolti, Serie I, pacco I/5: "Caeterum ut praedicta omnia ad communem singulorum notitiam deducantur, ne de eis ignorantia praetendi, seu etiam allegari possit, praesentes literas per Urbem publice praeconizari, et valvis Capitolii dictae Urbis mandavimus fecimusque affigi, decernentes auctoritate praefata, ut huiusmodi literae publicatae et ut praefertur affixae, omnes quos concernunt perinde arctent, ac si eis forent personaliter et praesentialiter intimatae. Cum non sit verisimile quod ad eos remanere incognitum, quod tam patenter omnibus extitit intimatum." Also ASR, Bandi, tom. 1, Bull of Innocent VIII, September 1, 1486, to be posted "in Curia Capitolii et aliis locis publicis dictae Urbis . . . sono tubae premisso."

67. Sixtus IV, Bulla, May 24, 1483, "Ad bonorum tutelam," printed in Rome by Johannes Philippus de Lignamine in 1483 (is00555030), fol. 10r: "Et cum ipse presentes littere Venetiis propter eorundem Venetorum potentiam nequeant tute publicari . . . volumus litteras ipsas in valvis basilice principis apostolorum de urbe affigi et per huiusmodi affixionem publicari, decernentes quod earundem litterarum publicatio sic facta perinde eosdem monitos et omnes quos littere ipse contingunt arctet ac si littere ipse die affixionis et publicationis huiusmodi eis personaliter lecte et insinuate forent. Cum non sit verisimilis coniectura, quod ea quae tam patenter fient debeant apud eos incognita remanere." Compare the wording in note 66.

68. In 1454, Nicholas V's bull granting Portugal rights over West Africa was posted on the doors of Lisbon Cathedral; Bilio et al., *Bullarum editio*, 5.115.

69. At the height of the Pazzi War, Sixtus IV had a bull denouncing Lorenzo de' Medici published in Siena, Bologna, and Perugia because he could not get it published in Florence itself. Sixtus IV, Bulla, August 17, 1479, "Decet Romanum pontificem" ([Rome: Georgius Lauer, after August 17, 1479]) (is00547800), fol. 4v.

70. Jütte, *Strait Gate*, 183; and *Corpus iuris canonici* (Rome, 1918), canons 1720 and 1721.

71. Julius II, *Bulla contra aspirantes ad papatum symoniace* ([Rome, s.n., after January 14, 1506]) (EDIT-16 54025), fol. 4r: "Ut autem Constitutionis decreti statuti ordinationis ac inhibitionis nostrae huiusmodi notitia deducatur, volumus presentes litteras nostras in valvis Basilicae principis Apostolorum necnon Cancellariae ac in acie campi florae affigi, nec aliam earundem litterarum publicationis solemnitatem requiri aut expectari debere, sed huiusmodi affixionem pro solemni publicatione et perpetuo robore sufficere."

72. See note 52.

73. William Brewyn, *A XVth Century Guide-Book to the Principal Churches of Rome*, trans. C. Eveleigh Woodruff (London, 1933), 67. See Burchard, 1.227 and 1.303, for accounts of this rite.

74. Frenz, *Die Kanzlei*, 150. See ASR, Bandi, 1/35 and 1/36 (both discussed below), 1/39 (April 13, 1502, to be posted on the doors of the chancery and the *Audientia*), and 1/42 (July 28, 1505, to be posted on the doors of the chancery); and Burchard, 2.220, 2.266, 2.270, for other bulls posted on the chancery doors.

75. Bilio et al., *Bullarum editio*, 5.251: Sixtus IV, March 21, 1478.

76. Fournier, "Affiches d'indulgence," 124–6.

77. See, e.g., Burchard, 1.130, 1.136, 1.272. An example of one such bull, from 1444, can be found in *Collectio bullarum brevium aliorumque diplomatum Sacrosanctae Basilicae Vaticanae* (Rome, 1750), 2.102.

78. Tiberio Alfarano identifies these doors, on the (liturgical) east end of the atrium, as *valvae aeneae*, while calling every other set of doors *portae*; see Tiberio Alfarano, *Almae urbi Divi Petri veteris noviqve templi descriptio* (Rome, 1590).

79. Ingersoll, "Ritual," 45–8, 143–51.

80. Ingersoll, "Ritual," 171–214, esp. 185–6; see also 122–9 on the Assumption Day procession, another incursion of the pope into secular civic space.

81. Ingersoll, "Ritual," 280–1, 408–19.

82. Evidence for this is circumstantial, and there are arguments to support the idea of nails. In favor of glue: the Roman Statutes use different words—*affigere* and *fodere*— to describe different kinds of posting, and only *fodere* can cause permanent damage to walls; papal documents invariably use *affigere*, *affixio*, and their cognates and never mention *fossura*; in the fifteenth century, the main doors of the basilicas at the Lateran and St. Peter's were bronze and could not have had anything nailed to them.

83. ASR, Bandi, 1/35 (Bull of Innocent VIII, May 2, 1492): "volumus quod literae ipsae in valvis basilicae principis Apostolorum de Urbe affigantur, et per decem dies continuos dum ibidem solemnis missa celebratur . . . teneantur, quodque illarum affixio et publicatio sic facta de illis perinde omnes quos concernunt arctent, ac si eis personaliter intimatae fuissent."

84. ASR, Bandi, 1/36 (Bull of Alexander VI, December 18, 1491): the bull should be posted on the doors of St. Peter's, the doors of the chancery, the doors of the *Audientia*, and at Campo de' Fiori, and it should be read out three times, on three different days, at sessions of the *Audientia*. The bull, which concerned the payment of feudal dues, would be considered in effect for those dwelling on this side of the Apennines two months after its publication, and for those living beyond the mountains, four months afterward.

85. ASR, Bandi, 1/35.

86. Innocent VIII, "Officii nostri debitum" (January 25, 1492), printed as *Bulla contra notarios et quascunque alias personas litteras apostolicas executioni demandari impedientes* ([Rome: Stephan Plannck, after January 31, 1492]) (ii00116400), targets parties to a dispute who try to intercept or suppress a papal judgment.

87. Infessura, 238. See 219–20 for the murder of another papal messenger.

88. Burchard, 2.303.

89. The inscription is in a style much later than the thirteenth century and describes the year 1238 as the sixth of Gregory's reign, while in 1238 Gregory had been ruling for eleven years. It was probably carved centuries after the issue of the original bull.

90. None of these epigraphic bulls are installed in their original locations, and we do not know how or where they were hung.

91. It is now installed in the cloister of the Lateran, near that of Gregory XI mentioned above.

92. Roberto Cannatò et al., *Umanesimo e primo Rinascimento in S. Maria del Popolo* (Rome, 1981), 26–7.

93. Armando Petrucci, *Public Lettering: Script, Power, and Culture*, trans. Linda Lappin (Chicago, 1993), 22.

94. Mitchell Merback, *The Thief, the Cross, and the Wheel: Pain and the Spectacle of Punishment in Medieval and Renaissance Europe* (Chicago, 1999); Nicholas Terpstra, ed., *The Art of Executing Well: Rituals of Execution in Renaissance Italy* (Kirksville, MO, 2008), 122–32; Terpstra, "Body Politics: The Criminal Body between Public and Private," *Journal of Medieval and Early Modern Studies* 45 (2015): 7–52; Esther Cohen, *The Modulated Scream: Pain in Late Medieval Culture* (Chicago, 2010), 44–7, 66–7. For Rome after 1527, see Adriano Prosperi, "Esecuzioni capitali e controllo sociale nella prima età moderna," in *La pena di morte nel mondo* (Casale Monferatto, 1983); Trevor Dean and K. J. P. Lowe, eds., *Crime, Society, and the Law in Early Modern Italy* (Cambridge, 2009); Guido Rebecchini, "Rituals of Justice and the Construction of Space in Sixteenth-Century Rome," *I Tatti Studies* 16 (2013): 153–79.

95. See, e.g., *The Diary of John Burchard*, trans. Arnold Harris Mathew (London, 1910), 107, 152–3, 262–4.

96. Infessura, 69–70: accused heretics "foro menati ad Araceli, dove fo fatto uno tavolato verso piazza di Campitoglio, et lì stettero colla mitria de carta in capo."

97. Infessura, 36.

98. Infessura, 65. See Terpstra, *Art*, 128, for similar cases in Bologna and Florence.

99. Infessura, 38.

100. Infessura, 107, 152–3, 292.

101. Burchard, 2.220: in May 1500, eighteen men were hanged on the bridge, nine on each side, arranged in such a way that the cardinals had to pass through them on their way back from consistory.

102. Burchard, 2.300.

103. After his death, Cola's body was dragged through the streets, hung upside down from a balcony on the via Lata, and left for two days. Then it was burned, the ashes thrown in the Tiber. Anonimo Romano, *Life of Cola*, 152; Musto, *Apocalypse*, 345–6.

104. Infessura, 36.

105. Infessura, 38.

106. Ingersoll, "Ritual," 409. Infessura, 255: a baron accused of colluding with the Turks was tied naked to a stake and dragged through the streets in a cart, his flesh torn with red-hot pincers, to the Campidoglio, where he was struck on the head with a wooden mallet and beaten about the chest with an iron rod till he died. Then his body was quartered and the parts nailed up on the gate of Castel Sant'Angelo and at Porta S. Paolo, Porta S. Giovanni, and Porta del Popolo. For the mutilation of condemned corpses, see Terpstra, *Art*, 129–30.

107. Infessura, 292: a Genoese man infuriated by gambling losses blasphemed at a crucifix painted on the wall of San Biagio, near Piazza Madama, striking the head and heart of the painted *corpus*. He was seized and his hands, with which he had struck the blows, were amputated and nailed or screwed ("cum clavis in dicto muro affixae") into the facade of the same church, while the rest of him was hanged in Piazza Navona.

108. Petrone, *La Mesticanza*, 51–2: a cardinal's servant, convicted of murdering and robbing the cardinal, was drawn through the streets in a cart, crossing Ponte Sant'Angelo to arrive at Campo de' Fiori, where his hands were cut off; they were then "chiavellate nello ditto ponte una da uno delli lati et la aitra dallo aitro." Then, he was hanged and his body quartered. The parts were displayed in Campo de' Fiori, at Porta San Giovanni, on Monte Mario, and on the Milvian Bridge; dello Mastro, *Diario*, 92.

109. Infessura, 38. The wolf was transferred to the Campidoglio in 1471.

110. Burchard, *Diary*, 12–3; Infessura, 140–2.

111. Ingersoll, "Ritual," 414. For examples, see Musto, *Apocalypse*, 253; Anonimo Romano, *Life of Cola*, 78, 93; Infessura, 38.

112. Infessura, 56; Terpstra, *Art*, 128.

113. Anonimo Romano, *Life of Cola*, 53 (two Senate scribes arrested and "mitred" as forgers), 94 (a nobleman convicted of forgery protests on account of his nobility but must wear a mitre nonetheless). See the comment of Infessura in note 95 above and at 38 (thieves hanged, wearing mitres), 69–70 (heretics made to wear mitres while they were preached at). See discussion earlier in this chapter of a mitre placed on a derelict process server.

114. Burchard, *Diary*, 152–3.

115. Burchard, 1.276–9; Infessura, 250–1.

116. Under Cola, a courier who killed another courier was buried alive beneath the body of his victim; Anonimo Romano, *Life of Cola*, 69.

117. Pastor, 3.125–6; Costanza Gislon Dopfel, "Reshaping Rome's Narrative: The Curious Case of Sigismondo Malatesta's Execution," in *Early Modern Rome 1341-1667*, ed. Portia Prebys (Rome, 2011), 682–90; Anthony F. D'Elia, *Pagan Virtues in a Christian World: Sigismondo Malatesta and the Italian Renaissance* (Cambridge, MA, 2016), 1–11, 253.

118. Compare the two ritual burnings of John XXII in effigy by Ludwig of Bavaria (discussed earlier in this chapter) and a case a few decades later in Forlì, when Fran-

cesco Ordelaffi, excommunicated by the pope, issued a declaration that he was excommunicating the pope in turn. He then "had stuffed paper effigies of the pope and cardinals burned in the piazza"; Anonimo Romano, *Life of Cola*, 114.

119. Infessura, 54: "folli letta la sentenza, o cascione perchè foro impiccati."

120. Infessura, 244: "aliqui nocte suspenduntur et mane suspensi reperiuntur apud turrim Nonae *sine nomine et sine causa*" (italics mine); Ingersoll, "Ritual," 417.

121. Infessura, 264: "capti fuerunt duo filii Francisci Buffali . . . ducti fuerunt in Castrum Sancti Angeli . . . mandato papae quodam brevi contento et ostenso, XVI hora noctis decapitati, et exhibiti in ecclesia Sancti Celsi. *Quid fecerint nescitur*" (italics mine).

122. Infessura, 247–8: "Et quod detestabilius est, secunda die post moretm, per famulum dicti domini Falconi, causa ipsius mortis in quadam cartula apud pedes eius apposita est falsa et mendax, nam gestsa per dictum famulum ei applicata sunt."

123. Infessura, 270–1.

124. Infessura, 242: "Istis temporibus nihil boni Romae actum est . . . "

125. Infessura, 243–4.

126. Infessura, 246.

127. Infessura, 244.

128. Infessura, 157.

129. Jütte, *Strait Gate*, 186; Jürgen Miethke, *Politiktheorie im Mittelalter: Von Thomas von Aquin bis Wilhelm Ockham* (Tübingen, 2008), 250–2.

130. Chris Schabel, "*Lucifer princeps tenebrarum* . . . The *Epistola luciferi* and Other Correspondence of the Cistercian Pierre Ceffons," *Vivarium* 56 (2018): 1–50. According to one chronicler, the letter was found posted on a door: "Affixa quoque fuit quedam clausa litera ostio cardinalis" (2 n. 1).

131. Mansi, *Sacrorum conciliorum collectio*, 31A, cols. 190–1: "has nostras literas in valvis ecclesiae cathedralis Gebennensi, ac in porta domus fratrum predicatorum extra muros Gebennenses, non in civitatibus et oppidis urbi Romae convicinis, si ad eundem Thomam tutus accessus non pateat, affigi ac simul publicari mandavimus."

132. Mansi, *Sacrorum conciliorum collectio*, 31B, col. 1897; 30, col. 1067, reproduces a conciliar complaint from 1436 about the posting of a bull.

133. Infessura, 158.

134. Infessura, 192.

135. Cecil H. Clough, "Pamphlets and Propaganda in the Italian Wars, 1494–1512," *Renaissance and Reformation* 3 (1967): 12–16; Burchard, 1.312–15.

136. Malipiero, 508.

137. Infessura, 293.

138. Cherubino Ghirardacci, *Historia di Bologna*, ed. Albano Sorbelli, *RIS*[2] 33.1.12.

139. Ghirardacci, *Historia di Bologna*, 354.

140. Ghirardacci, *Historia di Bologna*, 370.

141. Ghirardacci, *Historia di Bologna*, 375.

142. Burchard, 2.296.

143. Pastor, 1.280.

144. Pastor, 4.323.

145. See chapter four.

146. Pastor, 4.375–6.

147. Navagero, cols. 1182–3; Malipiero, 281–3.

148. In Malipiero's telling, the spy only managed to post on the door of San Celso.

149. Navagero, col. 1183: "debbano avere diligente cura, che sopra le porte delle chiese o in altro luogo della città non sia affissa scrittura alcuna."

150. Navagero, col. 1182–3. Sigismondo dei Conti notes that in 1481 Ercole d'Este forbade his heralds from publishing papal bulls or other documents; Sigismondo dei Conti, *Le storie de' suoi tempi dal 1475 al 1510*, 2 vols. (Rome, 1883; repr., Foligno, 2015), 1.115.

151. Pastor, 6.19; Christian, *Empire without End*, 189.

152. Sanudo, 8.187; Pastor, 6.312.

153. Pastor, 6.352, citing Sanudo, 12.250–4.

154. Margaret Meserve, "The Papacy, Power, and Print: The Publication of Papal Decrees in the First Fifty Years of Printing," in *Print and Power*, ed. Nina Lamal (Leiden, 2021), 259–99.

155. Sixtus IV, Bulla, May 24, 1493, "Ad bonorum tutelam" ([Rome: Johannes Philippus de Lignamine, after May 24, 1483]) (is00555030), fol. 10r.

156. Julius II, *Monitorium contra Venetos* ([Rome, s.n., 1509]), sig. a8v.

157. Julius II, *Bulla monitorii apostolici contra tres reverendissimos cardinales ut redeant ad obedientiam* ([Rome, s.n., 1511]), fol. 1v.

158. Fernanda Ascarelli, *Annali tipografici di Giacomo Mazzocchi*, 30–1 (no. 6).

159. Flavia Bruni, "In the Name of God: Governance, Public Order, and Theocracy in the Broadsheets of the Stamperia Camerale of Rome," in *Broadsheets: Single-Sheet Publishing in the First Age of Print*, ed. Andrew Pettegree (Leiden, 2017), 139–61.

Chapter 2 • Humanists, Printers, and Others

1. Paul II, Bulla, April 19, 1470, "Ineffabilis providentia" ([Rome: Sixtus Riessinger, after April 19, 1470]) (ip00156800).

2. Fifty-one editions of Sistine texts were printed in Rome between Sixtus's election in 1471 and his death in 1484; Margaret Meserve, "The Papacy, Power, and Print: The Publication of Papal Decrees in the First Fifty Years of Printing," in *Print and Power*, ed. Nina Lamal (Leiden, 2021), 259–99. ISTC counts 680 editions published in Rome in this period.

3. These were Sweynheym and Pannartz, who moved to Rome from Subiaco in 1467; Ulrich Han, who also started work in Rome in 1467; Giovanni Filippo de Lignamine (1470); Georg Lauer (1470); and the anonymous printer of the April 1471 edition of Silius Italicus. Adam Rot's first dated edition appeared a month after Sixtus's coronation, in September 1471. Sixtus Riessinger, who started printing in Rome in 1468, had departed for Naples by early 1471; *BMC* 4.vii–xii.

4. On Bussi, see the introduction, notes 18–21.

5. Franciscus della Rovere, *De futuris contingentibus* and *De sanguine Christi et de potentia Dei* (is00560500 and is00579000). On de Lignamine, see *BMC* 4.xi: "his fondness for dedications and small editions [suggests] that he looked for his profits rather to rewards from patrons than to ordinary sales." Laura Onofri, "Figure di potere e paradigmi culturali," in *Un pontificato ed una città: Sisto IV (1471–1484)*, ed. Massimo Miglio et al. (Vatican City, 1986), 76–7; Egmont Lee, *Sixtus IV and Men of Letters* (Rome, 1978), 99–105. De Lignamine would also print the obedience oration Giovanni d'Aragona presented Sixtus IV in 1471, discussed in chapter seven.

6. Gabriele Archetti, "Le forme della propaganda politica alla fine del medioevo," *Nuova rivista storica* 80 (1996): 681–706.

7. Margaret Meserve, "News from Negroponte: Politics, Popular Opinion, and Information Exchange in the First Decade of the Italian Press," *Renaissance Quarterly* 59 (2006): 440–80; Setton, 2.298–313.

8. Agostino Pertusi, *La Caduta di Costantinopoli*, 2 vols. (Verona, 1976); Antonio Carile, *Testi inediti e poco noti sulla caduta di Costantinopoli* (Bologna, 1983); Marios Philippides and Walter K. Hanak, *The Siege and Fall of Constantinople in 1453* (Farnham, UK, 2011); Noel Malcolm, *Useful Enemies: Islam and the Ottoman Empire in Western Political Thought* (Oxford, 2019), chap. 1.

9. Printers also set up in Foligno, Trevi, Treviso, Verona, Padua, and Perugia during this period.

10. Meserve, "News from Negroponte," appendix.

11. See Setton, 2.299, for Paul II's letters to Naples, Milan, and Florence.

12. Malipiero, 60. Venice advised its ambassadors abroad to press this point; Setton, 2.300 n. 108, 304.

13. See Pastor, 4.178–9, for letters from Paul II to Milan, Venice, Florence, Savoy, Brandenburg, Frankfurt, and Cologne on August 23–25, 1470. See Setton, 2.308, for papal correspondence with France.

14. Margaret Meserve, "Patronage and Propaganda at the First Paris Press: Guillaume Fichet and the First Edition of Bessarion's *Orations against the Turks*," *Papers of the Bibliographical Society of America* 97 (2003): 521–88.

15. Setton, 2.304–5. Michele Canensi, *De vita et pontificatu Pauli II*, ed. Giuseppe Zippel, *RIS*² 3.16.169, mentions legates from Naples, Venice, Milan, and Florence; see Pastor, 4.176, for correspondence among the Italian powers in August 1470.

16. Meserve, "News from Negroponte," appendix, nos. 1–5: *Piante di Negroponte*, printed once, in Venice (il00029500); *Lamento di Negroponte*, printed four times, in Milan (twice: il00029450 and il00029470), Naples (il00029400), and Florence (il00029350).

17. Marina Beer et al., eds., *Guerre in ottava rima*, 4 vols. (Ferrara, 1988–89); Antonio Medin and Ludovico Frati, *Lamenti storici dei secoli XIV, XV, e XVI*, 3 vols. (Bologna, 1969); Lauro Martines, *Strong Words: Writing and Social Strain in the Italian Renaissance* (Baltimore, 2001), 232–48.

18. Armando Petrucci, "Alle origini del libro moderno: Libri da banco, libri da bisaccia, libretti di mano," in *Libri, scrittura e pubblico nel Rinascimento* (Rome, 1970), 137–56; Rosa Salzberg, *Ephemeral City: Cheap Print and Urban Culture in Renaissance Venice* (Manchester, 2014); and Salzberg, *The Cantastorie in Renaissance Italy: Street Singers between Oral and Literate Cultures*, ed. Luca Degl'Innocenti et al., special issue of *Italian Studies* 71, no. 2 (2016). For Rome, see M. Adorisio, "Cultura in lingua volgare a Roma fra Quattro e Cinquecento," in *Studi di biblioteconomia e storia del libro in onore di Francesco Barberi*, ed. Giorgio de' Gregori and Maria Valenti (Rome, 1976), 19–36; Paola Farenga, "'Indoctis viris . . . mulierculis quoque ipsis.' Cultura in volgare nella stampa romana?," in *Scrittura, biblioteche e stampa a Roma*, ed. Concetta Bianca et al., 2 vols. (Vatican City, 1980), 1.403–1.

19. Martines, *Strong Words*, 232–48.

20. Arnoldo della Torre, *Paolo Marsi da Pescina: Contribuito alla storia dell'Accademia*

Pomponiana (Rocca San Casciano, 1903), 144–50, 169–73; Angela M. Fritsen, "Testing Auctoritas: The Travels of Paolo Marsi, 1468–69," *International Journal of the Classical Tradition* 6 (2000): 356–82.

21. Johannes Jacobus Canis composed a panegyric honoring Canal's commission as captain general in 1470, later printed as *Ad Nicolaum Canalem classem contra Turcos ducentem carmen* (ic00095400).

22. For a summary of the work, see della Torre, *Paolo Marsi*, 178–91.

23. Paulus Marsus, *Lamentatio de crudeli Eurapontinae urbis excidio* ([Venice: Federicus de Comitibus, ca. 1471]) (im00284000; Meserve, "News from Negroponte," appendix, no. 6). Other works attributed to Conti's first Venetian press are also humanistic: a grammar by Guarino, Galeotto Marzio's *De homine*, and the poetry of Propertius, Tibullus, Ovid, and Dante. After the Venetian printing bubble burst, Conti moved to Iesi, where he began printing more marketable texts: law books, proclamations, vernacular devotional works, and, in 1474, an *Esortazione ai Cristiani contro il Turco*; Victor Scholderer, "Federico de' Conti and the First Books Printed at Iesi," in *Fifty Essays in Fifteenth- and Sixteenth-Century Bibliography* (Amsterdam, 1966), 131–4; *BMC* 7:lvi.

24. Paulus Marsus, *Lamentatio de crudeli Eurapontinae urbis excidio* ([Rome: Printer of Silius Italicus, ca. 1471]) (im00284200; Meserve, "News from Negroponte," appendix, no. 7). Four editions are attributed to this press: Silius Italicus, edited by Leto and including Leto's life of the poet; Marsi's poem; and editions of Columella and Martial. Marsi's *Lamentatio*, dedicated to Paul II, must have been printed before the pope's death on July 26, 1471.

25. Georgius Fliscus, *Eubois* ([Naples: Sixtus Riessinger, ca. 1471]) (if00195000; Meserve, "News from Negroponte," appendix, no. 8; *BMC* 6:xl–xli, 855).

26. *BMC* 4 dates it to the final months of 1470. Recently, an edition of the *Lamentatio Nigropontis* printed in Riessinger's types and signed Naples, 1470, has been identified in Prague: *GWM* 2711250.

27. Johannes Aloysius Tuscanus, *Declamationes in Turcum* ([Rome: Ulrich Han, ca. 1470–71]) (it00557600); and Raphael Zovenzonius, *Carmen concitatorium ad principes Christianos in Turcum* ([Venice: Adam de Ambergau, ca. 1471]) (iz00029000). Meserve, "News from Negroponte," appendix, nos. 9–10.

28. Rodericus Zamorensis, *Epistola de expugnatione Nigropontis* ([Rome: Ulrich Han, ca. 1470]) (ir00212000). Reprinted shortly afterward by Ulrich Zel in Cologne (ir00213000). Meserve, "News from Negroponte," appendix, nos. 11–2. Roman intellectuals inhabited a small world. Sánchez, as governor of Castel Sant'Angelo, had been Pomponio Leto's jailer after the Roman Academy conspiracy; the two maintained a correspondence during Leto's detention. Cardinal Bessarion wrote to Sánchez to ask for clemency on Leto's behalf.

29. *Lamentatio Nigropontis* (Naples: [Sixtus Riessinger], 1470) (il00029039), not in Meserve, "News from Negroponte," appendix. The reprint (il00029040), assigned to Rome, ca. 1472, is detailed in Meserve, "News from Negroponte," appendix, no. 13.

30. *Lamentatio Nigropontis*, fol. 3r–v.

31. *Piante de Negroponte, terzina* 45, ed. Polidori, 408.

32. *Lamento di Negroponte*, Florentine edition (*Guerre in ottava rima*, 4:65–9), *ottave* 49–66. Only the first stanza appears in the Neapolitan edition (*Guerre in ottava*

rima, 4:100), *ottava* 77. The author of the *Piante* likewise addresses appeals to Paul II, Frederick III, and Cristoforo Moro for a new crusade against the Turks (*Piante de Negroponte, terzine* 30–40, ed. Polidori, 407–8).

33. Antonio Cornazzano, *La vita di Cristo* ([Venice?: Printer of Cornazzano (*GW* 7550)], 1472) (ic00912000; Meserve, "News from Negroponte," appendix, no. 14). See R. L. Bruni and Diego Zancani, *Antonio Cornazzano: La tradizione testuale* (Florence, 1992), 97–104; D. Bianchi, "*Vita e passione di Cristo* di Antonio Cornazzano," *Bolletino Storico Piacentino* 61 (1966): 1–16; A. Certui Burgio, "La *Vita e passione di Cristo* di Antonio Cornazzano," *Aurea Parma* 65 (1981): 56–71. The poem was composed between April and August 1471.

34. Cornazzano, *Vita di Cristo*, sig. [h]4v.

35. Cornazzano, *Vita di Cristo*, sig. [h]5r.

36. Fieschi, to Ferrante; Sánchez and the anonymous *Lamentatio* author, to Bessarion; Toscani and Marsi, to Paul II; Cornazzano, to the doge and senate of Venice.

37. Bianchi, "*Vita*," 1; Cornazzano had worked for the dukes of Milan and would later find patronage at the Este court.

38. Della Torre, *Paolo Marsi*, 191–2; Fritsen, "Testing Auctoritas," 360.

39. Zovenzoni also composed additional texts for de Spira editions of Appian, Boccaccio, Cicero, Martial and Terence, and Donatus's commentary on Terence.

40. Cornazzano corrected Jenson's 1470 Eusebius; Jenson published his *Vita della Vergine* in 1471; *BMC* 7.1147.

41. The anonymous press that produced Cornazzano's *Vita di Cristo*, known as the "printer of Cornazzano," later printed his *Vita della Vergine* (*GW* 7557). On contractual arrangements for editors, see Brian Richardson, *Print Culture in Renaissance Italy: The Editor and the Vernacular Text, 1460–1600* (Cambridge, 2002), 7–12.

42. Orietta Rossini, "Entusiasmi e riserve nei circoli umanistici," in *Gutenberg e Roma*, ed. Massimo Miglio and Orietta Rossini (Naples, 1997), 97–112, at 103–4.

43. Roberto Weiss, "Un umanista e curiale del Quattrocento: Giovanni Alvise Toscani," *Rivista della storia della chiesa in Italia* 12 (1958): 322–33. Toscani also edited classical texts for Roman presses.

44. Margaret Meserve, "Nestor Denied: Francesco Filelfo's Advice to Princes on the Crusade against the Turk," *Osiris* 25 (2010): 47–65.

45. Bussi selected, prepared, and corrected texts for the press with philological assistance from Gaza.

46. Francesco Filelfo, *Epistolae* (Venice, 1502), fol. 229v (letter to Giovanni Andrea Bussi, February 13, 1471).

47. This edition appeared in March 1471; Martin Davies, "Juan de Carvajal and Early Printing: The 42-Line Bible and the Sweynheym and Pannartz Aquinas," *The Library*, 6th ser., 18 (1996): 193–215, at 206.

48. Meserve, "News from Negroponte," appendix, nos. 15–6; Meserve, "Patronage and Propaganda"; Norman Housley, *Crusading and the Ottoman Threat, 1453–1505* (Oxford, 2012), 168–70.

49. For a different view of Bessarion's involvement, see Dan Ioan Muresan, "Bessarion's *Orations against the Turks* and Crusade Propaganda at the *Große Christentag* of Regensburg (1471)," in *Reconfiguring the Fifteenth-Century Crusade*, ed. Norman Housley (London, 2017), 207–44.

50. Cf. Martin Lowry, "Cristoforo Valdarfer tra politici veneziani e cortigiani estensi," in *Il Libro a Corte*, ed. Amadeo Quondam (Rome, 1994), 273–84; Lowry, "Diplomacy and the Spread of Printing," in *Bibliography and the Study of 15th-Century Civilisation*, ed. John Goldfinch and Lotte Hellinga (London, 1987), 124–37.

51. Meserve, "Patronage and Propaganda," 532–7.

52. Anna Esposito and Diego Quaglioni, eds., *Processi contro gli Ebrei di Trento (1475–1478)*, vol. 1, *I Processi del 1475* (Padua, 1990); Wolfgang Treue, *Der Trienter Judenprozess: Voraussetzungen, Abläufe, Auswirkungen (1475–1588)* (Hannover, 1996); R. Po-Chia Hsia, *Trent 1475: Stories of a Ritual Murder Trial* (New Haven, CT, 1992); and most recently, Stephen D. Bowd and J. Donald Cullington, *"On Everyone's Lips": Humanists, Jews, and the Tale of Simon of Trent* (Tempe, AZ, 2012).

53. R. Po-Chia Hsia, *The Myth of Ritual Murder: Jews and Magic in Reformation Germany* (New Haven, CT, 1988); Miri Rubin, *Gentile Tales: The Narrative Assault on Late Medieval Jews* (Philadelphia, 1999).

54. F. J. Worstbrock, "Simon von Trient," in *Die deutsche Literatur des Mittelalters: Verfasserlexicon*, 2nd ed., ed. W. Stammler et al. (Berlin, 1977–), 8.1260–75; Lamberto Donati, *L'inizio della stampa a Trento ed il Beato Simone* (Trent, 1968), esp. 3–13; F. Hamster, "Primärliteratur zu Simon von Trient: Drucke und Handschriften vom 1475 bis 1500 mit Standortnachweisen," in *Per Frumenzio Ghetta O.F.M.: Scritti di storia e cultura ladina, trentina, tirolese* (Trent, 1991), 311–9; Paul Oskar Kristeller, "The Alleged Ritual Murder of Simon of Trent (1475) and Its Literary Repercussions: A Bibliographical Study," *Proceedings of the American Academy for Jewish Research* 59 (1993): 103–35; Treue, *Der Trienter Judenprozess*, 285–308, 340–8; M. Korenjak, "Latin Poetry about Simon of Trento," in *Acta Conventus Neo-Latini Budapestensis*, ed. R. Schnur et al. (Tempe, AZ, 2010), 397–406 (focusing on Puscolo); David Areford, *The Viewer and the Printed Image in Late Medieval Europe* (Farnham, UK, 2010), 164–227; Bowd and Cullington, *"On Everyone's Lips,"* 18–34; Stephen D. Bowd, "Tales from Trent: The Construction of 'Saint' Simon in Manuscript and Print," in *The Saint between Manuscript and Print*, ed. Alison Frazier Knowles (Toronto, 2015), 183–218.

55. Diego Quaglioni, *Apologia Iudaeorum; Invectiva contra Platinam: Propaganda antiebraica e polemiche di curia durante il pontificato di Sisto IV (1471–1484)* (Rome, 1987), 22–3.

56. Esposito and Quaglioni, *Processi*, 66, 77; Anna Esposito, "Il culto del 'beato' Simonino e la sua prima diffusione in Italia," in *Il principe vescovo Johannes Hinderbach (1465–1486) fra tardo Medioevo e Umanesimo*, ed. Iginio Rogger and Marco Bellabarba (Bologna, 1992), 429–43, at 432.

57. Esposito and Quaglioni, *Processi*, 80–95.

58. Esposito, "Il culto del 'beato' Simonino," 431–2.

59. Bowd and Cullington, *"On Everyone's Lips,"* 6.

60. Text in Bowd and Cullington, *"On Everyone's Lips,"* 36–57, with notes at 24–6; see also Treue, *Der Trienter Judenprozess*, 287–93. The date of the letter is problematic. Treue, *Der Trienter Judenprozess*, 293–4, reports that the original manuscript letter bears the date "xv kal. Aprilis," or March 18, which must have been a slip of the pen. Tiberino likely meant "xv kal. Maii," or April 17, a date that appears in two early editions of the text (Guldinbeck, June and July 1475). In most other editions, the letter is dated "secundo nonas Aprilis," or April 4. As Treue points out, Tiberino refers to details

of the supposed plot that the Jewish detainees only admitted to during interrogations on April 8, so the letter cannot have been composed before then. The date that appears on the original letter in Brescia, though garbled, is probably authentic: April 17. News of the event arrived in Brescia on April 19 (see next note). Treuer suggests that Hinderbach changed the date to make it seem closer to the events it described and so more authoritative.

61. Esposito and Quaglioni, *Processi*, 74–7, 86–8.

62. Po-chia Hsia, *Trent 1475*, 50. The trial resumed on June 6 (61–2).

63. Bowd and Cullington, *"On Everyone's Lips,"* 14: ducal instructions to officials in Brescia from April 24, July 17, and November 5, 1475, and three more from 1476 and 1477; Po-chia Hsia, *Trent 1475*, 52, notes similar letters to officials in Padua (April 22) and Friuli (April 28).

64. Esposito and Quaglioni, *Processi*, 83: an inventory of the Observant convent of Poor Clares in Bologna records "Una sceleragine de' giudei tradotta in volgare et stampata in Verona l'anno 1475 alli 22 di maggio, mandata all badessa et composta per Feliciano antiquario." See Agostino Contò, "'Non scripto calamo': Felice Feliciano e la tipografia," in *L'Antiquario' Felice Feliciano veronese tra epigrafia antica, letteratura, e arti del libro*, ed. Contò and Leonardo Quacquarelli (Padua, 1995), 289–312, at 308–11, who suggests that the text was the *Tormenti*. Esposito and Quaglioni, *Processi*, 83, and Treue, *Der Trienter Judenzprozess*, 290, both suggest that it was an Italian translation of Tiberino's letter. Either, or neither, could be true.

65. See note 60 above.

66. Bowd and Cullington, *"On Everyone's Lips,"* 37 n. 2; Baccio Ziliotto, *Raffaele Zovenzoni, La vita, i carmi* (Trieste, 1950), 13–63; Giovanni Dellantonio, "Felice Feliciano e gli amici del principe vescovo di Trento Iohannes Hinderbach: Raffaele Zovenzoni e Giovanni Maria Tiberino," in *L'Antiquario' Felice Feliciano*, 43–8.

67. Bowd and Cullington, *"On Everyone's Lips,"* 39.

68. Tiberino's letter, dated April 17, was addressed "magnificis rectoribus senatui populoque Brixiano" and refers to "vos . . . magnifici rectores civesque preclarissimi." Hinderbach in his cover letter told Zovenzoni that he was enclosing a letter that Tiberino "praetoribus civitatis Brixiae . . . misit" (Bowd and Cullington, *"On Everyone's Lips,"* 36). Zovenzoni changed the salutation of the letter to "Raphaeli Zovenzonio p[oetae] cla[rissimo]" (40) and changed the valedictory to "Vale, Raphaele Tergestine, Musarum decus" (56).

69. Bowd and Cullington, *"On Everyone's Lips,"* 36.

70. Johannes Mathias Tuberinus, *Relatio de Simone puero Tridentino* (Treviso: G[erardus de Lisa, de] F[landria], June 20, 1475) (it00486500); the undated edition is it00482600.

71. Tuberinus, *Relatio* (Sant'Orso: Johannes de Reno, [after April 4, 1475]) (it00484000).

72. Tuberinus, *Relatio* (Mantua: [Johannes Schallus, after April 4, 1475]) (it00482500).

73. Tuberinus, *Relatio* ([Augsburg: Monastery of SS. Ulrich and Afra, after April 4, 1475]) (it00483000) and ([Nuremberg]: Friedrich Creussner, [after April 4, 1475]) (it00485000).

74. Tuberinus, *Die geschicht und legend von dem seyligen kind und marterer genannt*

Symon von den Iuden zu Trientt gemarteret und getoettet ([Nuremberg: Friedrich Creussner, after April 4, 1475]) (it00489100).

75. Tuberinus, *Die geschicht und legend von dem seyligen kind und marterer genannt Symon von den Iuden zu Trientt gemarteret und getoettet* ([Augsburg: Günther Zainer, [after April 4, 1475]) (it00489000).

76. "Ii nons aprilis, am vierden tag in dem aprill."

77. The Roman edition (it00488000) is assigned to Guldinbeck, ca. 1475–76; Treue, *Der Trienter Judenprozess*, 289 n. 22, notes an argument that this Guldinbeck edition was the first, printed before the dated editions of June and July 1475. Esposito and Quaglioni follow this claim, as does Po-chia Hsia, *Trent 1475*, 56. However, the two dated editions print the letter without its salutation and with the (correct) date of April 17, 1475. In the undated edition, the letter is addressed to the Brescians, it is dated April 4, and the *Miraculum* appears at its conclusion. It seems unlikely that Guldinbeck would have printed this redaction first and then switched to a different version for his edition of June 19. More likely, this redaction, which was widely printed in northern Italy, reached Guldinbeck and was printed by him sometime later (bibliographers have assigned it to as late as 1476). The Naples edition (it00488500) is assigned to 1477 and the Cologne one (it00482000) to 1478–81.

78. Tuberinus, *Relatio* (it00486000 and it00487000).

79. On Guldinbeck in Rome, see *BMC* 4.xiv, 48, 67.

80. Po-chia Hsia, *Trent 1475*, 69: in early summer 1475, Hinderbach wrote to correspondents in Vicenza, Innsbruck, and Rome, as well as Venice, and sent copies of Tiberino's letter to at least some of them.

81. Po-chia Hsia, *Trent 1475*, 69–70, 119–20.

82. "Quia multi et magni quidem viri submurmurare coeperunt," in Quaglioni, *Apologia*, 15.

83. "Sempre sia laude a te padre signore" ([Venice: Nel Beretin Convento della Ca Grande, ca. 1476]) (is00528450). See Martin Lowry, "'Nel Beretin Convento': The Franciscans and the Venetian Press (1474–1478)," *La Bibliofilia* 85 (1983): 39–54.

84. Johannes de Lübeck, *Prognosticon super Antichristi adventu Judaeorumque Messiae* ([Padua]: Bartholomaeus de Valdezoccho, [not before April 1474]) (ij00376000).

85. Martin Lowry, "Humanism and Anti-Semitism in Renaissance Venice: The Strange Story of 'Decor Puellarum,'" *La Bibliofilia* 87 (1986): 39–54.

86. Martin Lowry, *Nicholas Jenson and the Rise of Venetian Publishing in Renaissance Europe* (Oxford, 1991), 120–1.

87. Matheus Künig, *Gedicht von dem getöteten Knaben Simon von Trient* ([Sant'Orso?]: Johannes de Reno, 1475) (ik00040900).

88. *Tormenti del Beato Simone da Trento* (Sant'Orso: Johannes de Reno, June 24, 1475) (is00528500). The colophon mentions Giovanni da Conegliano in an indeterminate way; Dennis Rhodes, *La stampa a Treviso nel secolo XV* (Treviso, 1983), s.v., attributes the text to him (see also Contò, "'Non scripto calamo,'" 310 n. 3). The attribution to Feliciano is based on the inventory entry that mentions a book about Simon written by Feliciano and printed in Verona on May 22, 1475. No such edition survives, and in fact there were no printers active in Verona between about 1472 and 1478, so the entry is puzzling. The Treviso edition is is00528700.

89. Thomas Pratus, *De immanitate Judaeorum in Simonem infantem* (Treviso, [September 1475?]) (ip00957000).

90. Dennis E. Rhodes, *La tipografia nel secolo XV a Vicenza, Santorso e Torrebelvicino* (Vicenza, 1987).

91. On Kunne's press, see *BMC* 3.xxvii, 805 (mistakenly listed as the press of Hermann Schindeleyp); among his output were four Simon texts, an Italian *lamento* for the fall of Caffa to the Turks (the poem dated June 7, 1475), and a German lampoon of the Ottoman sultan, *Turckisher Kaiser bin ich*.

92. *Historie von Simon zu Trient* (Trent: Albrecht Kunne, September 6, 1475) (is00528800). Treue, *Der Trienter Judenprozess*, 290, confirms that there are no clues to the author's identity but asserts that it was not Tiberino, nor is it a German translation of Tiberino's work (cf. *Trent 1475*, 56).

93. Po-chia Hsia, *Trent 1475*, 69–70.

94. The same shield appears on the floor before the bishop in a painting of the Virgin and Child with various saints that he commissioned, and in which he appears as the donor, and on his tomb relief (*Il principe vescovo*, plates 2 and 3, after 380), both in the Diocesan Museum of Trent.

95. *Historie von Simon zu Trient*, fol. 10r: "Also stet sant Peter zu dem houpt des seligen kindes, und sant Vigilius zu den fuessen, der ain pischoff zu Trient was." The image is on fol. 9v. See Treue, *Der Trienter Judenprozess*, 359.

96. In the painting of Hinderbach with the Virgin and Child with SS. Jerome, John, Peter, and Paul (see above), the bishop kneels to the Virgin's left and in front of Peter and Paul; Peter bears a key but no papal regalia. Hannes Obermair and Michaela Schedl date the painting to the 1470s and suggest that it was associated with Hinderbach's restoration of the church of Peter and Paul and may have been the altarpiece; Hannes Obermair and Michaela Schedl, "Art in the Double Periphery: Commissions Ordered by the Bishops Johannes Hinderbach and Ulrich von Liechtenstein in Early Modern Trento," *Concilium medii aevi* 21 (2018): 53–73. That the woodcut artist of the *Historie* chose to represent the church's patron as devoted to Simon *and* as pope seems deliberate and connected to Hinderbach's concern to win Sixtus IV's support for the cult or to make it *seem* that he supported it.

97. Po-chia Hsia, *Trent 1475*, 78–9.

98. Po-chia Hsia, *Trent 1475*, 76.

99. Institoris, the future witch-hunter, baptised some of the Jewish captives before their executions and signed his name as a witness to the trial proceedings. He appears again in chapter four on the other side of an ecclesiastical dispute, publishing texts defending Sixtus against the conciliar agitator Andrija Jamometić in Basel. See Richard Kieckhefer, "The Office of Inquisition and Medieval Heresy: The Transition from Personal to Institutional Jurisdiction," *Journal of Ecclesiastical History* 46 (1995): 36–61, at 46. I am grateful to Prof. Kieckhefer for the reference.

100. On the Venetian order of August 12, 1475, see Ziliotto, *Raffaele Zovenzoni*, 50–1.

101. Bowd and Cullington, *"On Everyone's Lips,"* 15, 27–8.

102. Laura Dal Prà, "L'immagine di Simonino nell'arte trentina dal XV al XVIII secolo," in *Il principe vescovo*, 445–81 at 446–7: "Quam mihi laetitiam attulerit imago sanctissima quam ad me misisti, non possem vel coram aperire."

103. Boyd and Cullington, *"On Everyone's Lips,"* 14 n. 52; Po-chia Hsia, *Trent 1475*,

69; Quaglioni, *Apologia*, 21; E. Martène, *Veterum scriptorum*, 2 (Paris, 1724), cols. 1516–7: "accepimus nonnullos publice etiam in praedicationibus affirmare ipsum Simonem, quem beatum appellant . . . martyrem esse et plurimis miraculis clarescere . . . alios desuper imagines depingere, cartas in modum historiarum conscribere, et publice vendere, et venditioni exponere, et propterea Christi fideles contra Judaeos omnes eorumque bona incitare . . . Talia per bandimentum publicum in hac nostra urbe alma fieri amodo prohibuimus, et idcirco vobis tenore praesentium motu proprio et ex pastoralis officii debito committimus et districte etiam sub paena excommunicationis mandamus . . . eadem fieri."

104. Text in Bowd and Cullington, *"On Everyone's Lips,"* 14, 15 n. 52: "Scripsit summus pontifex omnibus dominis et potestatibus breve, per quod declarat et iubet sub pena excommunicationis ut nuper ille qui Tridenti ab Judeis interfectus dicitur, pingi non sinatur, neque illa res *ab scriptoribus imprimi*, aut a predicatoribus diffamari instigarique vulgus contra Judeos."

105. Bowd and Cullington, *"On Everyone's Lips,"* 14, 15 n. 54: "non dobia . . . epistole scrivere, ne scritte vendere."

106. Peter Parshall and Rainder Schoch, *Origins of European Printmaking: Fifteenth-Century Woodcuts and Their Public* (New Haven, CT, 2005), 208–10, Schreiber, 1967–70; Dal Prà, "L'immagine di Simonino"; and Dominique Rigaux, "L'immagine di Simone di Trento nell'arco alpino lungo il secolo XV," in *Il principe vescovo*, 445–82, 485–96. Areford, *Viewer*, 164–227, reproduces almost all known woodcut images of Simon.

107. Schreiber, nos. 1967 and 1969. A similar scene appears in the 1475 *Historie* and the Augsburg edition of Tiberino's letter in German. The illustration in the *Nuremberg Chronicle* derives from this tradition. Schreiber, no. 1967, is tipped into the back of a copy of Tiberino that was owned by Hartmann Schedel; Parshall and Schoch, *Origins of European Printmaking*, 211 n. 9.

108. Schreiber, 1969a.

109. Schreiber, 1967a, 1968. See Béatrice Hernad, ed., *Die Graphiksammlung des Humanisten Hartmann Schedel* (Munich, 1990), 214–5, cat. 53.

110. The 1476 broadside, *Epitaphium gloriosi pueri Simonis Tridentini novi martyris* ([Trent: Albrecht Kunne, ca. 1476]) (it00479500), shows him on a throne; the 1498 one, *Gedicht von dem Knaben Simon zu Trient* ([Ulm: Johann Zainer the Younger, ca. 1498]) (is00528750), shows him standing in triumph with a banner. The comparison with the resurrected Christ is made by Parshall and Schoch, *Origins of European Printmaking*, 212 n. 22. For the 1498 broadside, see Paul Heitz, ed., *Einblattdrucke des Fünfzehnten Jahrhunderts*, 100 vols. (Strasbourg, 1899–1942), vol. 31 (1912), no. 109; VE15 S-38.

111. Esposito, "Il culto del 'beato,'" 433: "Fecit ad se portare imaginem beati Symonis, quam emerat Venetiis."

112. Esposito and Quaglioni, *Processi*, 85, esp. n. 94.

113. Dal Prà, "L'Immagine," 449–50.

114. Biblioteca Classense, inventory no. 22, extracted from MS 374, fol. 144r.; Fiora Bellini, *Xilografie italiane del Quattrocento da Ravenna e da altri luoghi* (Ravenna, 1987), no. 46; Areford, *Viewer*, 120–2.

115. Quaglioni, *Apologia*, 29, 109.

116. Po-chia Hsia, *Trent 1475*, 79–80.

117. Sylvester de Balneoregio, *Conclusiones cum earum declarationibus super*

canonizatione Simonis Tridentini ([Trent: Albrecht Kunne, after December 6, 1475])
(is00880000), fol. 9v.

118. Esposito and Quaglioni, *Processi*, 90–1; Silvestro's letter is edited on 447–8.

119. Johannes Mathias Tuberinus, *Historia completa de passione et obitu pueri Simonis* (Trent: [Albrecht Kunne for] Hermann Schindeleyp, February 9, 1476) (it00481000); see Bowd and Cullington, *"On Everyone's Lips,"* 28. Zovenzoni also believed that his daughter was saved from death after he made a vow to Simon.

120. Bowd and Cullington, *"On Everyone's Lips,"* 121–3.

121. Bowd and Cullington, *"On Everyone's Lips,"* 117–31.

122. Johannes Franciscus de Pavinis, *Responsum de jure super controversia de puero Tridentino a Judaeis interfecto* (Rome: Apud Sanctum Marcum (Vitus Puecher), 1478) (io00062370).

123. Bowd and Cullington, *"On Everyone's Lips,"* 28–32, with the text of the letter, dated June 29, 1481, on 120–1. Puscolo's text was not printed till 1511; Bowd suggests that it was too long to be printed earlier. On Hinderbach's suggestion, Puscolo prepared a deluxe manuscript copy for Maximilian of Austria sometime in 1481–82. Hinderbach himself wrote a *Historia* of the episode, which was never printed; he also commissioned deluxe manuscript copies of the trial records and sent them to elite recipients. See Esposito and Quaglioni, *Processi*, 66, 76–7.

124. Johannes Calphurnius, *Mors et apotheosis Simonis infantis novi martyris* ([Trent]: Z. L. (Giovanni Leonardo Longo), [ca. 1481]) (ic00062000), with poems by Zovenzoni. Longo operated presses in Vicenza, Torrebelvicino, and Trent, where he began printing around 1481 (*BMC* 3.805–6). Calfurnio's poem was also printed at the back of his edition of Catullus, Tibullus, Propertius, and Statius, printed at Vicenza by Johannes de Reno ca. 1481 (ic00323000). On September 5, 1482, Longo completed an edition of Tiberino's *Epigrammata* on Simon (it00479000). Giorgio Sommariva, *Martirio di Simone da Trento*, a verse lament with poems by other humanists, was first printed in Verona in 1478 (is00679300) and republished twice in Treviso in the next decade (May 12, 1480, and July 14, 1480, is00628000 and is00629000). Another of his poems on Simon was printed by a different Treviso printer in September 1484 (is00630000). Around 1493, Quintus Aemilianus's *Historia beati Simonis Tridentini* appeared from a press in Vienna (ia00096500); in this same year Hartmann Schedel devoted a page of the *Nuremberg Chronicle* to Simon's martyrdom and death.

Chapter 3 • Sixtus IV and his Pamphlet Wars

1. Sixtus IV, Bulla, August 29, 1473, "Quemadmodum operosi vigilisque pastoris solertia." Three Roman editions, all unsigned, are attributed to Georg Lauer, Johannes Gensberg, and the Printer of the *Mercuriales quaestiones* (is00537800; is00537830; is00537850).

2. Three more editions were printed in Germany, and there were at least two more reprints in Rome after his death.

3. Bulle, a native of Bremen, was only active in Rome in 1478 and 1479, but he published other Sistine documents, including a bull concerning benefices, a *formularium* for ecclesiastical trials, and an edition of the *Taxae cancellariae apostolicae*; *BMC* 4.78–9.

4. Pastor, 4.300–30; Lauro Martines, *April Blood: Florence and the Plot against the Medici* (Oxford, 2003), esp. 178–85; Marcello Simonetta, *The Montefeltro Conspiracy: A*

Renaissance Mystery Uncoded (New York, 2008); Simonetta, *La congiura della verità* (Naples, 2012); Tobias Daniels, "Die Pazzi-Verschwörung, der Buchdruck und die Rezeption in Deutschland. Zur politischen Propaganda in der Renaissance," *Gutenberg-Jahrbuch* 87 (2012): 109–20; and Daniels, *La congiura dei Pazzi: I documenti del conflitto fra Lorenzo de' Medici e Sisto IV* (Florence, 2013).

5. George Holmes, "The Pazzi Conspiracy as Seen from the Apostolic Chamber," in *Mosaics of Friendship: Studies in Art and History for Eve Borsook,* ed. Ornella Francisci Osti (Florence, 1999), 163–73.

6. Martines, *April Blood,* 96–110.

7. Martines, *April Blood,* 164–73.

8. Martines, *April Blood,* 158, quoting Montesecco's confession of May 4, 1478; on the question of the pope's words, whether he said them, and whether he meant them, see 159–61.

9. Samuel Edgerton, *Pictures and Punishment: Art and Criminal Prosecution during the Florentine Renaissance* (Ithaca, 1985).

10. Lorenzo de' Medici to René of Anjou, June 19, 1478, in Lorenzo de' Medici, *Lettere,* ed. Nicolai Rubinstein (Florence, 1977), 3.72–4, at 73; Lorenzo to Guillaume d'Estouteville, June 1, 1478, in Lorenzo, *Lettere,* 3.42.

11. Sixtus IV, Bulla, June 1, 1478, "Ineffabilis et summi patris providentia," in Daniels, *La congiura,* 105–12, with analysis at 23–8; Pastor, 4.316–7; Martines, *April Blood,* 177.

12. Lorenzo had supported local lords in Umbria and northern Lazio against their overlord, the pope.

13. This was a group of German merchants with whom the Apostolic Camera did regular business, who had been caught in a customs dispute at the Florentine border; Sixtus, invoking the terms of the bull *In coena Domini,* declared that Florence had violated the immunities due "provisioners" of the papal court and then gilded the lily by declaring them *Romipeti* as well.

14. The Milanese ambassadors in Rome wrote on June 5 that the promulgation had occurred the previous day: Lorenzo, *Lettere,* 3.46. For June 10, see Daniels, *La congiura,* 23, 27, but cf. Luca Landucci, *A Florentine Diary from 1450–1516,* trans. Alice de Rosen Jervis (New York, 1969), 20: June 7 was the day the Florentines learned that "the pope excommunicated us."

15. Quoted by Rubinstein in Lorenzo, *Lettere,* 3.48. The bull was published in Naples on June 14; see 3.75 n. 2.

16. Bulla, June 1, 1478, "Ineffabilis et summi patris providentia" ([Rome: Johannes Bulle, after June 1, 1478]) (is00545000).

17. Bulla, June 1, 1478, "Ineffabilis et summi patri providentia" ([Rome:] Johannes de Monteferrato and Rolandus de Burgundia, [after June 1, 1478]) (is00545500); Albano Sorbelli, "La scomunica di Lorenzo de' Medici in un raro incunabolo romano," *L'Archiginnasio* 31 (1937): 331–5. The first page is blank; the text fills fols. 1v–2v. Sorbelli suggests that the recto of the first leaf would have been glued to a wall or door, with the second leaf floating free, allowing the public to view all three pages of text (333), but this seems an impractical method of display.

18. Martines, *April Blood,* 177; Alison Brown, *Bartolommeo Scala, 1430–1497, Chancellor of Florence: The Humanist as Bureaucrat* (Princeton, NJ, 1979), 159.

19. Sorbelli, "La scomunica," 335; Pastor, 4.327.

20. The "Christian princes" were an elastic group, which could stretch to several dozen recipients depending on the gravity of the message a pope desired to send. The list usually included the emperor and the kings of France, Naples, Spain, and England, and sometimes those of Poland, Portugal, Denmark, and Scotland; major dukes (Burgundy, Brittany, Bavaria, Savoy, Milan, etc.); imperial electors and counts palatine; and the communal governments of republics like Florence and Venice. Other authors could address the same group—for example, Doge Cristoforo Moro and Cardinal Bessarion sent open letters to "the Christian princes" after the fall of Negroponte.

21. Sixtus IV to Federico da Montefeltro, July 5, 1478, quoted in Brown, *Bartolomeo Scala*, 159 n. 68: "Al re di Francia habbiamo mandato multi nuncii con nostre justificationi, similiter alli altri Principi, come lo è l'Imperatore, Re di Ungheria, di Spagna, e tutti li altri. Praeterea la Bolla nostra, *quale è stata posta in stampa a tutto il mondo*, dimostrerà nostra justificatione" (italics mine).

22. Becchi's text is discussed below. See also a letter of L. Botta quoted by Brown, *Bartolomeo Scala*, 159 n. 68: "Perchè non sequendo pace, me pare essere certo ch'el pontefice con le consuete justificationi sue, le faria mettere in stampa et le mandaria ad sua justificatione ad tutti li potentati christiani."

23. Richard C. Trexler, *The Spiritual Power: Florence under the Interdict* (Leiden, 1974), 167, 170–3.

24. Bulls of June 22, 1478, "Ad apostolicae dignitatis auctoritatem" and "Inter cetera quorum nos cura sollicitat," in Daniels, *La congiura*, 115–21. The printed edition is is00545550.

25. Martines, *April Blood*, 86.

26. Landucci, *Florentine Diary*, 22.

27. Bulls of excommunication or privation printed outside Rome before this time were almost certainly printed without the pope's knowledge or instruction. In 1476, a Roman printer published Sixtus IV's bull *In coena Domini* (April 11, 1476: is00538500), but this contains only the usual formulas published annually on Maundy Thursday. The Sistine bulls of June 1478 were the first bulls printed in Rome to target a particular political opponent.

28. Brief "Si qui sunt," addressed "prioribus et populo nobis dilecto Communitatis Florentiae." F. Di Benedetto, "Un breve di Sisto IV contro Lorenzo," *ASI* 159 (1992): 371–84, at 374–6. Grifo was Francesco della Rovere's private secretary from 1467 and apostolic secretary throughout his pontificate; Marcello Simonetta, "Leonardo Griffi," *DBI* 59.360–3.

29. Di Benedetto, "Un breve," 374: "Si qui sunt qui existiment nos defecisse a desiderio iuvandae reipublicae Christianae, et arma adversus civitatem istam movere, errant quidem vehementer."

30. Lorenzo, *Lettere*, 3.131 n. 13. See also Di Benedetto, "Un breve," 381.

31. *Litterae directae universitati et hominibus Florentiae* ([Naples: Sixtus Riessinger or Francesco del Tuppo, not after July 1478]) (is00574500). The text survives in a unique copy in the Biblioteca Casanatense consisting of a single sheet folded into two leaves, the second blank. The printer may have intended to print the text again on the second leaf and cut the sheet to make two printed broadsides with text on recto and verso.

32. Di Benedetto, "Un breve," 382–3.

33. Paolo Farenga, "Francesco del Tuppo," *DBI* 38.317–21.

34. Del Tuppo and Riessinger printed legal books for the kingdom later in the 1480s, and del Tuppo would print indictments against Neapolitan noblemen after the Conspiracy of the Barons.

35. Lorenzo, *Lettere*, 3.131–2 n. 13.

36. "Mirati primum sumus," in Di Benedetto, "Un breve," 376–8; Bartolomeo Scala, *Humanistic and Political Writings*, ed. Alison Brown (Tempe, AZ, 1997), 195–8; Daniels, *La congiura*, 37.

37. Di Benedetto, "Un breve," 372, records six; Daniels, *La congiura*, 37 n. 43, identifies two more.

38. Daniels, *La congiura*, 39.

39. "Si qui sunt, frater Francisce (postquam pater esse non vis), qui existiment nos semper tyrannidem non odisse, vehementer falluntur." Compare with the text in note 29 above.

40. See Di Benedetto, "Un breve," 379–81. The letter levels charges that would reappear in drafts of Becchi's *Florentina synodus*. Daniels suggests that Becchi is the likely author of the letter. For the charges of luxury and sexual license, see Paola Farenga, "Monumenta memoriae: Pietro Riario fra mito e storia," in *Un pontificato ed una città*, 179–216, at 212 n. 92. On Teresa, see Simonetta, *La congiura della verità*, 93 n. 3; and Gerald Stanley Davies, *Renascence: The Sculptured Tombs of the Fifteenth Century in Rome* (London, 1910), 103–5.

41. Di Benedetto, "Un breve," 381.

42. Bartolomeo Scala, *Excusatio Florentinorum* (without the transcript of Montesecco's confession), in *Humanistic and Political Writings*, 199–202; see also Brown, *Bartolomeo Scala*, 84–7, 158–9; Martines, *April Blood*, 181–2.

43. Scala, *Excusatio*, in *Humanistic and Political Writings*, 199.

44. Cf. Daniels, *La congiura*, 61: the confession is not an appendix to the work but an integral—indeed, the major—part of it. In the printed edition (Bartholomaeus Scala, *Excusatio Florentinorum ob poenas de sociis Pactianae in Medices coniurationis* [Florence: Nicolaus Laurentii, Alamanus, after August 11, 1478]; is00300700), Montesecco's confession starts at the bottom of the first folio and continues for another six; the book is only seven and a half folios long.

45. Martines, *April Blood*, 158.

46. Scala, *Excusatio Florentinorum*, fols. 5v–7v.

47. Daniels, *La congiura*, 30 and 40–1, esp. n. 59.

48. Daniels, *La congiura*, 202, a deliberate echo of Lorenzo's words to René of Anjou (above).

49. Daniels, *La congiura*, 202: "est doctissimorum iurisconsultorum et collegiorum declaratum testimonio, et publicis eorum scriptis in aperto positum."

50. Daniels, *La congiura*, 199 n. b.

51. Roberto Ridolfi, "Contributi sopra Niccolò Todesco," *La Bibliofilia*, 58 (1956): 1–14; Ridolfi, "Le ultime imprese tipografiche di Niccolò Todesco," *La Bibliofilia* 67 (1965): 140–51; L. Böninger, "Ricerche sugli inizi della *stampa* fiorentina," *La Bibliofilia* 105 (2003): 225–48; Concetta Bianca, "Le dediche a Lorenzo nell'editoria fiorentina,"

in *Laurentia laurus, per Mario Martelli*, ed. F. Bausi and V. Fera (Messina, 2004), 51–89; Dennis Rhodes, *Gli annali tipografici fiorentini*, 117.

52. Scala, *Excusatio Florentinorum*, fol. 1v:

SINGULIS ATQUE UNIVERSIS IN QUOS HAEC SCRIPTA
inciderunt Priores libertatis et vexillifer iustitiae et Populus Florentinus
.SALUTEM.

53. Scala, *Excusatio*, in *Humanistic and Political Writings*, 199.

54. Scala, *Excusatio*, in *Humanistic and Political Writings*, 200.

55. Text edited by Simonetta, *La congiura della verità*, 90–168, and Daniels, *La congiura*, 122–60, with analysis at 45–80. See also Martines, *April Blood*, 179–81. The printed edition is *Florentina synodus ad veritatis testimonium et sixtianae caliginis dissipationem* ([Florence: Nicolaus Laurentii, Alamanus, after July 20, 1478]) (if00207300).

56. Simonetta, *La congiura della verità*, 90.

57. Simonetta, *La congiura della verità*, 90.

58. Simonetta, *La congiura della* verità, 92.

59. Simonetta, *La congiura della* verità, 92.

60. Simonetta, *La congiura della verità*, 150.

61. Simonetta, *La congiura della verità*, 158: "Hinc illam [bullam] imprimi fecit, non contentus calamo; illam vendi in campo Flore, non contentus valvis ecclesiarum."

62. One manuscript of the text is dated July 20 and another July 23, but both dates are fictive.

63. Daniels, *La congiura*, 51–60, 79–80.

64. His changes were incorporated into the printed edition; Daniels, *La congiura*, 47, 57, 79.

65. On August 19, Lorenzo told Becchi to show the text to "one or two jurists" for their review; Lorenzo, *Lettere*, 3.181–2.

66. Lorenzo, *Lettere*, 3.51–52: "et guardate non s'adoperi per altro che allo impressore." See also the letter quoted on 57.

67. Simonetta, *La congiura della verità*, 21; Daniels, *La congiura*, 51–4.

68. The jurists included Francesco Accolti (then teaching law at Siena), Bartolomeo Sozzini (Pisa), and an anonymous team described as Florence's *collegium doctorum legium*. Their *consilia* examine the legality of Sixtus's bulls of excommunication and interdict. Among their conclusions: Sixtus denied Lorenzo due process by excommunicating him without issuing a summons or letting him defend himself; it was impossible to prove who had killed Salviati; because the archbishop had dressed in layman's clothing, he had forfeited his ecclesiastical immunities, anyway; the Florentines had not *arrested* the cardinal but placed him in protective custody at his own request; there were no grounds for punishing the entire city with an interdict because some in the government had been accused (not convicted) of violating ecclesiastical laws. See Kenneth Pennington, *The Prince and the Law, 1200–1600: Sovereignty and Rights in the Western Legal Tradition* (Berkeley, 1993), 238–68. Several of the *consilia* were printed later in the fifteenth century (241 n. 16, 242 n. 20, 248 n. 46, 268), but it seems that none were put into print in 1478. Rather, Lorenzo drew on them in his private correspondence with allies and supplied them to the Florentine authors charged with composing

publica scripta in his defense. Enrico Spagnesi, "In difesa del Magnifico: A proposito di alcuni *Consilia* legali al tempo della Congiura dei Pazzi," in *La Toscana al tempo di Lorenzo il Magnifico: Politica, Economia, Cultura, Arte* (Pisa, 1996), 3.1233–53.

69. Brown, *Bartolomeo Scala*, 202; Daniels, *La congiura*, 29, 42–3, 61–2.

70. Pastor, 4.320–4; Aldo Landi, *Concilio e papato nel Rinascimento (1449–1516): Un problema irrisolto* (Turin, 1997), 114–9; Daniels, *La congiura*, 33. The council, which was supposed to convene in Lyons in May 1476, never met.

71. Landi, *Concilio e papato*, 123–9; Landi, "Prolungamenti del movimento conciliare e riflessi nella politica di Firenze in età Laurenziana," in *La Toscana al tempo di Lorenzo il Magnifico*, 3.1255–73, at 1265–7; Daniels, *La congiura*, 33–5, 51.

72. Landi, *Concilio e papato*, 126.

73. Daniels, *La congiura*, 51–7; J. Blanchard, *Commynes et les italiens: Lettres inédits* (Klincksiek, 1993); on July 16, Cardinal Ammannati warned Sixtus that Louis planned to call a council; Daniels, *La congiura*, 35.

74. Daniels, *La congiura*, 39–42, 60–2.

75. Daniels, *La congiura*, 61.

76. Daniels, *La congiura*, 41.

77. Angelo Poliziano, *Della congiura dei Pazzi*, ed. Angelo Perosa (Florence, 1958); Simonetta, *La congiura della verità*, 55–85.

78. Simonetta, *La congiura*, 16.

79. Angelus Politianus, *Pactianae coniurationis commentarium* ([Florence: Nicolaus Laurentii, Alamanus], 1478) (ip00892500).

80. *Tradimento per la morte di Giuliano de' Medici* ([Florence: Apud Sanctum Jacobum de Ripoli, 1478]) (it00421600); Francesco Flamini, "Versi in morte di Giuliano de' Medici, 1478," *Il Propugnatore*, n.s., 22 (1889): 315–34; Martines, *April Blood*, 183–4; Melissa Conway, *The "Diario" of the Printing Press of San Jacopo di Ripoli, 1476–84* (Florence, 1999), 32 n. 76, 162.

81. Flamini, "Versi," 330–4, edits a poem by Luigi Pulci, "Morale . . . per la morte di Giuliano," which did not appear in print.

82. *Dissentio inter papam et Florentinos suborta*, in Daniels, *La congiura*, 162–74, with commentary at 81–97. The text addresses "all beloved Christians" and shares damaging information about Lorenzo that "every Christian should know" (162).

83. Daniels suggests that the author may have been the papal librarian, Bartolomeo Platina (*La congiura*, 84–7), but see the review by M. G. Blasio in *Roma nel Rinascimento* (2014), 53–8.

84. Daniels, *La congiura*, 164: "nullo sanctissimi domini nostri, neque dignitatis sue . . . habito respectu."

85. Daniels, *La* congiura, 166: "[Sixtus] quem Florentina sodomitica hereticaque sinodus iudicare non veretur." This could be a reference to Becchi's reputation for homosexuality (166), but there was a long tradition of impugning Florence as a whole as a city of sodomites.

86. Daniels, *La congiura*, 162: "Ad . . . rei ipsius elucentem veritatem quam, ob falsos relatores florentinos schismaticos hereticos excommunicatos anathematizatos ac sancte matris ecclesie persecutores et publicos hostes, plurimos Christi fideles ignorare con[i]cio, hec in publicum quorumcumque fidelium prodire cogit virtus." I depart from the punctuation Daniels imposes on the text here.

87. Daniels, *La congiura*, 173: "ob tam execrandam iniuriam non modo ecclesie Romane private, sed universis Christiano nomine illustratis per Florentinos illatam."

88. Sixtus IV, *Dissentio inter papam et Florentinos suborta* ([Basel: Johann Amerbach, after July 1, 1478]) (is00558000).

89. Barbara C. Halporn, *The Correspondence of Johann Amberbach: Early Printing in Its Social Context* (Ann Arbor, 2000), 2–5; *BMC* 6.742.

90. Sigismondo dei Conti, *Le storie de' suoi tempi dal 1475 al 1510*, 2 vols. (Rome, 1883; repr., Foligno, 2015), 1.35–41.

91. Conti, *Storie*, 36: "Rem paullo altius repetemus, palamque faciemus necessario hoc bello gerendum esse, si pace Italiae frui volumus; nec posse Italiam pace frui, si hoc bellum omissum est."

92. Conti, *Storie*, 40.

93. Conti, *Storie*, 41.

94. Also relevant: Cola Montanus's oration to the people of Lucca encouraging them to join Sixtus and Ferrante in an alliance against the Medici; delivered in November 1478, it was printed by Bulle in Rome the next year. Tobias Daniels, *Umanesimo, congiure e propaganda politica. Cola Montano e l'*Oratio ad Lucenses (Rome, 2015). It was reprinted twice in Rome in the next decade by Plannck and Silber.

95. Pastor, 4.319.

96. Martines, *April Blood*, 184–5, 189; Alison Brown, "Lorenzo and Public Opinion in Florence," in *Medicean and Savonarolan Florence: The Interplay of Politics, Humanism, and Religion* (Turnhout, 2011), 87–111, at 103 n. 56: "Ogni notte sonno apichate scripte ali cantoni di Firenze contro Lorenzo."

97. Bulla, August 16, 1479, "Et si natura omnibus in rebus" ([Rome: Georgius Lauer, after August 16, 1479]) (is00547700), excommunicating Roberto Malatesta of Rimini, Costanzo Sforza of Pesaro, Antonello Zampeschi of Forlì, and Galeazzo Sanseverino of Sarsina; confiscating their goods; and declaring their titles void.

98. Sixtus IV, Bulla, August 17, 1479, "Decet Romanum pontificem" ([Rome: Georgius Lauer, after August 17, 1479]) (is00547800), fols. 1v–2r: "affirmavimus nos promptos et paratos ad pacem semper fuisse, et esse nec quicquam ullo tempore magis desiderasse quam Italie et universorum Christianorum tranquillitatem et expeditionem ipsam contra Thurcos." Another bull issued on August 17 excommunicated Ercole d'Este, duke of Ferrara and commander of the Florentine forces, but if it was printed, no copy survives.

99. The first bull specifies that posting in the two places in Rome and just two or three other places in Romagna would be sufficient to render the four men "notified" of their sentences; Bulla, August 16, 1479, "Et si natura omnibus in rebus," fol. 4v.

100. Bulla, August 17, 1479, "Decet Romanum pontificem," fol. 4v.

101. See *BMC* 4.xi–xii.

102. Martines, *April Blood*, 196.

103. Pastor, 4.348–9; Conti, *Storie*, 1.114; Setton, 2.375.

104. Pastor, 4.350. For the larger dispute, see Pastor, 4.348–88; Setton, 2.364–80; Michael Mallett, "Diplomacy and War in Later Fifteenth-Century Italy," *Proceedings of the British Academy* 67 (1981): 267–88; Mallett, "Venice and the War of Ferrara, 1482–84," in *War, Culture, and Society in Renaissance Venice: Essays in Honour of John Hale*, ed. David S. Chambers, Cecil H. Clough, and Michael Mallett (London, 1993), 57–72.

105. Pastor, 4.349–57.

106. Jacopo Gherardi, *Diarium Romanum*, in *RIS*² 23.3.111.

107. Pastor, 4.364–71.

108. *Transactio pacis factae Romae MCCCCLXXXII die XII decembris inter Sixtum IV Pontificem Maximum, Ferdinandum Siciliae Regem, Johannem Galeatium Sfortiam Mediolani Ducem, Rempublicam Florentinam et Herculem Estensem Ferrariae Ducem* ([Rome: Eucharius Silber, after December 12, 1482]) (it00425500).

109. Malipiero, 269: "Questa novità ha dà gran travaglio a la Signoria."

110. Conti, *Storie*, 1.158–76; see Ignazio Campi, "Dei *Libri historiarum sui temporis* di Sigismondo de' Conti da Foligno, *Archivio storico italiano*, 4th ser., 1 (1878): 71–89; R. Ricciardi, "Sigismondo de' Conti," *DBI* 28.470–5. Conti, appointed chancellor of Foligno by Sixtus in 1471, wrote in 1482 that he had been occupied "dictandis scribendisque epistolis pontificis" for nearly twelve years.

111. Conti, *Storie*, 1.158.

112. Conti, *Storie*, 1.163–5.

113. Conti, *Storie*, 1.165.

114. Bernardus Justinianius, *Epistolae et Orationes* (Venice: Bernardus de Benalibus, [1492?]) (ij00611000), sigs. h1r–h2v. On Giustiniani's contributions to the debate, see Patricia Labalme, *Bernardo Giustiniani, a Venetian of the Quattrocento* (Rome, 1969). Sigismondo includes copies of all four letters in his history: the letter from Sixtus and the two Venetian replies are fairly accurate copies, but the text presented as that of the cardinals is only vaguely related to the text preserved in every other witness; Bullen (see note 129 below) suggests that Sigismondo did not have a copy of this text at hand and reconstructed its sense in a new composition of his own devising. Malipiero copied down the texts of all four letters as they appear in contemporary printed sources (269–78), as did Marin Sanudo, *Le Vite dei dogi (1474–94)*, ed. Angela Caracciolo Aricò, 2 vols. (Padua, 1997), 1.315–24. Some also appear in Odorico Rinaldi, *Annales ecclesiastici ab anno MCXCVIII ubi desinit Cardinalis Baronius*, 15 vols. (Lucca, 1747–56), 11.23–4.

115. Bernardino Zambotti, *Diario Ferrarese dall'anno 1476 sino al 1504*, in *RIS*² 24.7.123–4.

116. Rinaldi, *Annales*, 11.36–7. The recipients would understand how devoted Sixtus was to the cause of peace "ex exemplo litterarum nostrarum ad ipsos Venetos, quod tibi his inclusum mittimus."

117. Pastor, 4.373, mentions a "state paper," dating from later in the spring of 1483, which also seeks to refute Venetian protests against the new peace.

118. Sanudo, *Vite dei dogi*, 1.325: "le qual letere scrite al Papa et Collegio d'I Cardinali, fonno butate a stampa insieme con ditti brievi."

119. The bibliographic literature on these editions is confused, with the same texts identified by different titles and assigned conflicting dates. The Paduan edition is listed as *Argumentum belli inter Venetos ducemque Ferrariensem exorti causas exponens* ([Padua: Matthaeus Cerdonis, after January 12, 1482/83]) in *GW* 2326 and ia00955000 (the title comes from the text that heads the preface on fol. 1r), while the Venetian edition is listed as Sixtus IV, *Breve ad Johannen Mocenigum de bello ferrariensi* ([Venice: n.p., after January 11, 1482]) (is00555500). Both editions present the same preface and the same four letters, although the letters are ordered differently in each. With the

prefatory *Argumentum* appearing in both, the editions are clearly related, but neither was printed from the other. *BMC* 7.920 dates the Paduan edition to [after February 22, 1483] on the basis of its style of foliation. The Paduan edition, including the *Argumentum*, was later transcribed by Flaminio Corner and printed as an appendix to Laurentius de Monacis, *Chronicon de rebus Venetis* (Venice, 1758), 343–52.

120. In the Paduan edition, the letters alternate, with Sixtus's letter followed by the doge's reply, and the letter from the College of Cardinals followed by that from the Senate. In the Venetian edition, the two papal letters appear together first, followed by the two letters from the doge; this is how the letters appear in Malipiero's *Annali Veneti* and Conti's own *Storie*.

121. *Argumentum*, fol. 1r.

122. *Argumentum*, fol. 1r–v.

123. Sixtus IV, *Epistola ad Joannem Mocenigum* ([Rome: Stephan Plannck, after February 13, 1482/83]) (is00558500); *BMC* 4.102 dates it "after 14 February 1483." The text is also printed (but misdated February 17) in an appendix to Conti, *Storie*, 1.413–9. The Bodleian incunabula catalogue (followed by Sheehan and others) identifies this letter as the same as the sixth in the Caxton edition (see below), but it is in fact the fifth.

124. Justinianius, *Epistolae et Orationes* (Venice: Bernardus de Benalibus, [1492?]), sigs. h2v–h3r; this edition also includes his two previous letters of January 1483 (sigs. h1r–h2v), as well as later correspondence with the pope from late 1483 and early 1484.

125. Rinaldi, *Annales*, 11.38–42. Cf. Gherardi, *Diarium Romanum*, 118, who says that the bull was drawn up on Saturday, May 24, and posted on the doors of St. Peter's the following day, Trinity Sunday.

126. Rinaldi, *Annales*, 11.42–3: "Bullam censurarum adversus Venetos ipsos edidimus, quam ad nobilitatem tuam mittimus, hortantes ut per omnia dominii tui loca publicari, et observari eam facias."

127. Sixtus IV, *Bulla* "Ad bonorum tutelam" ([Rome: Johannes Philippus de Lignamine, ca. 1483]) (is00555030).

128. Pastor, 4.375–6; Navagero, cols. 1182–3; Malipiero, 281–3. This episode is described in more detail in chapter one.

129. Malipiero, 281–3; David Chambers and Brian Pullan, *Venice: A Documentary History 1450–1630* (Toronto, 2001), 219–20.

130. The official diplomatic correspondence continued with (1) Giustiniani's letter of March 15; (2) a letter from the cardinals to Doge Mocenigo, warning of the impending interdict, dated April 24; (3) the bull of excommunication of May 23; and (4) another letter from Giustiniani, responding to the bull and addressed to the cardinals, dated May 28. (Giustiniani's letters are in his *Orationes et Epistolae* of 1492 [see above]; the text of *Ad bonorum tutelam* is in Odorico Rinaldi, *Annales ecclesiastici ab anno MCXCVIII ubi desinit Cardinalis Baronius*, vol. 19 [Cologne, 1694], 319–22.) These letters all respond to one another; there is no room in the exchange for another text, which further suggests that the letter eventually printed as the sixth in the exchange was a fictive text, perhaps composed by Giustiniani, not for dispatch to Rome but for public circulation.

131. *Sex quam elegantissimae epistolae* (Westminster: William Caxton, [1484]) (is00558550), which survives in a single copy in the British Library; *BMC* 11.157–8 supplies the date of [1484] from the paper stock. Facsimile with translation: *Sex quam*

elegantissimae epistolae, Printed by William Caxton in 1483, ed. James Hyatt and George Bullen (London, 1892). *BMC,* following Bullen, attributes the *Argumentum* to Carmeliano, but that short text had already appeared in print twice in the Veneto. The entry also misdates Sixtus's third letter to February 14, 1483. The correct date (xvii kalendiis Martii) is February 13.

132. J. B. Trapp, "Pietro Carmeliano," *Oxford Dictionary of National Biography,* 60 vols. (Oxford, 2004), 10.165–7; Massimo Firpo, "Pietro Carmeliano," *DBI* 20.410–3. Carmeliano weathered the political storms of the 1480s, dedicating texts to Edward IV, Richard III, and Henry VII in turn.

133. Cicero, *De oratore* (Venice: Andreas de Paltasichis, August 20, 1478) (ic00660000).

134. George Painter, *William Caxton: A Quincentenary Biography of England's First Printer* (London, 1976), 135–6; and David R. Carlson, *English Humanist Books: Writers and Patrons, Manuscript and Print, 1475-1525* (Toronto, 1993), 134–5. Carmeliano later corrected texts for the Oxford printer Theodoric Rood; it is possible he offered Caxton help with corrections in exchange for Caxton's printing the Venetian-papal correspondence, which may also have been underwritten with Venetian funds.

135. Trapp, "Pietro Carmeliano," 166.

136. *Sex quam elegantissimae epistolae,* sig. c8r.

137. "Interpretacio magnarum litterarum punctuatarum parvarumque."

138. Margaret Meserve, *Empires of Islam in Renaissance Historical Thought* (Cambridge, MA, 2008), 228–9.

139. Carmeliano also copied the fictive epistolary attributed to the ancient tyrant Phalaris, another collection of letters that circulated for their rhetorical as well as (pseudo-)historical interest; Anna Modigliani, "Un nuovo manoscritto di Pietro Carmeliano: Le 'Epistole' dello pseudo-Falaride nella Trinity College Library di Dublino," *Humanistica Lovaniensia* 33 (1984): 86–102.

140. Johannes Nicolaus Faela, *De origine et laudibus Maffeorum* ([Rome: Stephan Plannck, after December 12, 1482]) and ([Rome: Johann Schömberger, after April 16, 1484]) (if00042800 and if00043000). The composition of these editions is complex: Sheehan F-2 includes two letters by Benedetto Maffei dated August 23 and December 12, 1482 (the latter entitled "De laudibus pacis"); Sheehan F-3 includes another letter by the same author, addressed to Antonello Petrucci and dated August 13, 1483, and a reply from Petrucci dated April 26, 1484; but many copies contain fewer of the letters. See *BMC* 4.101, *GW* 9702 and 9703.

141. Benedetto Maffei to Antonello Petrucci, August 13, 1483. The letter to Petrucci in fact forms a preface to another letter, which Maffei addressed to Arcamone (and this is what is dated August 13, 1483). The fourth and final letter is a brief reply from Petrucci, dated April 26, 1484.

Chapter 4 • Broadsides in Basel

1. Johann Heinrich Hottinger, *Historiae ecclesiasticae novi testamenti seculi XVI,* 9 vols. (Zurich, 1651–57), 4.347–604; Joseph Schlecht, *Andrea Zamometić und der Basler Konzilsversuch vom Jahre 1482* (Paderborn, 1902); Alfred Stoecklin, *Der Basler Konzilsversuch des Andrea Zamometić vom Jarhe 1482 (Genesis und Wende)* (Basel, 1938); Stoecklin, "Das Ende des Basler Konzilsversuchs von 1482," *Zeitschrift für*

Schweizerische Kirchengeschichte 79 (1985): 3–118; Jürgen Petersohn, *Reichsrecht versus Kirchenrecht: Kaiser Friedrich III. im Ringen mit Papst Sixtus IV. um die Strafgewalt über den Basler Konzilspronuntiator Andreas Jamometić 1482–1484* (Cologne, 2015). This episode, thoroughly studied in German historiography, has attracted only brief notices in English; see Mandell Creighton, *A History of the Papacy from the Great Schism to the Sack of Rome*, 6 vols. (London, 1897), 4.106–9; Pastor, 4.358–63; Hubert Jedin, *A History of the Council of Trent*, trans. Ernst Graf, 2 vols. (London, 1957), 1.101–6; Ronald C. Finucane, *Contested Canonizations: The Last Medieval Saints, 1482–1523* (Washington, DC, 2011), 94–5, 103–5, 115–6; Tamar Herzig, *Christ Transformed into a Virgin Woman: Lucia Brocadelli, Heinrich Institoris, and the Defense of the Faith* (Rome, 2013), 17–22; Norman Housley, *Crusading and the Ottoman Threat, 1453–1503* (Oxford, 2012), 52–3, 144.

2. Tobias Daniels, "Die italienischen Mächte und der Basler Konzilsversuch des Andreas Jamometić," *Zeitschrift der Savigny-Stiftung für Rechtsgeschichte* 131 (2014): 339–67.

3. Sigismondo dei Conti, *Le storie de' suoi tempi dal 1475 al 1510*, 2 vols. (Rome, 1883; repr., Foligno, 2015), 1.157–8; Infessura, 98; Jacopo Gherardi, *Diarium Romanum*, in *RIS*² 23.3.57, 102, 113, 115.

4. Many of these are preserved in unique copies in the Basel city archive; Carl Christoph Bernoulli, "Die Incunabeln des Basler Staatsarchivs," *Basler Zeitschrift für Geschichte und Altertumskunde* 9 (1910): 1–35.

5. Contemporary sources identify him with variants of "Andreas archiepiscopus Crainensis" or "Craynensis" but also as Andreas de Cranea, Andreas Dalmata, and Andreas de Udine. Modern historians call him Andrea Zuccalmaglio, Andrea Za-mometić, or Andreas or Andrija Jamometić, the last of which now seems the most widely used form. See Jürgen Petersohn, "Zum Personalakt eines Kirchenrebellen. Name, Herkunft, und Amtssprengel des Basler Konzilsinitiators Andreas Jamometić (+1484)," *Zeitschrift für historische Forschung* 13 (1986): 1–14. For his career up to his flight from Rome, see Petersohn, *Kaiserliche Gesandter und Kurienbischof: Andreas Jamometić am Hof Papst Sixtus IV (1478–1481)* (Hannover, 2004).

6. He had already formed a dim view of the pope during the Pazzi War, when he blamed Sixtus for provoking Florence and fomenting discord across Italy; Petersohn, *Kaiserliche Gesandter*, 147; Tobias Daniels, *La congiura dei Pazzi: I documenti del conflitto fra Lorenzo de' Medici e Sisto IV* (Florence, 2013), 32.

7. Schlecht, *Andrea Zamometić*, 42–4, 52–8.

8. Ten printed texts relating to the dispute survive: five imperial broadsides (VE15 F-64–F-68) in favor of Otto, all in German and printed in either Constance or Ulm; four more texts by Otto himself, all short, folio leaflets printed in Constance, three in German and one in Latin (io00125400; io00125300; io00125450; io00125460); and a quarto edition of a Sistine bull in favor of Ludwig, in Latin and also printed in Constance (is00538200). They are dated between January 1475 and August 1480.

9. Peter F. Kramml, *Kaiser Friedrich III. und die Reichsstadt Konstanz (1440–1493): Die Bodenseemetropole am Ausgang des Mittelalters* (Sigmaringen, 1985), 223–9.

10. Renée Weis-Müller, *Die Reform des Klosters Klingental und ihr Personenkreis* (Basel, 1956); Anne Winston-Allen, *Convent Chronicles: Women Writing about Women and Reform in the Late Middle Ages* (University Park, PA, 2004), 148–50.

11. February 13, 1480: Sixtus IV, Bull for the reformation of Klingenthal, printed twice, once in Latin and once in German (VE15 S-77 and S-78, is00550500 and is00550510).

12. Sixtus IV, Bulla, April 7, 1480, "Regimen universalis," for the excommunication of nuns resisting the reform, printed twice, once in Latin and once in German (VE15 S-79 and S-80, is00551000 and is00551100). Copies of both are in Basel SA, Klosterarchiv Klingental HH 4, documents 59 and 69. The two broadsides are identical in format, but the German text is longer, so much so that the text has been printed on two sheets that overlap at the short ends to form a longer single broadside.

13. Frederick III, exhortation to the estates to support and protect the reformed convent, April 13, 1480 (VE15 F-75, if00318453).

14. Guillermus de Pereriis, *Instrumentum citationis contra moniales*, October 29, 1481 (VE15 P-201, ip00269500).

15. Winston-Allen, *Convent Chronicles*, 149–50; Weis-Müller, *Die Reform*, 53–4, 182–3, 188–9; Herzig, *Christ Transformed*, 24.

16. But cf. Johannes Savageti, *Oratio lamentabilis ad Sixtum IV* ([Rome: Wolf Han, after December 12, 1476]) (is00165500). This volume contains an "oration" by Savageti deploring the controversy over the bishopric and a separate treatise (*Tractatus super controversia ecclesiae Constantiensis*) outlining arguments to back Sixtus's position. That the tract was published in Rome could indicate some official or semiofficial effort to circulate his arguments in Roman circles.

17. Andreas, Archiepiscopus Crainensis, [*Call for a General Church Council at Basel*] ([Basel: Johannes Besicken, after March 25, 1482]) (*GW* 1857, VE15 A-490, Bernouilli, "Incunabeln," 32, ia00650900); Basel SA, Politisches H 2, document 18. The text of the appeal is also edited by Schlecht, *Andrea Zamometić*, 96–101, with the revised version of April 1482.

18. Charlotte de Lusignan, exiled from Cyprus by her half-brother James, who by his marriage to Caterina Cornaro was allied to Venice, sought support from the Mamluks to oust him but was unsuccessful.

19. Schlecht, *Andrea Zamometić*, 79.

20. On Besicken's work in Basel, see *BMC* 3.xxxv. He would move to Rome in the next decade and become one of the most prolific printers of texts for the 1500 Jubilee.

21. The two copies are Basel SA, Politisches H 2, document 18 (the only printed text in an 11.5-meter-long roll of manuscript documents relating to the Jamometić affair; I am grateful to Brigitte Heiz of the Staatsarchiv, who helped me examine the roll in July 2017); and Chur, Bischöflichen Archiv, Mappe 50, on which see Ingrid Heike Ringel, "Ein bisher unbekanntes Exemplar der Konzilsproklamation des Andreas Jamometić von 1482 im Bischöflichen Archiv Chur," *Gutenberg-Jahrbuch* 64 (1989): 101–5, with facsimile on 102. It is a large sheet (43 × 31 cm).

22. The text of this version, addressed to Sixtus and dated April 11, 1482, is printed as Document XX ("Die Konzilsproklamation") in Schlecht, *Andrea Zamometić*, 36–41. Schlecht derives the text from a manuscript of the *Memorial* of Konrad Stolle in Jena, with readings from the copy made by Schedel (MS Munich, BSB, Clm 414, fols. 230–4). Sanudo's version is in Marin Sanudo, *Le vite dei dogi (1474–1484)*, ed. Angela Caracciolo Aricò, 2 vols. (Padua, 1989), 1.227–31.

23. Schlecht, *Andrea Zamometić*, 78, 81; Stoecklin, *Der Basler Konzilsversuch*, 40.

Stolle probably got the text from the Erfurt cleric Hugo von Landenberg, who arrived in Basel in spring 1482 as one of Sixtus's earliest envoys.

24. Schlecht, *Andrea Zamometić*, 66–8; Jürgen Petersohn, *Diplomatische Berichte und Denkschriften des päpstlichen Legaten Angelo Geraldini aus der Zeit seiner Basel-Legation (1482–1483)* (Stuttgart, 1987), 78–9.

25. Ringel, "Ein bisher unbekanntes Exemplar," 103.

26. Schlecht, *Andrea Zamometić*, 42–3, at 43.

27. Schlecht, *Andrea Zamometić*, 42.

28. Schlecht, *Andrea Zamometić*, 42–3. See also Ringel, "Ein bisher unbekanntes Exemplar," 105.

29. For earlier cases, see John Monfasani, "The First Call for Press Censorship: Niccolò Perotti, Giovanni Andrea Bussi, Antonio Moreto, and the Editing of Pliny's *Natural History*," *Renaissance Quarterly* 41 (1988): 1–31.

30. Text in Schlecht, *Andrea Zamometić*, 70–1. It was read out in the Basel churches on June 9; Ringel, "Ein bisher unbekanntes Exemplar," 104–5.

31. Ringel, "Ein bisher unbekanntes Exemplar," 105.

32. In an undated reply to Rudolf von Scheerenberg (Schlecht, *Andrea Zamometić*, 43–4), Piccolomini explains that he did not send the *libellus* itself to the pope—not just because he had bitten it to pieces but also because he did not want to aggravate him when Rome was under siege. In April 1482, Ferrante of Naples sent troops into the papal states to besiege Marino, south of Rome; his son Alfonso, duke of Calabria, led troops to the walls of Rome in late May (Pastor, 4.352–5).

33. Gherardi, *Diarium Romanum*, 102.

34. Schlecht, *Andrea Zamometić*, 45–6; Conti, *Storie*, 1.410–2, esp. 411 n. 2.

35. Herzig, *Christ Transformed*, 18, 24.

36. Petersohn, *Diplomatische Berichte*, 122–3; Schlecht, *Andrea Zamometić*, 47–8.

37. Letters to Salvo Casetta, master general of the Dominicans and Emerich von Kemel, papal commissary; Schlecht, *Andrea Zamometić*, 49–51.

38. Andreas, Archiepiscopus Crainensis, [*Invective against Sixtus IV*] ([Basel: Johann Besicken, after July 21, 1482]) (VE15 A-491, Bernouilli, "Incunabeln," 33, ia00650950). See Stoecklin, *Basler Konzilsverzuch*, 43–9. The text is in Hottinger, *Historia ecclesiastica*, 368–94. The unique copy is Basel SA, Politisches H 1 I, documents 18 and 19. It is almost exactly twice as long as the March 25 "Proclamation." Later in 1483, Geraldini told the pope that Jamometić's initial appeal had been reworked, had doubled in size (from around forty articles to one hundred), and had been both posted (*affixi*) and distributed widely throughout Germany.

39. Matt. 18:15–7.

40. Basel SA, Politisches H 1 I, documents 18 and 19.

41. See, e.g., the Strassburg broadside edition of Heinrich Institoris's *Epistola* and Geraldini's two publications of February and April 1483, discussed below. See also the German-language version of Sixtus's bull excommunicating the nuns of Klingental mentioned in note 12.

42. Jürgen Petersohn, "Kirchenrecht und Primatstheologie bei der Verurteilung des Konzilsinitiators Andreas Jamometić durch Papst Sixtus IV: Die Bulle 'Grave gerimus' vom 16. Juli 1482 und Botticellis Fresko 'Bestrafung der Rotte Korah' (mit Edition des Quellentextes)," in *Proceedings of the Twelfth International Congress of*

Medieval Canon Law, ed. Ute-Renate Blumenthal (Vatican City, 2008), 667–98. Jamometić's printing activity is described at 688–9.

43. Jürgen Petersohn, *Ein Diplomat des Quattrocento. Angelo Geraldini (1422–1486)* (Tübingen, 1985); Petersohn, *Diplomatische Berichte*; Petersohn, "Geraldini, Angelo," in *DBI* 53.316–21; Petersohn, "Azioni di governo e missioni diplomatiche di Angelo Geraldini (1422–1486)," in *I Geraldini di Amelia nell'Europa del Rinascimento* (Terni-Amelia, 2004), 19–24.

44. Infessura, 97–8. In fact, Geraldini's credentials were drawn up on July 21 (Conti, *Storie*, 1.411 n. 2).

45. Geraldini to Sixtus IV, September 22, 1482, in Petersohn, *Diplomatische Berichte*, 49–53, at 49–50. Geraldini traveled through Venetian territory, avoiding the duchy of Milan, which was still allied with Florence and Naples against Rome.

46. Petersohn, *Diplomatische Berichte*, 50. Jamometić was staying in the Basel suburb of Kleinbasel, across the Rhine from the city proper and therefore in the diocese of Constance; Geraldini meant to have the bull published in both dioceses so that the sentence would be valid no matter which side of the Rhine Jamometić was on.

47. Petersohn, *Diplomatische Berichte*, 50–1.

48. Geraldini to Sixtus IV, February 3, 1483, in Petersohn, *Diplomatische Berichte*, 71.

49. Von Kemel carried a secret letter from the pope dated July 22, 1482, which authorized him to overrule Geraldini; Stoecklin, *Basler Konzilsverzuch*, 116; Petersohn, *Diplomatische Berichte*, 50 n. 5, 71 n. 40.

50. Weis-Müller, *Die Reform*, 7, 53–4, 182–3.

51. See letters from Sixtus to the Swiss Confederation, Sigismund of Austria, and the bishop of Constance, as well as his general letter of July 2; Schlecht, *Andrea Zamometić*, 88–94.

52. Stoecklin, *Basler Konzilsverzuch*, 23–4.

53. Stoecklin, *Basler Konzilsverzuch*, 134–9; Petersohn, *Diplomatische Berichte*, 57 n. 22.

54. Basel SA, Politisches H 1 I, document 45: [*Appeal of the City of Basel to Sixtus IV against the Impositions of the Papal Nuncio, the Prior of Veltpach*] ([Basel: Johann Besicken, ca. 1482]) (VE15 B-27; Bernouilli, "Incunabeln," 35, ib00270200); Stoecklin, *Basler Konzilsverzuch*, 136–9, 156; Petersohn, *Diplomatische Berichte*, 59.

55. Basel SA, Politisches H 1 I, document 45.

56. VE15 B-27, quoting account books for the year June 1482 to June 1483: "Item XII lb die appellaciones se trucken"; "Item II lb von den appellaciones wider den legaten ze trucken." An entry in the weekly journal for October 19, 1482, also refers to payments disbursed to cover printing costs.

57. That autumn, the Florentine envoy Ugolini complained to Lorenzo that his interviews with Jamometić were growing awkward, since matters were at a standstill and he had no new instructions to give.

58. Stoecklin, *Basler Konzilsverzuch*, 134–8.

59. Salvo Casetta, letter to Sixtus IV, October 28, 1482, in Petersohn, *Diplomatische Berichte*, 122–3; Creighton, *History of the Papacy*, 4.108.

60. Stoecklin, *Basler Konzilsverzuch*, 24, 158.

61. Stoecklin, *Basler Konzilsverzuch*, 167.

62. Geraldini to Sixtus IV, November 27, 1482, in Petersohn, *Diplomatische Berichte*,

54–65, at 56. Later, Geraldini reported that one of his agents had been attacked as he tried to publish the bull; Geraldini to Sixtus IV, June 4, 1483, in Petersohn, *Diplomatische Berichte*, 105.

63. Geraldini to Sixtus IV, November 27, 1482, in Petersohn, *Diplomatische Berichte*, 57.

64. Petersohn was unaware that the bull was printed: Sixtus IV, Bulla [1482] ([Strassburg: Printer of the 1483 Jordanus de Quedlinburg (Georg Husner), ca. 1482]) (VE15 S-101; is000555000). It survives in a single, fragmentary copy in the Library of Congress, Incun. X. C3536. Katie Labor kindly sent me a photo.

65. Stoecklin, *Basler Konzilsverzuch*, 99–105; Herzig, *Christ Transformed*, 3–27.

66. The broadside is Henricus Institoris, *Epistola contra quendam conciliistam archiepiscopum videlicet Crainensem* ([Strassburg: Heinrich Eggestein, after August 10, 1482]) (ii00160000; VE15 I-69; Bernouilli, "Incunabeln," 34); the quarto edition is ii00160050. Copies of the broadside in Basel SA and the Pierpont Morgan Library, New York, show further signs of creative cutting and pasting: to correct an error in the typesetting of the first few lines, a portion of a reprinted sheet has been pasted rather clumsily across the top. I examined the Basel copy myself, and I am grateful to John McQuillen of the Morgan Library for sending me photos and his description of the Morgan copy.

67. The Nuremberg broadside is ii00160080; VE15 I-70; the Nuremberg quarto is ii00160100. Both editions were formerly attributed to Michael Greyff of Reutlingen, and VE15 I-70 still attributes the broadside to Greyff, but an attribution to Nuremberg is more likely. On Greyff and the printer of the *Rochus Legende*, see *BMC* 2.466.

68. Stoecklin, *Basler Konzilsverzuch*, 104.

69. Henricus Institoris, *Epistola contra quendam conciliistam archiepiscopum videlicet Crainensem* ([Rome: Stephan Plannck, after August 10, 1482]) (ii00160200). The humanist Pietro Ursuleo (d. 1483) made a manuscript copy of Institoris's *Epistola* in Rome around this time, preserved in BAV Ottob. lat. 736.

70. Leonellus Chieregatus, *Epistola ad Georgium praepositum Basiliensem data* ([Rome]: Bartholomaeus Guldinbeck, [after October 18, 1482]) (ic00451000); text in Creighton, *History of the Papacy*, 4.331–7. See also Herzig, *Christ Transformed*, 21–2.

71. *Verteidigung des Konzils zu Basel* ([Mainz: Printer of the Darmstadt *Prognostication*, ca. 1480]) (ib00270400); the text has been attributed to the Erfurt theologian Heinrich Toke; see *Gutenberg: Aventur und Kunst, von Geheim unternehmen zur ersten Medienrevolution* (Mainz, 2000), 387.

72. Aldo Landi, *Concilio e papato nel Rinascimento (1449–1516): Un problema irrisolto* (Turin, 1997), 151–2, raises the possibility that the *editio princeps* of the conciliarist Jean Gerson's *Opera*, which appeared in Cologne in 1483 (ig00185000), also reflects contemporary interest in Jamometić's enterprise.

73. Geraldini to Sixtus IV, June 4, 1483, in Petersohn, *Diplomatische Berichte*, 102.

74. Petersohn, *Diplomatische Berichte*, 102.

75. Basel SA, Politisches H 1 I, document 81, letter of Sixtus IV to Angelo Gerardini: Geraldini should give the provost a few days to negotiate and only then, if his efforts have failed, may Geraldini act "prout res postulabit."

76. In a letter to the pope, the Basel councilors later praised Gratiadei, saying they would never have accomplished anything without him; the discord among the papal

ambassadors had made negotiations impossible. Petersohn, *Reichsrecht versus Kirchenrecht*, 176–7: "propter diversitatem oratorum sanctitatis vestrae atque divisionem inter eos et animorum varietatem."

77. News of his arrest reached Rome at the very end of the year. Gherardi notes it between entries for December 30 and January 1; see Gherardi, *Diarium Romanum*, 113.

78. Andreas, Archiepiscopus Crainensis, *Retractatio eius a notario publico d. 2 Jan. 1483 testata* ([Basel: Johann Besicken, after January 2, 1483]) (ia00651000; VE15 A-492).

79. Andreas Crainensis, *Retractatio*.

80. Andreas Crainensis, *Retractatio*.

81. VE15 A-492: the council's account books record payments of 3 pounds, 9 shillings to de la Woestine for drawing up the document and 5 pounds, 3 shillings for printing it.

82. VE15 A-492 lists four copies with de la Woestine's manuscript authentication ("Ita est attestor ego Iohannis notarius prenominatus manu propria"); a fifth, Chicago, Newberry Library, Inc., +7651, also contains his manuscript signature and notarial sign.

83. Andreas, Archiepiscopus Crainensis, *Retractatio eius a notario publico d. 2 Jan. 1483 testata* ([Mainz: Peter Schoeffer, after January 2, 1483]) (ia0651100; VE15 A-493). A single, fragmentary copy of this edition survives in Giessen.

84. Geraldini to Sixtus IV, February 3, 1483, in Petersohn, *Diplomatische Berichte*, 71.

85. Petersohn, *Diplomatische Berichte*, 72–3. ISTC records eight editions of Emeric's 1482 indulgence, by printers in Nuremberg, Reutlingen, and Mainz. Norman Housley, "Crusading Responses to the Turkish Threat in Visual Culture, 1453–1519," in *The Crusades and Visual Culture*, ed. Elizabeth Lapina et al. (Farnham, UK, 2015), 201–19, at 209–11, notes that von Kemel was ordered by Sixtus in May 1482 to stop selling his indulgences, probably in response to Jamometić's call for a council. In July of that year, Sixtus sent von Kemel a brief indicating that his call for a crusade was still in force and that sales could continue. Someone, presumably von Kemel, had this brief printed by Peter Schoeffer in Mainz (two editions survive: is00537780 and is00537781). This short text says nothing about the Basel interdict. If von Kemel did have a document printed that declared the interdict lifted, it does not survive.

86. Petersohn, *Diplomatische Berichte*, 71–2.

87. Geraldini to Sixtus IV, May 30, 1483, in Petersohn, *Diplomatische Berichte*, 83–4.

88. Petersohn, *Diplomatische Berichte*, 85.

89. Against this, Geraldini does seem to distinguish between the document that de le Woestine allegedly contrived and the one published by his master, Gratiadei; he says that the former was longer and more elaborate than the latter.

90. Angelus Geraldinus, *[Publication of a Bull of Sixtus IV against Basel]* ([Strassburg: Heinrich Eggestein, before January 31, 1483]) (ig00158100; VE15 A-507). Eggestein had already printed Heinrich Institoris's pro-papal tract. Institoris may have served as Geraldini's middleman, conveying to Eggestein both the crusade bull of January 1483 and Sixtus's briefs regarding the interdict that Geraldini published with him in April.

91. Details in VE15 A-507.

92. The text of the bull was read out in the bishoprics of Strassburg, Constance, Speyer, and Basel.

93. Frederick III to Sixtus IV, March 10, 1483, in Petersohn, *Reichsrecht versus Kirchenrecht*, 192–3.

94. Frederick III to Geraldini, March 20, 1483, in Petersohn, *Reichsrecht versus Kirchenrecht*, 196–8.

95. He issued a Latin version on March 20 (Petersohn, *Reichsrecht versus Kirchenrecht*, 199–203) and a German version the following day (204–7).

96. Frederick III, *Ausschreiben an alle Stände des Reiches mit der Weisung, Angelus (Geraldini), Bischof von Sessa, nicht zu unterstützen und den über Basel verhängten Bann als ungültig anzusehen*, issued Vienna, March 21, 1483 ([Basel: Johannes Besicken, after March 21, 1483]) (if00318500, if00318510, if00318515; VE15 F-76–78: three slightly variant settings, all broadsides).

97. Fiscal records for the year June 1482–June 1483 (quoted in VE15 F-76) include the following: "Item III lb II ss die keiserlichen mandatt ze trucken. Item XII ss einem knecht lonsz die keiserlichen mandat helffen verkunden. Item XII ss Jorgen Spittelschreiber das keiserlich mandat mim herren von Costentz ze verkunden." Thanks to Claire Taylor Jones for help with these records.

98. Geraldini to Sixtus IV, June 4, 1483, in Petersohn, *Diplomatische Berichte*, 105.

99. Sixtus IV to Angelo Geraldini, March 17, 1483, printed on the April broadside: "Venerabilis frater: salutem et apostolicam benedictionem. Volumus et fraternitati tue mandamus ut omnino supersedeas in publicatione bulle cruciate contra Basilienses donec aliud a nobis habueris in mandatis, quia speramus Craynensem olim archiepiscopum in manibus nos habituros absque alio tumultu; faciemus postea te certiorem quid agere debeas. Datum Rome."

100. Geraldini to Sixtus IV, May 30, 1483, in Petersohn, *Diplomatische Berichte*, 86–7.

101. *GW* 1949: "Bekanntmachung zweier Breve Sixtus' IV. betr. die Anhänger des Andreas Erzbischof von Kraina." See also ig00158200; VE15 A-508.

102. The abridged text omits more than half of the original bull text; in all, some forty-six lines of text from the February broadside are missing from the April republication.

103. Geraldini to Sixtus IV, May 30, 1483, in Petersohn, *Diplomatische Berichte*, 86.

104. Petersohn, *Diplomatische Berichte*, 86. In his letter to the pope, Geraldini said he had drawn up another letter for use in the territory immediately surrounding Basel, warning the local clergy to observe the interdict, but he does not say whether this was printed.

105. Petersohn, *Diplomatische Berichte*, 86.

106. Petersohn, *Diplomatische Berichte*, 78. He repeatedly mentions guards who accompanied the printed texts as they traveled.

107. Petersohn, *Diplomatische Berichte*, 78. Geraldini means Martin Flach, who is also mentioned in the city's account books as the person paid for the printing of Frederick III's decrees in 1483. However, the types used in all the Basel broadsides are those of Johannes Besicken; while Flach printed books in the late 1470s, by this point he was operating more as a publisher commissioning work on the city's behalf. See *BMC* 3.xxxiv–xxxv and the discussion in VE15 F-76. Flach was elected to the city council in 1483, but Geraldini's accusation of *quid pro quo* seems unfair.

108. Petersohn, *Diplomatische Berichte*, 78–9.

109. Petersohn, *Diplomatische Berichte*, 79.

110. Petersohn, *Diplomatische Berichte*, 80–1.

111. VE15 F-76.

112. Geraldini, or Institoris, or Eggestein himself may have got the idea for this from Jamometić, since his July 21 broadside *seems* to have been printed so that it could be assembled into an extralong sheet in just this way, although the only copy that survives consists of two printed sheets preserved separately. More immediately, Institoris's August 10 broadside against Jamometić, also printed by Eggestein, is constructed exactly this way, from two printed sheets glued together.

113. Conti, *Storie*, 1.157–8.

114. A postscript: around 1488, the bishop of Basel, Caspar zu Rhein, published a text of his own, derived indirectly, at least, from the controversies of 1482–83. It was not another tract for or against Jamometić; rather, the bishop offered advice to his fellow clerics, presumably derived from his own experience, on how one might still celebrate mass during an interdict. *Wie man messen haben mag in interdicto* ([Basel: Printer of the Form der Copyen, ca. 1488]) (ic002360600).

Chapter 5 • The Holy Face, Imprinted and in Print

1. *In dem puchlin stet geschreiben wie Rom gepauet wart* ([Southern Germany]: Hans Aurl, September 28, 1481) (im00607200).

2. *In dem puchlin stet geschreiben wie Rom gepauet wart* (Munich: Johann Schaur, June 28, 1482) (im00607300).

3. *Reliquie rhomane urbis et indulgentiae* ([Basel: Peter Kollicker, ca. 1483]) (ir00140000).

4. *Oratio Salve Sancta Facies* ([Ulm: Johann Zainer, ca. 1482]) (io00065950); *Gebet zu dem heiligen Antlitz Christi* ([Ulm: Conrad Dinckmut, ca. 1482]) (io00065970).

5. Debra J. Birch, *Pilgrimage to Rome in the Middle Ages: Continuity and Change* (Woodbridge, 1998); Claudio Strinati et al., eds., *I Giubilei: Roma e il sogno dei pellegrini* (Prato, 1999); Mario D'Onofrio, ed., *Romei e Giubilei: Il pellegrinaggio medievale a San Pietro (350–1350)* (Milan, 1999); Arsenio Frugoni, *Pellegrini a Roma nel 1300* (Casale Monferrato, 1999); Massimo Miglio, ed., *Pellegrinaggi a Roma* (Rome, 1999); Herbert Kessler and Elizabeth Zacharias, *Rome 1300: On the Path of the Pilgrim* (New Haven, CT, 2000); Raissa Teodori and Alessandra Casamassima, eds., *Antiquorum habet: I giubilei nella storia di Roma attraverso le raccolte librarie e documentarie del Senato* (Rome, 2016).

6. "Veronica" was thought to derive from the cloth's status as a *vera icone*, or "true image," of Christ; tradition transferred the name to the woman who held it, so the relic was sometimes called the Veronica and sometimes the veil of Veronica; it was sometimes depicted on its own and sometimes held by St. Veronica.

7. Hans Belting, *Likeness and Presence: A History of the Image before the Era of Art* (Chicago, 1994), 59–63; Erik Thunø, *Image and Relic: Mediating the Sacred in Early Medieval Rome* (Rome, 2002), 13–6.

8. On the Assumption Eve procession, first attested in 847 during the reign of Leo IV (847–55), see Richard Ingersoll, "The Ritual Use of Public Space in Renaissance Rome" (PhD diss., University of California, Berkeley, 1978), 34–5, 224–52; Belting, *Likeness and Presence*, 64–73, 498–502; Gerhard Wolf, *Salus populi Romani: Die*

Geschichte römischer Kultbilder im Mittelalter (Weinheim, 1990), 37–78; Kessler and Zacharias, *Rome 1300*, chaps. 3–5; Enrico Parlato, "La processione di Ferragosto e l'acheropita del Sancta Sanctorum," in *Il Volto di Cristo*, ed. Giovanni Morelli and Gerhard Wolf (Milan, 2000), 51–2; Parlato, "Le icone in processione," in *Arte e Iconografia a Roma*, ed. Maria Andaloro and Serena Romano (Milan, 2002), 55–72; Kirstin Noreen, "Serving Christ: The Assumption Procession in Sixteenth-Century Rome," in *Remembering the Middle Ages in Early Modern Italy*, ed. Lorenzo Pericolo and Jessica N. Richardson (Turnhout, 2015), 231–45; Carla Keyvanian, *Hospitals and Urbanism in Rome, 1200–1500* (Leiden, 2015), 133–4, 282–3, 326–7.

9. Kirsten Noreen, "The Icon of Santa Maria Maggiore, Rome: An Image and Its Afterlife," *Renaissance Studies* 19 (2005): 660–72.

10. Parlato, "Le icone in processione," 61. The first mention of an encounter between the two icons dates to 1170; Noreen, "Icon," 660. By the fifteenth century, elaborate stage machinery was used for the encounter between the two icons, which were made to move together and even bow to one another; Barbara Wisch, "Keys to Success: Propriety and Promotion of Miraculous Images in Roman Confraternities," in *The Miraculous Image in the Late Middle Ages and Renaissance*, ed. Erik Thunø and Gerhard Wolf (Rome, 2004), 161–84, esp. 174–6.

11. Hans Belting, "Icons and Roman Society," in *Italian Church Decoration of the Middle Ages and Early Renaissance*, ed. William Tronzo (Bologna, 1989), 27–41, at 40–1. A long tradition linked the pope's protective custody of Rome with the icons he could deploy: Gregory I saved the city from plague by taking the *Salus populi Romani* in procession; Stephen II and Leo IV processed with the Salvatore to ward off attacks by Lombards and the plague.

12. Gerald of Wales considered the images pendants to one another, calling them "Veronica" and "Uronica"; B. M. Bolton, "Advertise the Message: Images in Rome at the Turn of the Twelfth Century," in *The Church and the Arts*, ed. D. Wood (Oxford, 1992), 117–30; Gerhard Wolf, "Christ in His Beauty and Pain: Concepts of Body and Image in an Age of Transition," in *The Art of Interpreting*, ed. Susan C. Scott (University Park, PA, 1995), 164–97; Keyvanian, *Hospitals*, 129–37.

13. Keyvanian, *Hospitals*, esp. chap. 5.

14. Wisch, "Keys to Success," 165–6; Anna Esposito, "Gli ospedali romani tra iniziative laicali e politica pontificia," in *Ospedali e città: L'Italia del Centro-Nord, XIII–XVI secolo*, ed. A. J. Grieco and L. Sandri (Florence, 1997), 233–51; Andreas Rehberg, "Die Römer und ihre Hospitäler: Beobachtungen zu den Trägergruppen der Spitalsgründungen in Rom," in *Hospitäler in Mittelalter und Früher Neuzeit: Frankreich, Deutschland, Italien: Eine vergleichende Geschichte*, ed. Gisela Drosselbach (Munich, 2007), 225–60; Keyvanian, *Hospitals*, 293–309.

15. Parlato, "Le icone in processione," 71. Before the establishment of the confraternity in 1288, a hereditary group of *ostiarii*, drawn from baronial families, were the main custodians of the image; they were gradually eclipsed by the confraternity but did not disappear till the end of the fifteenth century. Members of the butcher's guild also played a role in the procession; Kirsten Noreen, "Sacred Memory and Confraternal Space: The Insignia of the Confraternity of the Santissimo Salvatore (Rome)," in *Roma Felix: Formation and Reflections of Medieval Rome*, ed. Éamonn Ó Carragáin and Carol Neuman de Vegvar (Aldershot, 2007), 164–6; Wisch, "Keys to Success," 166–7.

16. Ronald Musto, *Apocalypse in Rome: Cola di Rienzo and the Politics of the New Age* (Berkeley, 2003), 184–5.

17. A sixteenth-century tablet in the Musei Capitolini records the order of the Roman guilds in the procession; Vincenzo Forcella, *Iscrizioni delle chiese e d'altri edificii di Roma dal secolo XI fino ai giorni nostri*, 14 vols. (Rome, 1869–84), 1.37 n. 60.

18. Noreen, "Sacred Memory," 159–87. Antoniazzo Romano made several copies of the icon of the Salvatore for lay patrons in Rome; Anna Cavallaro, *Antoniazzo Romano e le confraternite del Quattrocento a Roma* (Rome, 1984), 335–65.

19. Keyvanian, *Hospitals*, 293–309, 315–21.

20. Andre Chastel, "La Veronica," *Revue d'Art* 40–1 (1978): 71–81; Marcello Fagiolo and Maria Luisa Madonna, eds., *L'Arte degli Anni Santi: Roma 1300–1875* (Milan, 1984), 41–130; Herbert L. Kessler and Gerhard Wolf, eds., *The Holy Face and the Paradox of Representation* (Bologna, 1998), 129–80; Herbert Kessler, " 'Pinta della nostra effigie,' La Veronica come richiamo dei Romei," in D'Onofrio, *Romei e Giubilei*, 211–18; Kessler, " 'Or, fu sì fatta la semblanza vostra?' Sguardi alla 'vera icona' e alle sue copie artistiche," in *Il volto di Cristo*, ed. Giovanni Morello and Gerhard Wolf (Milan, 2000), 103–14; Amanda Murphy et al., eds., *The European Fortune of the Roman Veronica in the Middle Ages* (Brno, 2017).

21. Keyvanian, *Hospitals*, 129–30.

22. Wolf, "Christ in His Beauty and Pain," 168.

23. Kessler and Zacharias, *Rome 1300*, 211–2. Celestine installed the ciborium.

24. Christoph Eggler, "Papst Innocent III. und die Veronica. Geschichte, Theologie, Liturgie und Seelsorge," in Kessler and Wolf, *Holy Face*, 181–204; Rebecca Rist, "Innocent III and the Roman Veronica," in Murphy et al., *European Fortune*, 115–24.

25. Gisela Drossbach, "The Roman Hospital of Santo Spirito in Sassia and the Cult of the Vera Icon," in Murphy et al., *European Fortune*, 159–66.

26. The procession, which replaced the earlier station of Sant'Eustachio, was to take place on the feast of the Marriage at Cana, the first Sunday after the Octave of Epiphany. Innocent's bull is in *PL* 215, cols. 1270–1: just as Jesus met his mother at the wedding feast, so the *effigies Christi* in the Vatican should go to Santa Maria in Sassia on the day that commemorates the Cana miracle. See Robert Brentano, *Rome before Avignon: A Social History of Thirteenth Century Rome*, 2nd ed. (Berkeley, 1990), 19–20; Kessler and Zaccharias, *Rome 1300*, 187–8; Keyvanian, *Hospitals*, 78–80, 129–37; Drossbach, "Roman Hospital." Smaller towns in Lazio like Sutri and Tivoli also staged processions in which an image of Christ was taken to an image of his mother; Filippo Caraffa, "La processione del SS. Salvatore a Roma e nel Lazio nella notte dell'Assunta," *Lunario Romano: Feste e cerimonie nella tradizione romana e laziale* 5 (1976): 127–51.

27. Brentano, *Rome before Avignon*, 20; Keyvanian, *Hospitals*, 129–37.

28. Keyvanian, *Hospitals*, 80; Rist, "Innocent III.," 118–21.

29. Wolf, "Christ in His Beauty and Pain," 169–71, esp. 169: the Veronica was "a spiritual presence, a transitory foreshadowing of the face-to-face encounter with God" and "the first universal image of Christ in the Western Church." See also Aden Kumler, "*Signatis . . . tui vultus*: (Re)impressing the Holy Face before and after the Veronica," in Murphy et al., *European Fortune*, 103–12.

30. Uwe-Michael Lang, "Origins of the Liturgical Veneration of the Roman Veronica," in Murphy et al., *European Fortune*, 144–57.

31. See Gerhard Wolf, "From Mandylion to Veronica," in Kessler and Wolf, *Holy Face*, 168, for the uncertain authorship of this prayer; see 170–4 for painted copies of the Holy Face. See also Flora Lewis, "Rewarding Devotion: Indulgences and the Promotion of Images," in Wood, *Church and the Arts*, 179–94, at 179–81. In 1289, Nicholas IV issued further indulgences for those pilgrims who came to Rome to view the original; Étienne Doublier, "*Sui pretiosissimi vultus Imago*: Veronica e prassi indulgenziale nel XIII e all'inizio del XIV secolo," in Murphy et al., *European Fortune*, 181–92.

32. S. Corbin, "Les Offices de la sainte face," *Bulletin des Études Portugaises*, n.s., 11 (1947): 37–8.

33. Kessler and Zaccharias, *Rome 1300*, 210–1; Giovanni Villani, *Cronica*, 5 vols. (Rome, 1980), 8.36; Dante, *Vita nuova*, 40.1; *Paradiso*, 31.103–8.

34. Chiara Di Fruscia, "*Datum Avenioni*: The Avignon Papacy and the Custody of the Veronica," in Murphy et al., *European Fortune*, 219–28, suggests that the Avignon popes promoted the cult of the Veronica as a way of reasserting their authority in Rome, from a distance, with the face of Christ in St. Peter's standing in for the *vicarius Christi* away in Avignon.

35. Keyvanian, *Hospitals*, 327. In his 1478 bull *Tandem ascensurus*, Sixtus IV allowed that the Veronica should be exposed (within St. Peter's) as part of the annual liturgy for Santa Maria in Sassia, but only other, less prestigious relics were now sent in procession to the church.

36. Wolf, "Christ in His Beauty and Pain," 168. Innocent III gave the canons of St. Peter's a monopoly on the production and sale of metal badges of the Veronica, which they sold inside the Atrium of the basilica; beyond, various merchants hawked paper and cloth reproductions of the image.

37. Keyvanian, *Hospitals*, 169–70: the very ephemerality of the Veronica assisted in this process—it was not venerated as a cult object, like the Salvatore, but rather as a universal image that could be reproduced indefinitely, even as it remained "bound conceptually" to the original in St. Peter's. See also Noreen, "Sacred Memory," 181.

38. Jörg Bölling, "Face to Face with Christ in Late Medieval Rome," in Murphy et al., *European Fortune*, 137–42.

39. MS Cortona, Biblioteca del Comune e dell'Accademia Etrusca di Cortona, 101.6, written by the papal *scriptor* Sylvester, and dated February 22, 1300. Reproduced in Murphy et al., *European Fortune*, 187.

40. *Il Volto di Cristo*, 123, cat. IV.9 and cat. IV.10.

41. Gerhard Wolf, "'Pinta della nostra effigie': La Veronica come richiamo dei Romei," in D'Onofrio, *Romei e Giubilei*, 211–8, at 217; on the coin, see William R. Day, "Antiquity, Rome, and Florence: Coinage and Transmissions across Time and Space," in *Rome across Time and Space: Cultural Transmission and the Exchange of Ideas*, ed. Claudia Bolgia (New York, 2011), 237–61, at 251, 256–7; D'Onofrio, *Romei e Giubilei*, 418; Teodori and Casamassima, *Antiquorum habet*, 209. Philip the Good's collection of paintings of the Veronica, assembled in the middle decades of the fifteenth century, includes one image of the Veronica flanked by Peter and Paul and another of the Veronica surmounted by a papal tiara and pallium; Murphy et al., *European Fortune*, 171–2.

42. Wolf, "Christ in His Beauty and Pain," 168–9. In his "From Mandylion to Veronica," 175–6, Wolf reiterates how the Veronica differed from other sacred images belonging to other cities, calling it the "Vexillum Ecclesiae."

43. The *Stationes* is included in numerous fifteenth-century editions of the *Indulgentiae*. See, e.g., *Indulgentiae ecclesiarum principalium urbis Romae* ([Rome: Stephan Plannck, 1485–87]) (im00594000), fols. 62v–66v.

44. J. R. Hulbert, "Some Medieval Advertisements of Rome," *Modern Philology* 20 (1923): 403–24; and Christian Hülsen, *Le chiese di Roma nel Medio Evo* (Florence, 1927), both with transcriptions. See also the English translation, C. Eveleigh Woodruff, ed., *A XVth Century Guide-Book to the Principal Churches of Rome, Compiled c. 1470 by William Brewyn* (London, 1933). Nine Robijntje Miedema documents the manuscript and printed traditions of the text in *Die römische Kirchen im Spätmittelalter nach den Indulgentiae ecclesiarum urbis Romae* (Tübingen, 2001) and *Rompilgerführer in Spätmittelalter und Früher Neuzeit: Die Indulgentiae ecclesiarum urbis Romae (deutsch-niederlandisch)* (Tübingen, 2003), with further bibliography at 475–6. See Miedema, *Rompilgerführer*, 22–3, for the dating of the text; see also Birch, *Pilgrimage*, 179–82.

45. Different redactions of the *Indulgentiae* vary in length, although the section on the seven churches remains relatively stable; here I quote from Stephan Plannck's first Latin edition, printed in Rome around 1485–87 (im00594000), in which the seven churches are described over fols. 22v–32v. Of these twenty pages, the description of the Lateran fills about seven and St. Peter's about four and a half.

46. *Indulgentiae*, fol. 22v.

47. *Indulgentiae*, fols. 22v–23r. The author attributes this statement to Boniface VIII. The statement was a formulaic one that appeared in many thirteenth- and fourteenth-century guides, usually to compare the whole of Rome to other pilgrim sites; Birch, *Pilgrimage*, 179–82. Giovanni Rucellai, visiting Rome for the 1450 Jubilee, made a similar comment on the Sancta Sanctorum; Giovanni Rucellai, "Della bellezza e anticaglia di Roma," in *Codice topografico della città di Roma* (Rome, 1953), 4.406–7.

48. *Indulgentiae*, fol. 23v.

49. *Indulgentiae*, fol. 25r–v.

50. The Vatican, like the Lateran, offers forty-eight years + forty-eight quarantines in indulgences every day. The Scala Sancta offers nine years off for each of its twenty-eight steps; the steps of St. Peter's are not counted, but the pilgrim can gain seven years off for each one; *Indulgentiae*, fol. 26v.

51. *Indulgentiae*, fol. 27r–v.

52. *Reliquie*, sig. a3r.

53. *Reliquie*, sig. a6r.

54. See note 47 above; see also Germano Buccilli, "Viaggiatori a Roma nella seconda metà del Quattrocento," *La Cultura* 26 (1988): 39–62.

55. Rucellai, "Della bellezza e anticaglia di Roma," 4.407; his comments on the Veronica are at 4.403.

56. Quoted by Buccilli, "Viaggiatori," 43.

57. Flavio Biondo, *Roma instaurata*, 3.94, ed. Anne Raffarin-Dupuis, 2 vols. (Paris, 2005), 2.215.

58. Previous catalogues of this complicated corpus include Ludwig Schudt, *Le guide di Roma: Materialen zu einer Geschichte der Römischen Topographie* (Vienna, 1930); Nine Robijntje Miedema, *Die Mirabilia Romae: Untersuchungen zu ihrer Überlieferung mit Edition der deutschen und niederlandischen Texte* (Tübingen, 1996), 173–251, as well as her *Die römische Kirchen* and *Rompilgerführer*; Sergio Rossetti, *Rome: A*

Bibliography from the Invention of Printing through 1899, 2 vols. (Florence, 2000–2004), vol. 1, *The Guide Books*, esp. 1–26 (i.e., books printed 1471–1530). See also Paolo Veneziani, "I Mirabilia Romae: Le edizioni a stampa," in *Roma 1300–1875*, 216–9, reprinted in his *Tracce sul foglio* (Rome, 2007), 37–9; Veneziani, "Il *Mirabilia* scomparso," *Gutenberg Jahrbuch* 75 (2000): 145–51; Lorenzo Amato, "I *Mirabilia urbis Romae*: trasformazioni tipologiche dal manoscritto alla stampa," in *Dal Libro manoscritto al libro stampato*, ed. Outi Merisalo and Caterina Tristano (Spoleto, 2010), 109–32. The project has been vexed by the different ways in which bibliographers have classified these texts: some describe them all as variants of the *Mirabilia*, some consider the *Indulgentiae* a subsidiary section of the *Historia* (and some vice versa), and some consider the *Oratio de Sancta Veronica* to be a subsidiary part of the *Indulgentiae*. I consider these four texts as distinct and independent, each with its own tradition of publication and illustration.

59. The *Mirabilia urbis Romae* proper, with its catalogue of ancient monuments, is recorded in forty-three different printed editions in ISTC (forty-two in Latin and one in French; thirty-seven printed in Rome, and five elsewhere). The *Historia* is recorded in some sixty-three editions in ISTC, in Latin, German, Italian, French, and Dutch versions, fifty-four of them printed in Rome, nine printed abroad.

60. A German translation of both texts was published in an earlier block book, discussed below.

61. *BMC* 1.7; L. Donati, "Del *Mirabilia Romae* xilografico," *La Bibliofilia* 64 (1962): 1–36; Nine Miedema, "Überlegungen zu den *Mirabilia Romae*," in *Blockbücher des Mittelalters* (Mainz, 1991), 329–40; Miedema, *Die* Mirabilia Romae, 204–6 (item d1); facsimile of the Gotha copy in Rudolph Ehwald, *Mirabilia Romae* (Berlin, 1903).

62. Miedema, *Die* Mirabilia Romae, 205–6. Printed on Italian paper, the book is often attributed to the early Roman printer Ulrich Han. Donati, "Del *Mirabilia Romae*," argued for the earlier date of ca. 1450.

63. Ehwald, *Mirabilia Romae*, 10–1: "als eine Art bildlichen Kolophons."

64. This image appears at the very start of the book, on fol. 1v, in Ehwald's facsimile of the Gotha copy; in the British Library copy, it appears, as in the Munich copy, on fol. 27v. Ehwald, *Mirabilia Romae*, 10–1, discusses these variants.

65. Jean George Théodore Graesse, *Trésor de livres rares et précieux*, 7 vols. (Dresden, 1859–69), 4.535–6. Paolo Veneziani, "I *Mirabilia urbis Romae*," in Fagiolo and Madonna, *L'Arte degli anni santi*, 212–9, at 216, says that this block book shows repairs to the woodcuts, suggesting they had perhaps been used before for the Jubilee of 1450.

66. This redaction of the text includes indulgences for the seven major basilicas but no other churches. The editions were printed by Adam Rot (three), Theodore Schencbecher (three), Ulrich Han (one), and Arnold Pannartz (one). A ninth edition was printed in Ferrara in 1475.

67. Schencbecher's second edition of the *Indulgentiae* (ii00068450) was described by Giuseppina Li Calsi, *Catalogo degli incunabuli della Biblioteca Comunale di Palermo* (Palermo, 1978), 197, as including the text of the *Mirabilia Romae* (inc. "Murus habet . . .") before the *Indulgentiae*. GWM 1203550 and 2354750 cast doubt on this claim.

68. John Oliver Hand, "*Salve sancta facies*: Some Thoughts on the Iconography of the Head of Christ by Petrus Christus," *Metropolitan Museum Journal* 27 (1992): 7–18, at 10 and fig. 4.

69. Printed once each by Johannes Bulle, Bartolomaeus Guldinbeck, Eucharius Silber, and Stephan Plannck.

70. *Oratio Salve Sancta Facies* ([Ulm: Johann Zainer, ca. 1482]) (io00065950); *Gebet zu dem heiligen Antlitz Christi* ([Ulm: Conrad Dinckmut, ca. 1482]) (ii00065970), reproduced as *The Holy Face on a Cloth,* woodcut on paper, ca. 1482, in Peter Parshall and Rainder Schoch, *Origins of European Printmaking: Fifteenth-Century Woodcuts and Their Public* (New Haven, CT, 2005), no. 71. Later, but similar: *Lied von der Heilige Veronika* (Nuremberg: Peter Wagner, 1497) (iv00234800), fol. 1r, a different German text expressing devotion to the Holy Face.

71. Editions of the *Indulgentiae* and *Historia* in German translation never include the *Oratio,* in Latin or in German; when the same texts appeared in Italian translation, slightly later, those editions omitted the *Oratio* as well.

72. Germano Bucilli, "L'Aggiornamento riguardante reliquie ed indulgenze in alcune edizioni romane di *Libri indulgentiarum* a stampa del secolo XV," *Quellen und Forschungen aus italienischen Archiven und Bibliotheken* 70 (1990): 328–47, at 339–40; *BMC* 4.144.

73. ISTC describes this composite edition as (*Mirabilia Romae* vel potius) *Historia et descriptio urbis Romae.* Add: ps-Aegidius Romanus: *Oratio de S. Veronica. Indulgentiae ecclesiarum principalium urbis Romae* ([Rome: Andreas Freitag, not after 1489]) (im00596000). One copy contains a rubricator's date of 1489, providing a *terminus ante quem.* In the following discussion of subsequent editions containing these texts, I cite only the ISTC number. However, some ISTC entries for composite editions are incomplete and list only some of the constituent texts. I have independently checked the bibliographical literature, consulted facsimile reproductions, or corresponded with librarians concerning sixty-three editions listed in ISTC containing the *Indulgentiae* either with or without the *Historia* and *Oratio.* My tallies and accounts of these editions derive from this exercise rather than just the ISTC records.

74. im00607500. In Guldinbeck's edition, the woodcut of the Veronica exposition appears, as it does in the block book, just before the start of the text of the *Indulgentiae* (in this case on fol. 17r).

75. im00607400. Bibliothèque Nationale, *Catalogue des incunables,* 2 vols. (Paris, 1981), M-381 derives the date from the accession of Innocent VIII in September 1484, not noticing the reference to August 1485 in the text.

76. im00594000. This edition, at sixty-eight folios, is significantly longer than that described in the next note.

77. im00595000, described by Herman R. Mead in "Bibliograhical Notes," *Papers of the Bibliographical Society of America* 36 (1942): 229, as fifty-two folios long, with four full-page woodcuts. I have not seen a copy of this edition.

78. The Nuremberg printer Peter Wagner included a woodcut of the same scene, copied from Plannck's, in two German-language editions of the *Historia* and *Indulgentiae* (April, 13, 1491, im00608500; and February 28, 1500, im00610000). I am grateful to Steven Tabor of the Henry Huntington Library, who sent me a photo of Wagner's woodcut in the latter edition.

79. Solange Corbin, "Le Culte de voile de Véronique à Rome et en occident: L'office milanais de la Sainte Face," *Bulletin des Études Portugaises,* n.s., 11 (1947): 15–65.

80. im00596000, London, BL, IA.19480.

81. im00613500 and im00603400.

82. Paolo Veneziani, "Xilografie in edizioni romane della fine del Quattrocento," *La Bibliofila* 82 (1980): 1–21, at 12–20.

83. Hulbert, "Some Medieval Advertisements."

84. Presented as an *epistola generalis* addressed by Constantine to all Christian bishops and prelates.

85. Freitag, in his first edition, used the cut of the angels holding the Veronica over three shields, taken from the original block book. In his 1492 edition, Freitag replaced this with a cut of the Virgin and Child standing on the crescent moon.

86. im00609000, fol. 1v.

87. im00613500, sig. c4r.

88. *Figuren van die seven kercken van Romen* (Gouda: Collaciebroeders, [not before September 12, 1494]) (im00606700), fol. 1v.

89. Giuliano Dati, *Statione, indulgentie, e reliquie quadragesimale de l'alma ciptà di Roma* ([Rome: Andreas Freitag, 1492–93]) (id00048300); reprinted once by Freitag in Rome (id00048700) and twice in Florence (id00048600 and id00049000). See Nerida Newbigin, "Giuliano Dati, *Statione, indulgentie, e reliquie quadragesimale de l'alma ciptà di Roma,*" *Letteratura Italiana Antica* 5 (2004): 227–57.

90. Dati, *Statione*, fols. 3r and 4v.

91. Alexander VI, Bulla, April 12, 1498, "Consueverunt Romani pontifices" ([Rome: Eucharius Silber, not before April 12, 1498]) (ia00371200).

92. British Museum 1855,0625.32; compare the wax on parchment triptych of Veronica with Peter and Paul in the Staatliches Museum, Schwerin.

93. *Weg und Meilen von Erfurt bis Rom* (Erfurt: [Hans Sporer], 1500) (iw00010850).

94. Reproduced in Fagiolo and Madonna, *L'Arte degli anni santi*, 106–7.

95. Max Geisberg, *The German Single-Leaf Woodcut, 1500–1550*, rev. and ed. Walter L. Strauss, 4 vols. (New York, 1974), 2.423; Fagiolo and Madonna, *L'Arte degli anni santi*, 120, fig. II.8(a). See also the 1514 print of Peter, Paul, and the Veronica by Luca da Leida; Fagiolo and Madonna, *L'Arte degli anni santi*, 120, fig. II.8(b).

96. Geisberg, *German Single-Leaf Woodcut*, 2.431.

97. Miedema makes this point about Boniface VIII and the earliest redactions of the *Indulgentiae* (*Rompilgerführer*, 22).

98. Infessura, 72–3; Margaret Meserve, "News from Negroponte: Politics, Popular Opinion, and Information Exchange in the First Decade of the Italian Press," *Renaissance Quarterly* 59 (2006): 440–80, at 448.

Chapter 6 • *Refugee Relics*

1. Patrick J. Geary, *Furta Sacra: Thefts of Relics in the Central Middle Ages*, rev. ed. (Princeton, NJ, 1991).

2. *Reliquie rhomane urbis et indulgentiae* ([Basel: Peter Kollicker, ca. 1483]) (ir00140000), sig. a3v.

3. See stanzas 97–102 in the edition by Nerida Newbigin, "Giuliano Dati, *Statione, indulgentie, e reliquie quadragesimale de l'alma ciptà di Roma,*" *Letteratura Italiana Antica* 5 (2004): 227–57.

4. Jacopus de Voragine, *The Golden Legend: Readings on the Saints*, trans. William Granger Ryan, rev. ed. (Princeton, NJ, 2012), 554–9; Stephan Borgehammar, "Heraclius

Learns Humility: Two Early Latin Accounts Composed for the Celebration of *Exaltio Crucis*," *Millenium: Jahrbuch zu Kultur und Geschichte des ersten Jahrtausends n. Chr.* 6 (2009): 145–201.

5. Margaret Meserve, *Empires of Islam in Renaissance Historical Thought* (Cambridge, MA, 2008), chap. 2.

6. Completed between 1492 and 1495. See Francesca Cappelletti, "L'affresco nel catino absidale di Santa Croce in Gerusalemme a Roma," *Storia dell'Arte* 66 (1989): 119–26; Meredith J. Gill, "Antoniazzo Romano and the Recovery of Jerusalem in Late Fifteenth-Century Rome," *Storia dell'Arte* 83 (1995): 28–47.

7. Germano Bucilli, "L'Aggiornamento riguardante reliquie ed indulgenze in alcune edizioni romane di *Libri indulgentiarum* a stampa del secolo XV," *Quellen und Forschungen aus italienischen Archiven und Bibliotheken* 70 (1990): 328–47, at 339–40.

8. Later known as the Madonna del Perpetuo Soccorso; Fabriciano Ferrero, *The Story of an Icon* (Chawton, 2001), 108–9.

9. Pius II, *Commentarii*, ed. Adrianus van Heck, 2 vols. (Vatican City, 1984), 2.467–90; see also the brief notice in Infessura, 66. See also Ruth Olitsky Rubinstein, "Pius II's Piazza S. Pietro and St. Andrew's Head," in *Essays in the History of Architecture Presented to Rudolf Wittkower*, ed. Douglas Fraser (London, 1967); Arianna Antoniutti, *Pio II e Sant'Andrea Apostolo: Le ragioni della devozione* (Rome, 2004); Claudia Barsanti, "In memoria del reliquiario del Sacro Capo dell'apostolo Andrea," in *Enea Silvio Piccolomini: Pius secundus, poeta laureatus, pontifex maximus*, ed. M. Sodi and A. Antoniutti (Rome, 2007), 319–40; Peter Godman, "Pius II in the Bath: Papal Ceremony and Cultural History," *English Historical Review* 129 (2014): 808–29; Maya Maskarinec, "Mobilizing Sanctity: Pius II and the Head of St. Andrew in Rome," in *Authority and Spectacle in Medieval and Early Modern Europe*, ed. Yuen-Gen Liang et al. (London, 2017), 186–202.

10. Paolo Prodi, *Papal Prince: One Body and Two Souls: The Papal Monarchy in Early Modern Europe*, trans. Susan Haskins (Cambridge, 1988), 176–7.

11. Pius II, *Orations "Adveniisti tandem/Si loqui possent" of Pope Pius II (12–13 April 1462, Rome)*, ed. and trans. Michael von Cotta-Schönberg, 4th version (2019), https://hal.archives-ouvertes.fr/hal-01234345.

12. Pius II, *Commentarii*, 2.477–9.

13. Cf. Godman, "Pius II in the Bath," 827–9, on gestures of humility the pope made during the ceremony.

14. According to Pius II, *Commentarii*, 2.480–2, the procession went south from Piazza del Popolo toward the Mausoleum of Augustus, but there, where it might have turned right on the via Recta and made for Ponte Sant'Angelo, it went south and east toward the Pantheon; the procession picked up the via Papalis through Campo de' Fiori, past cardinals' palaces specially decorated for the occasion, to Ponte Sant'Angelo, the Borgo, and St. Peter's.

15. Andrew was thought to have founded the church of Constantinople (F. Dvornik, *The Idea of Apostolicity in Byzantium and the Legend of the Apostle Andrew* [Cambridge, MA, 1958]), so having Bessarion, titular patriarch of Constantinople, speak for him was especially apt. For Pius's oration at St. Peter's ("Si loqui possent"), see note 11 above.

16. Pius II, *Commentarii*, 2.489–90.

17. Margaret Meserve, "Pomponio Leto's *Life of Muhammad*," in *For the Sake of*

Learning: Essays in Honor of Anthony Grafton, ed. Ann Blair and Anja-Silvia Goeing, 2 vols. (Leiden, 2016), 1.215–31, esp. 225–7.

18. Rubinstein, "Pius II's Piazza S. Pietro," 31–2.

19. For examples, see *The Diary of John Burchard*, trans. Arnold Harris Mathew (London, 1910), 121, 163, 199, 233, 269, and 310 (expositions on the Feast of St. Andrew, November 30).

20. *Indulgentiae ecclesiarum principalium urbis Romae* ([Rome: Stephan Plannck, 1485–87]) (im00594000), fol. 27r.

21. William Brewyn, *A XVth Century Guide-Book to the Principal Churches of Rome*, trans. C. Eveleigh Woodruff (London, 1933), 8.

22. *Reliquie rhomane*, sig. a5v.

23. Arnold von Harff, *The Pilgrimage of Arnold von Harff, Knight, from Cologne: Through Italy, Syria, Egypt, Arabia, Ethiopia, Nubia, Palestine, Turkey, France, and Spain Which He Accomplished in the Years 1496 to 1499*, trans. Malcolm Letts (London, 1946), 26.

24. Dati, *Statione*, stanza 58, in Newbigin, "Giuliano Dati."

25. Pastor, 5.311–3 316–8; John Freely, *Jem Sultan: The Adventures of a Captive Turkish Prince in Renaissance Europe* (London, 2004), 203–5; Sigismondo dei Conti, *Le storie de' suoi tempi dal 1475 al 1510*, 2 vols. (Rome, 1883; repr., Foligno, 2015), 2.28–9; Infessura, 274. Antonio de Vascho understood the reasons for the gift very well: "Lo Gran Turco la manda al papa per graduirlo perché tenera prigione il fratello del detto gran Turco" (*Diario, RIS*² 23.3.545).

26. Burchard, 2.364: an ampule of Christ's blood from the Lateran and the Titulus Christi from Santa Croce were thought particularly appropriate items to meet the lance.

27. Peter Parshall and Rainder Schoch, *Origins of European Printmaking: Fifteenth-Century Woodcuts and Their Public* (New Haven, CT, 2005), 212–4; David S. Areford, "Multiplying the Sacred: The Fifteenth-Century Woodcut as Reproduction, Surrogate, Simulation," in *The Woodcut in Fifteenth-Century Europe*, ed. Peter Parshall (New Haven, CT, 2009), 119–54, at 131–5 and fig. 11. The pamphlet is *Heiltum zu Nurnberg* (Nuremberg: Peter Vischer, 1487) (in00278410), fols. 4r (exposition) and 5v (lance); the text was reprinted in 1493 (in00278415).

28. Burchard, 2.356–68.

29. Burchard, 2.356–7; another discussion of its authenticity is at 2.364–5.

30. Burchard, 2.357.

31. Burchard, 2.365.

32. Infessura, 274; see also the comments of the Florentine ambassador Filippo Valori, quoted at 274 n. 2: "pare certamente cosa molta antiqua."

33. Conti, *Storie*, 28.

34. Conti, *Storie*, 29.

35. Dati, *Stazione*, stanza 57.

36. Von Harff, *Pilgrimage*, 27.

37. Timothy B. Smith, "Up in Arms: The Knights of Rhodes, the Cult of Relics, and the Chapel of St. John the Baptist in Siena Cathedral," in *Images, Relics, and Devotional Practices in Medieval and Renaissance Italy*, ed. Sally J. Cornelison and Scott B. Montgomery (Tempe, AZ, 2006), 213–38.

38. Luca da Monterado, *Storia del culto e del pellegrinaggio a Loreto: Secoli XIV–XV*, 2nd ed. (Loreto, 1979); Bernard Hamilton, "The Ottomans, the Humanists, and the Holy House of Loreto," *Culture, Theory and Critique* 31 (1987): 1–19; Floriano Grimaldi, *La Historia della Chiesa di Santa Maria de Loreto* (Loreto, 1993); Floriano Grimaldi and Katy Sordi, eds., *L'Iconografia della Vergine di Loreto nell'Arte* (Loreto, 1995); Ferdinando Citterio and Luciano Vaccaro, eds., *Loreto: Crocevia religioso tra Italia, Europa e Oriente* (Brescia, 1997); Floriano Grimaldi, *Pellegrini e pellegrinaggi a Loreto nei secoli XIV–XVIII* (Loreto, 2001); Alexander Nagel and Christopher S. Wood, *Anachronic Renaissance* (New York, 2010), 195–215; Karin Vélez, *The Miraculous Flying House of Loreto: Spreading Catholicism in the Early Modern World* (Princeton, NJ, 2019).

39. Grimaldi, *Historia*, 15–7, with a transcription of Tolomei's text. For his biography and dates, see 15.

40. Pietro di Giorgio Tolomei, *Translatio miraculosa ecclesie beate Virginis Mariae de Loreto* ([Rome: Eucharius Silber, 1486–90]) (it00425700), fol. 2r. All quotations from Tolomei in this chapter are taken from this edition.

41. Giuseppe Santarelli, "La Madonna di Loreto nei testi poetici del secolo XV," in Citterio and Vaccaro, *Loreto*, 491–519, at 504.

42. Giacomo Ricci, *Virginis Mariae Loretae historia*, ed. Giuseppe Santarelli (Loreto, 1987), 116.

43. Santarelli, "Madonna di Loreto," 509: "non sum passa hostes inter adesse meos."

44. Baptista Mantuanus (Battista Spagnoli), *Redemptoris mundi matris Ecclesiae Lauretanae historia* ([Bologna: Bazalerius or Caligula de Bazaleriis, ca. 1495]) (ib00084400), sig. [a]3r.

45. Text in Pietro Valerio Martorelli, *Teatro istorico della Santa Casa Nazarena della Vergine Maria e sua ammirabile Traslazaione in Loreto*, 2 vols. (Rome, 1722–23), 1.515.

46. Nanni Monelli and Giuseppe Santarelli, *La basilica di Loreto e la sua reliquia* (Loreto, 1999).

47. Kathleen Weil-Garris, "The Santa Casa di Loreto: Problems in Cinquecento Sculpture" (PhD diss., Harvard University, 1965; repr., New York, 1977); Floriano Grimaldi and Katy Sordi, *L'Ornamento marmoreo della Santa Cappella di Loreto* (Loreto, 1999).

48. Fabio Bisogni, "Iconografia lauretana: Prototipi e sviluppi," in Citterio and Vaccaro, *Loreto*, 346–7.

49. Ricci must have completed his work by 1477, since he presented it to Cardinal Latino Orsini, who died that year; Ricci, *Virginis Mariae Loretae historia*, 19.

50. Tolomei, *Translatio miraculosa*, 3v: "et in una pariete ibi prope est scriptum et sculptum in muro quomodo ista ecclesia fuit ibi et postea recessit."

51. Tolomei, *Translatio miraculosa*, 4r.

52. Bartolomeo da Vallombrosa, *La dichiaratione della chiesa di sancta Maria delloreto et come ella venne tutta intera* ([Florence: Francesco di Dino, after May 20, 1483]) (it00427080), fol. 4v: "tucta questa scriptura fu copiata da uno originale autentico scripto nella decta chiesa di sancta Maria delloreto a di xx del mese di magio, MCCCCLXXXIII." Spagnoli, *Redemptoris*, sig. [a]2r: "Et ecce mihi sese offert tabella situ et vetustate corrosa, in qua unde et quonam pacto locus ille tantam sibi vindicasset auctoritatem, conscripta erat historia." Giuliano Dati, *Traslazione della sacrata camera di Nostra Donna di Nazareth in Italia appresso la città di Recanati* ([Rome: Andreas

Freitag, ca. 1492–93]) (id00049400), fol. 3r, suggests that the sign was made after the delegation to Nazareth: "e piu anchora per dichiaratione / era scritto in un muro li appresso / el caso aperto chiaro e molto spresso." Nicola da Manopello visited Loreto in 1478 and copied the story from a parchment sign attached to a column in the church, "coscripta litteris grossis patentibus in carta pergamena"; Grimaldi, *Historia*, 79.

53. The *Rosarium* of Caterina de' Vigri, recording a vision she had in 1440, includes six lines on the flight of the holy house from Nazareth to Italy. Some scholars date the text itself to 1440 and call this the earliest known reference to the legend, but Vigri could have composed the lines on Loreto at any point up to her death in 1463, and Tolomei's account could have been in circulation by then.

54. Gian Ludovico Masetti Zannini, "I papi e Loreto," in Citterio and Vaccaro, *Loreto*, 246.

55. Flavio Biondo, *Italy Illuminated*, trans. Jeffrey A. White (Cambridge, MA, 2005), 1.263. For other fourteenth- and fifteenth-century notices, see Giuseppe Santarelli, "Il pellegrinaggio loretano attraverso i secoli," in *I pellegrini alla Santa Casa di Loreto: Indagine socio-religiosa* (Loreto, 1992), 51–72, at 55–8. Ricci, *Virginis Mariae Loretae historia*, also mentions silver votive items hanging on the walls of the shrine.

56. Mentioned in a document of 1315; Bisogni, "Iconografia lauretana," 331–3.

57. The pope refers to the church, "in qua, sicut fidedignorum habebat assertio, ipsius Virginis gloriose ymago angelico comitante cetu mira Dei clementia collocata existebat." Quoted by Hamilton, "Ottomans," 1, from U. Chevalier, *Notre-Dame de Lorette. Étude historique sur l'authenticité de la Santa Casa* (Paris, 1906), 141–3.

58. Corrado Ricci, "Per l'iconografia lauretana," *Rassegna d'Arte* 16 (1916): 266–7; text in Santarelli, "Madonna di Loreto," 492–3.

59. Mark J. Zucker, "The Madonna of Loreto: A Newly Discovered Work by the Master of the Vienna Passion," *Print Quarterly* 6 (1989): 149–60.

60. *The Illustrated Bartsch*, 166 vols. (New York, 1978–2012), 2402.023, 55–7; present whereabouts unknown.

61. *Illustrated Bartsch*, 2405.068, 234–6; Zucker, "Madonna of Loreto," 160; present whereabouts unknown.

62. *Miracoli della Vergine Maria* (Brescia: Baptista Farfengus, March 2, 1490) (im00618000), fol. 1v.

63. Ricci, "Per l'iconografia lauretana," 266–7, with reproduction; M. Falocci Pulignani, *La Santa Casa di Loreto secondo un affresco di Gubbio* (Rome, 1907), 67; Grimaldi, *Historia*, 116.

64. Ricci, "Per l'iconografia lauretana"; Grimaldi, *Historia*, 287–342; Grimaldi, "L'iconografia della Vergine Lauretana nell'arte: I prototipici iconografici," in Grimaldi and Sordi, *L'iconografia della Vergine di Loreto*, 15–30, at 23–4; Grimaldi, *La chiesa di Santa Maria di Loreto nei documenti dei secoli XII–XV* (Ancona, 1984), 83–5.

65. Marche (Tolentino, Matelica, and Fabriano), Umbria (Morra and Bevagna), Abruzzo (Atri), and Lazio (Velletri); two more were in collections in Rome and Caserta.

66. Two frescoes in Ripatransone, dated 1403 and 1426 (Giuseppe Crocetti, "Il culto mariano-lauretano in area farfense, in Sabina, e nel Piceno," in Citterio and Vaccaro, *Loreto*, 416–7 and figs. 78 and 79); another in Loro Piceno from the middle of the fifteenth century (417–8 and fig. 80); a fresco by Bernardino Campilio, formerly in the church of Santa Maria di Loreto and now in Sant'Agata in Spelonga near Arquata del

Tronto (Grimaldi and Sordi, *L'iconografia della Vergine di Loreto*, 21–2); a panel painting in Serra San Quirico (74–7); a fresco from Foligno, possibly dated 1448 (80–3), with mention of another in Serravalle del Chienti, near Macerata; and a fresco in Ussita, in the same region (86–7).

67. Cristina Giorgetti, "Andrea de Litio," in *Dalla testa ai piedi: Costume e moda in età gotica*, ed. Laura Dal Prà and Paolo Peri (Trent, 2006), 435–63; *Atri: una storia per immagini* (Atri, 2011); *Il Duomo di Atri e il Museo* (Capitolare, 2011).

68. A fifteenth-century marble relief in the basilica at Loreto shows the church moving over waves; Giuseppe Santarelli, *Loreto: History and Art* (Ancona, 1998), 85–6 and fig. 109. A panel painting dated to 1506–13 and now in a Sicilian collection shows the Madonna and Child atop a church sailing, not flying, over the sea with angelic help.

69. Nagel and Wood, *Anachronic Renaissance*, 197. A Neapolitan poem written in the 1490s also attributes construction of the church to St. Helen; Santarelli, "Madonna di Loreto," 499.

70. Giovanni Antonio Campano, *Vita Pii II pontificis Maximi*, ed. Giulio C. Zimolo, *RIS*² 3.3.85 and n. 6; Pastor, 3.357.

71. Spagnoli, *Redemptoris*, sig. b1r–v.

72. Hamilton, "Ottomans," 1.

73. Cf. Alberti's model of the Holy Sepulcher in the Rucellai chapel in San Pancrazio, Florence, or the Portiuncula chapel outside Assisi, enshrined in the later basilica of Santa Maria degli Angeli.

74. Grimaldi, *Historia*, 156–7; Eva Renzulli, "Tales of Flying Houses and Paved Roads: Loreto, an Early Modern Town of Pilgrimage," *Città e Storia* 7 (2012): 27–41, at 37–8.

75. The nave, the crossing over the shrine, several lobed chapels, and the four towers of the basilica were started under Paul II, whose marble stemma is embedded in the exterior masonry; Sixtus's arms appear on adjacent areas of exterior construction, indicating that construction continued after 1471.

76. Masetti Zannini, "I papi e Loreto," 247.

77. Eva Renzulli, "Loreto, Leo X and the Fortifications on the Adriatic Coast against the Infidel," in *Italy and the European Powers: The Impact of War, 1500–1530*, ed. Christine Shaw (Leiden, 2006), 57–73, at 64: Giuliano da Maiano was appointed architect in March 1486 but left the site in 1487. Pontelli was appointed in 1488.

78. Masetti Zannini, "I papi e Loreto," 247.

79. Renzulli, "Loreto, Leo X."

80. Details of this edition are given in note 44.

81. Giovanni Sabbadino degli Arienti, *Hystoria del sacro templo de Laureto* ([Bologna: Bazalerius de Bazaleriis, ca. 1489]) (ib00084500).

82. Ricci presented his manuscript to Cardinal Latino Orsini, and Spagnoli presented his to Cardinal Girolamo Basso della Rovere; Sabbadino dedicated his translation to Ginevra Sforza, wife of Giovanni Bentivoglio of Bologna.

83. See note 52.

84. This is disingenuous: Tolomei's narrative was known to Ricci and would soon begin to circulate in print; it is possible Spagnoli had no knowledge of it beyond the placard version hanging on the wall of the shrine, but he may have exaggerated how old and faded the placard was (and suppressed any mention of other copies circulating

in manuscript) in order to present himself as rescuing a narrative otherwise lost to posterity.

85. Spagnoli, *Redemptoris*, sigs. [a]2r–4v.

86. Spagnoli, *Redemptoris*, sig. [a]3r.

87. Spagnoli, *Redemptoris*, sig. b1r–v.

88. Spagnoli, *Redemptoris*, sig. b1v.

89. Spagnoli, *Redemptoris*, sig. b1v.

90. The Carmelites formed their first community atop Mount Carmel in the twelfth century and believed they were joining a community of Christian hermits that had dwelt on the mountain since antiquity. The notion that the order continued this Christian presence in the Holy Land since the time of Christ was central to Carmelite identity.

91. Spagnoli, *Redemptoris*, sig. b2r. This claim makes little sense, since Spagnoli claims that the house departed Nazareth in the seventh century, while the Carmelites were only expelled from the Holy Land with the Crusaders in 1291.

92. Spagnoli, *Redemptoris*, sig. b2v.

93. Dati, *Traslazione*, fol. 1v.

94. Dati, *Traslazione*, fol. 2r.

95. Dati, *Traslazione*, fol. 2v. This detail, unattested in any other source, is not easily explained.

96. Dati, *Traslazione*, fol. 3v.

97. Dati, *Traslazione*, fol. 4v.

98. Tolomei, *Translatio miraculosa*, fols. 1v–2r.

99. Tolomei's text only fell out of favor with the publication of Girolamo Angelita's version around 1525.

100. Despite the fact that many are listed in ISTC, very few have been assigned dates before 1500.

101. Sander, nos. 4280–4303, lists twenty-four editions of Tolomei's text, some lost and others with too little information to identify them. ISTC also lists twenty-four editions, a group overlapping Sander's list only partially, and includes several editions dated after 1500. *LL* lists thirty-two editions, including many, but not all, listed by Sander. Grimaldi's entries are not always complete; he does include reproductions of twenty-two different woodcuts used to illustrate different editions. Nestore Zanandrea, "Postilla al libro lauretano," *L'Esopo* 49 (1979): 58–64, identifies three more editions unknown to Grimaldi in the first edition of *LL*; two of these are included in the second edition of the census, but not a (lost) Ripoli broadside, a reference to which Zanandrea found in the Ripoli *Diario* and which probably predates all surviving editions. Relying on Sander, Grimaldi, Zanandrea, ISTC, EDIT-16, VD16, and advice from librarians and colleagues at distant collections, I have identified forty-three editions of Tolomei's *Translatio* printed between ca. 1483 and ca. 1525. A few isolated editions of the text appear later in the sixteenth century but do not figure in my tally.

102. *Translatio miraculosa* ([Rome: Eucharius Silber?, ca. 1486–90]) (it00425700 = *LL* 13).

103. *Translatio miraculosa* ([Rome: Eucharius Silber, ca. 1495]) (it00425900 = *LL* 29).

104. These are: (1) [Stephan Plannck, ca. 1491–1500] (it00425800 = *LL* 8); (2) [Rome: Stephan Plannck, after 1500?] (it00425820, not in *LL*); (3) [Rome: Stephan Plannck, ca.

1500] (not in ISTC; *LL* 8A = Sander, 4289, known in a single copy, once held in a princely library in Germany and now lost, with the same woodcut of the Virgin).

105. These are: (1) [Rome: Johannes Besicken, after 1500] (it00426200, not in *LL*); (2) [Rome: Johannes Besicken, after 1500] (*LL* 14, not in ISTC; a copy is Venice, Fondazione Cini, 1115); (3) [Rome: Johannes Besicken, ca. 1504] (it00426800 = *LL* 17). This is the cut missing from Besicken's Jubilee broadside of 1500 (ia00372400), where the printer used three other cuts from his *Indulgentiae* series: St. Peter, St. Paul, and St. John, with a different cut of the Virgin to represent Santa Maria Maggiore.

106. [Rome: Eucharius Silber, ca. 1500?] (it0042600 = *LL* 18).

107. [Rome: Etienne Guillery, ca. 1513]; not in ISTC or *LL*; a copy is Cambridge UL F.151.e.2.8. Thanks to Mary Laven for an identifying photo.

108. [Rome: Etienne Guillery, ca. 1524?]; also not in ISTC or *LL*. See Francesco Barberi, *Tipografi romani del Cinquecento* (Florence, 1983), 47: "l'immagine della Vergine col Bambino in gloria."

109. [Rome: Eucharius Silber, not after 1509]; Sander, 4299, and Tinto, 56, not in ISTC or *LL*; a copy is BSB, Rar. 1240. ISTC identifies it00427000 as identical to Sander, 4299, but it is not. The copy of T-427 in the Library of Congress, Rosenwald 355, is a different edition, discussed below. Marianna Stell at the Library of Congress confirmed this.

110. [Rome: Johannes Besicken, after 1500] (it00426100 = *LL* 16); I examined Rome, Biblioteca Casanatense, Rari 246.

111. These are: (1) [Rome: Johannes Besicken (and Martinus de Amsterdam?), ca. 1504] (it00426900 = *LL* 6); (2) [Rome: Johannes Besicken, ca. 1500], *LL* 7, not in ISTC.

112. All identified as [Rome: Eucharius Silber, after 1500]: (1) it00426300 = *LL* 30; (2) it00426400 = *LL* 9; (3) it00426500, also listed under *LL* 9 but a different edition; (4) it00426600, not in *LL*. Mary Laven supplied photos of the copies of versions 1, 2, and 4 in the Cambridge UL, which helped with identifications. Marcellus Silber would reuse this cut in 1515.

113. [Rome: Eucharius Silber, ca. 1505] (*LL* 24, not in ISTC).

114. [Rome: Eucharius Silber, before 1509] (it00427020 = *LL* 24).

115. [Rome: Marcellus Silber, not before 1510]: (1) it00427030; (2) it00427040. Both are listed under *LL* 23 but are in fact separate editions.

116. [Rome: Marcellus Silber, after 1513] (it00427060 = *LL* 25).

117. Santarelli, "Madonna di Loreto," 496–7.

118. *LL* 30 (not in ISTC), a copy of which is Fondazione Cini 1114.

119. *LL* 33 (= Sander, 4293, not in ISTC). A copy is Munich, BSB, Rar. 1242(6), dated "not before 1521."

120. [Italy: n.p., after 1500] (it00427000); the two other editions are *LL* 11 and *LL* 12. The woodcut was used again in a Venetian edition of the vernacular text, discussed below in note 127.

121. Zanandrea, "Postilla," 61; *LL* 37 and fig. 21.

122. *LL* 32 = Sander, 4303, a copy of which is Tübingen Universitätsbibliothek, Gb 534b. Thanks to Nicole Domka for help with identification. *GW* M17576 refers to this copy, but the description does not match it; it describes an unknown edition of twelve leaves. *LL* 32A and *LL* 32B are two more Blado editions with this same cut, dated September 16, 1524, and November 24, 1524.

123. Zanandrea, "Postilla," 59–60.

124. *La dichiaratione della chiesa di Sancta Maria del Loreto et come venne tucta intera*, trans. Bartolomeo da Vallombrosa. Editions: (1) [Florence: Francesco di Dino, after May 20, 1483] (it00427080 = *LL* 5); (2) [Venice: Simon Bevilaqua, ca. 1500] (it00427150); (3) [Italy: n.p., after 1500] (it00427300 = *LL* 15); (4) [Venice: Georgius de Rusconibus, ca. 1515] (it00427200 = *LL* 31).

125. These are: (1) [Florence: Bartolommeo di Libri, ca. 1497] (it00427100 = *LL* 1); (2) *LL* 31A = Sander, 4282, the only copy of which is now lost. The Italian translation appeared once more in our period in Ancona in 1513 (*LL* 24, the only copy of which is missing the first page, so its illustration is unknown).

126. *Translatio miraculosa* (German) ([Augsburg: Johann Schönsperger, ca. 1500]) (it00427320) and ([Augsburg, ca. 1510]), VD16 T-1485.

127. Zucker, "Madonna of Loreto," 158–9.

128. Sander, 2338 = *LL* 36.

129. Charles Hope, "Religious Narrative in Renaissance Art," *Journal of the Royal Society of Arts* 134 (1986): 804–18; Hope, "Altarpieces and the Requirements of Patrons," in *Christianity and the Renaissance*, ed. Timothy Verdon and John Henderson (Syracuse, 1990), 535–71.

130. Giovanni Battista Rosito, *La Vergine di Loreto sotto il Baldacchino*, Velletri, Museo Capitolare, reproduced in Grimaldi and Sordi, *L'iconografia della Vergine di Loreto*, 23.

131. Mario Salmi, "Luca Signorelli a Morra," *Rivista d'arte* 26–7 (1950): 131–47; Ricci, "Per l'iconografia lauretana," 269. Signorelli's work has been dated to 1507–10.

132. Carol Plazzotta et al., "The Madonna di Loreto: An Altarpiece by Perugino for Santa Maria dei Servi, Perugia," *National Gallery Technical Bulletin* 27 (2006): 72–95, at 72.

133. Christoph Jobst, *Die Planungen Antonios da Sangallo des Jüngeren für die Kirche S. Maria di Loreto in Rom* (Worms, 1992).

134. Attributed to Perugino, Marco Palmezzano, Antoniazzo Romano, and even Raphael. In the same year, 1507, Julius II issued a bull granting Agostino Chigi the right to construct a funeral chapel in Santa Maria del Popolo, dedicated to both St. Augustine and the Virgin of Loreto. Raphael's program for the chapel made no reference to the Loreto legend. J. Shearman doubts that the altarpiece represented the Madonna di Loreto in any way; "The Chigi Chapel in Santa Maria del Popolo," *JWCI* 24 (1961): 129–60, at 143.

135. E.g., in the church of San Lorenzo in Fiastra and in the Chiesa della Maddalena in Fabriano; Ricci, "Per l'iconografia lauretana," 268–9.

136. Caravaggio later painted the Madonna di Loreto in even starker isolation, greeting a pair of kneeling pilgrims from a darkened doorway.

137. Don Skemer, "Magic Writ: Textual Amulets Worn on the Body for Protection," in *Schriftträger, Textträger: Zur materialen Präsenz des Geschriebenen in frühen Gesellschaften*, ed. Annette Kehnel and Diamantis Panagiotopoulos (Berlin, 2015), 127–49, at 144–5; Nine Miedema, "Die *Oratio ad sanctam crucem* des Johannes Mercurius Corrigiensis: Ein Einblattdruck als Apotropäum?," in *Einblattdrucke des 15. und frühen 16. Jahrhunderts: Probleme, Perspektiven, Fallstudien*, ed. Volker Honneman et al. (Tübingen, 2000), 325–48. Johannes Mercurius's broadside *Oratio ad sanctam crucem* (Rome:

[Eucharius Silber], 1499) (ij00378050) is a prayer typeset in the shape of a cross, sprinkled with typographic crosses printed in red; the single surviving copy was folded numerous times into a tiny square.

138. See, e.g., Johannes Franciscus de Pavinis, *Oratio in laudem Leopoldi Marchionis Austriae* ([Rome: Eucharius Silber, after November 20, 1484]) (ip00243000) and ([Rome: Eucharius Silber, ca. 1485]) (ip00243500); *Relatio circa canonizationem Beati Bonaventurae* ([Rome: Eucharius Silber, 1486–87]) (ip00245300); Ronald C. Finucane, *Contested Canonizations: The Last Medieval Saints, 1482–1523* (Washington, DC, 2011), 33–116; Roberto Cobianchi, "Printing a New Saint: Woodcut Production and the Canonization of Saints in Late Medieval Italy," in *The Saint between Manuscript and Print: Italy 1400–1600*, ed. Alison K. Frazier (Toronto, 2015), 73–98.

Chapter 7 • Kissing the Papal Foot

1. *Cronaca di Bologna, RIS*² 33.1, 232.

2. *Cronaca di Bologna*, 232. Cecilia M. Ady, *The Bentivoglio of Bologna: A Study in Despotism* (Oxford, 1937), 82–3; Giovanni Gozzadini, *Memorie per la vita di Giovanni II Bentivoglio* (Bologna, 1839), 40.

3. Battista Spagnoli's tract in praise of the Madonna di Loreto and its Italian translation, dedicated to Bentivoglio's wife, Ginevra Sforza, would be published in Bologna within a few years of Giovanni's trip.

4. The brief Innocent sent the Bolognese on April 22 is in Gozzadini, *Memorie*, appendix, xli–xlii.

5. Gaetano Moroni, "Ubbidienza," *Dizionario di erudizione storico-ecclesiastica* (Venice, 1857), 82.3–30; see also *Catholic Encyclopedia*, s.v. "Kiss."

6. Michel de Montaigne, *Travel Journal*, trans. Donald Frame (San Francisco, 1983), 74–5.

7. Burchard, 1.100; translations with some changes from *The Diary of Johann Burchard*, trans. Arnold Harris Mathew (London, 1910), 94.

8. Burchard, 1.116; *Diary*, 108.

9. Burchard, 1.118; *Diary*, 109.

10. There is no comprehensive study of the history of obedience embassies or obedience orations beyond Christian Gottlieb Buder, *De legationibus oboedientiae Romam missis* (Jena, 1737). Philipp Stenzig, *Botschafterzeremoniell am Papsthof der Renaissance: Der Tractatus de oratoribus des Paris de Grassi*, 2 vols. (Frankfurt, 2013), deals extensively with questions of diplomatic ceremonial at the papal court and touches on obedience orations at several points, esp. 1.215–27 (de Grassi on obedience ceremonies) and 2.1281–1363 (anthology of sample orations). Individual studies include Cesare Guasti, *Due legazioni al Sommo Pontefice per il Comune di Firenze presedete da Sant'Antonino Arcivescovo* (Florence, 1857); Francis M. Rogers, *The Obedience of a King of Portugal* (Minneapolis, 1958), esp. 3–18; Alfred A Strnad, "Johannes Hinderbachs Obedienz-Ansprach vor Papst Pius II," *Römische Historische Mitteilungen* 10 (1967): 43–183; Franco Martignone, "Diplomazia e politica della Republica di Genova nella 'Oratio de obedientia' ad Innocenzo VIII," in *Atti del III Convegno Internazionale di Studi Colombiani* (Genoa, 1979), 101–50; Martignone, "L'orazione di Giacomo Spinola per l'obbedienza genovese al pontefice Alessandro VI," in *La Storia dei Genovesi* (Genoa, 1987), 7.390–409; Martignone, I *Turchi e l'Europa nelle orazioni d'obbedienza ai pontefici*

412   Notes to Pages 263–264

del secondo Quattrocento (Genoa, 2002); Martignone, "Le 'orazione d'obbedienza' ad Alessandro VI: Immagine e propaganda," in *Principato ecclesiastico e riuso dei classici: Gli umanisti e Alessandro VI* (2002), 237–54; Adrie van der Laan, "Rudolph Agricola's Address to Innocent VIII," in *Christian Humanism: Essays in Honor of Arjo Vanderjagt*, ed. A. A. MacDonald and Jan Veenstra (Leiden, 2009), 431–43; José C. Miralles Maldonado, "Discurso de obediencia de Antonio Agustín y Siscar al Papa Julio II en nombre del Rey Fernando el Católico (1507), *Humanistica Lovaniensia* 65 (2016): 131–63. Florentine obedience embassies are described by Brian Jeffery Maxson, *The Humanist World of Renaissance Florence* (Cambridge, 2014), 85–106; Isabella Lazzarini, *Communication and Conflict: Italian Diplomacy in the Early Renaissance 1350–1520* (Oxford, 2015), 158–60. Tessa Beverley, "Venetian Ambassadors 1454–1494: An Italian Elite" (PhD diss., University of Warwick, 1999), includes Venetian obedience embassies to Rome from Nicholas V to Julius II.

11. Concetta Bianca, "Le orazioni a stampa," in *Roma di fronte all'Europa al tempo di Alessandro VI*, 3 vols. (Rome, 2001), 2.441–67; Holger Nickel, "Orations Crossing the Alps," in *Incunabula and Their Readers: Printing, Selling, and Using Books in the Fifteenth Century*, ed. Kristian Jensen (London, 2003), 153–8.

12. John O'Malley, *Praise and Blame in Renaissance Rome: Rhetoric, Doctrine, and Reform in the Sacred Orators of the Papal Court, c. 1450–1521* (Durham, NC, 1979); John M. McManamon, *Funeral Oratory and the Cultural Ideals of Renaissance Humanism* (Chapel Hill, NC, 1989). For an equally important study of Counter-Reformation rhetoric, see Frederick J. McGinness, *Right Thinking and Sacred Oratory in Counter-Reformation Rome* (Princeton, NJ, 1995).

13. Catherine Fletcher, *Diplomacy in Renaissance Rome: The Rise of the Resident Ambassador* (Cambridge, 2015), and Gianvittorio Signorotto and Maria Antonietta Visceglia, eds., *Court and Politics in Papal Rome, 1492–1700* (Cambridge, 2002), landmark studies on diplomacy and court culture in early modern Rome, do not treat the question of diplomatic oratory. See Brian Jeffery Maxson, "Diplomatic Oratory," and Monica Azzolini and Isabella Lazzarini, "Diplomacy and the Papacy," in *Italian Renaissance Diplomacy: A Sourcebook*, ed. Monica Azzolini and Isabella Lazzarini (Durham, NC, 2017), 27–41, 96–115.

14. Maxson, *Humanist World*, 85–106. Karsten Plöger, *England and the Avignon Popes: The Practice of Diplomacy in Medieval Europe* (London, 2005), 209–18, describes the (physical) gifts ambassadors gave popes in the fourteenth century (bolts of cloth, vestments, gilded dishes and plates) as tokens in a larger exchange of honor, prestige, and favor.

15. Subject cities sent embassies to congratulate newly elected Venetian doges in these same years; the ambassadors would deliver an oration in praise of the doge and his city. See Luka Špoljarić, "Power and Subversion in the Ducal Palace: Dalmatian Patrician Humanists and Congratulatory Orations to Newly Elected Doges," in *Neo-Latin Contexts in Croatia and Tyrol: Challenges, Prospects, Case Studies* (Göttingen, 2018), 81–106; see also Monique O'Connell, "The Multiple Meanings of Ritual: Orations and the Tensions of Empire," in *Rituals of Politics and Culture in Early Modern Europe*, ed. Mark Jurdjevic and Rolf Strøm Olsen (Toronto, 2016), 91–109, esp. 98–103.

16. Hector Fliscus, *Oratio ad Innocentium VIII pontificem maximum* ([Rome: Stephan Plannck, after April 27, 1485]) (if00196000), fol. 1r.

17. Burchard, 1.374–5, 1.384, 1.425: whether an envoy could visit the pope before formally delivering his obedience; 1.118: Johannes Dalberg's poor accent; 1.330: the bishop of Glasgow appearing in the wrong vestments ("quod inhonestissimum fuit"); 1.381: the Venetian ambassador's letter does not mention "obedience" explicitly; 1.568: an argument about lunch for the embassy of Charles VIII; 2.450: the English ambassador "fecit orationem brevem et bonam"; 2.461, the orator for the Knights of Rhodes "fecit orationem nimis longam."

18. In May 1485, after Innocent fell ill, four different obedience embassies were stacked up in lodgings in Rome waiting to deliver their orations; Burchard had them all come in and orate, one after another; the pope replied to the group collectively.

19. Arnold Harris Mathew dismisses one as "not at all remarkable" (Burchard, *Diary*, 106 n. 1); Setton calls Caoursin's oration to Innocent VIII "unctuous" and "turgid" (Setton, 2.392–3).

20. Pius II, *Commentaries*, vol. 1, ed. Margaret Meserve and Marcello Simonetta (Cambridge, MA, 2004), 1.29.4: "pro vetusta consuetudine"; 1.24.1: "pro veteri more."

21. Aeneas Sylvius Piccolomini, *Oratio ad Calixtum papam offerendo obedientiam imperatoris* [1455], in Pius II, *Opera Omnia* (Basel, 1551), 923–8, at 923: "Solent plerique omnes, beatissime pater, maxime pontifex, qui ad tradendam obedientiam huic sacrosanctae Romanae sedi mittuntur . . ."

22. See, e.g., Bartholomaeus Scala, *Oratio ad Innocentium VIII* ([Rome: Stephan Plannck, after December 15, 1484]) (is00302000), fol. 1r: "Florentinus populus, qui modo nos sex oratores ad te *de more* gratulatum misit"; Andreoccius de Shinucciis, *Oratio pro Senensibus ad Innocentium VIII* ([Rome: Johannes Schoemberger, after October 30, 1484]) (is00488000), fol. 2v: "pollicemur atque offerimus congruam, *solitam*, sanctamque obedientiam"; Robertus Guiba, *Oratio ad Innocentium VIII in obedientia praestanda* ([Rome: Stephan Plannck, after June 10, 1485]) (ig00550000), fol. 2r–v: "Gratulatur sanctitati tue illustrissimus dux . . . obedientiam veram et integram *per eum et predecessores suos prestari solitam* summa cum veneratione exhibet et pollicetur" (emphasis mine).

23. Walter Ullmann, *The Growth of Papal Government in the Middle Ages* (London, 1962), 253–61, esp. 257 and 339 n. 3: "Juro . . . fidelitatem tuisque successoribus."

24. See I. S. Robinson, *The Papacy, 1073–1198: Continuity and Innovation* (New York, 1990), 371 and 394, for the oath Robert Guiscard made to Gregory VII (written by Gregory himself) in 1080.

25. Ullmann, *Growth of Papal Government*, 339 n. 3; David Abulafia, *The New Cambridge Medieval History*, vol. 5, *c. 1198–c. 1300* (Cambridge, 1999), 131. The emperor did not promise to *obey* the pope, but to protect and defend him. Ullmann stresses the difference between this imperial oath (*iuramentum subditi*, the oath of an officer the pope had *created*) and the *iuramentum fidelitatis*, or fealty offered by a vassal to the pope. An oath of fealty—such as the one the king of Sicily made—included an explicit promise of obedience in exchange for the pope's and St. Peter's protection. This feudal use of "obedience" was different from what would obtain during and after the Schism.

26. Charles du Cange, *Glossarium ad scriptores mediae et infimae latinitatis* (Niort, 1883–87), s.v., notes earlier medieval meanings of *oboedientia*: the deference that monks owe an abbot; then by extension the lands and dependencies of an abbey; more

generally, lands or people subject to a particular ruler; and finally, and only during the Schism, the faction or following of a particular pope or antipope.

27. Philip H. Stump, "The Council of Constance and the End of the Schism," in *A Companion to the Great Western Schism (1378–1417)*, ed. Thomas Izbicki and Joëlle Rollo-Koster (Leiden, 2009), 395–442, at 399. In 1396, the French theologian Simon de Cramaud circulated a tract entitled *De subtractione obedientie* for the use of a synod of French prelates who withdrew their obedience from Benedict XIII; Howard Kaminsky, *Simon de Cramaud and the Great Schism* (New Brunswick, 1983); Hélène Millet and Emmanuel Poule, *Le vote de la soustraction d'obedience en 1398* (Paris, 1988).

28. Thomas M. Izbicki, "The Missing Antipope: The Rejection of Felix V and the Council of Basel in the Writings of Aeneas Sylvius Piccolomini and the Piccolomini Library," *Viator* 41 (2010): 301–14.

29. One of the clearest accounts of this complicated episode is in John B. Toews, "Pope Eugenius IV and the Concordat of Viena (1448)—an Interpretation," *Church History* 34 (1965): 178–94, esp. 184–6. Toews characterizes the diplomatic agreements of 1447 and 1448 as more favorable to the empire than the papacy, in that they reduced Rome to the status of one state among many that must negotiate for privileges and rights.

30. Pius II, *Commentaries*, 1.16.1: "nationis nomine Sanctae Sedi Apostolicae oboedientiam restituerunt."

31. Letter 67 in *Reject Aeneas, Accept Pius: Selected Letters of Aeneas Sylvius Piccolomini (Pope Pius II)*, trans. Thomas M. Izbicki, Gerald Christenson, and Philip Krey (Washington, DC, 2006), 243–73, with the delivery of the obedience at 254; Pius II, *Commentaries*, 1.16.2–6. I quote here from my translation in Pius II, *Commentaries*, vol. 1, with minor changes.

32. Pius II, *Commentaries*, 1.16.5.

33. Pius II, *Commentaries*, 1.16.6; including the crown of Pope Sylvester and the head of John the Baptist.

34. Maxson, *Humanist World*, 92–95.

35. Quoted by Maxson, *Humanist World*, 94; Heinz Willi Wittschier, *Gianozzo Manetti: Das Corpus der Orationes* (Cologne, 1968), 79–84. The Florentine Signoria's instructions to Manetti's embassy make no reference to "obedience" at all. The ambassadors were to congratulate the pope and express how much joy the Florentines felt at his election and their hopes for good relations with Rome; Stefano Baldassari and Bruno Figliuolo, *Manettiana* (Rome, 2010), 21–2. The text of the speech itself, edited in J. B. Mittarelli, *Bibliotheca codicum manuscriptorum monasterii S. Michaelis Venetiarum prope Murianum* (Venice, 1779), cols. 716–21, likewise offers congratulations but makes no mention of obedience.

36. Stenzig, *Botschafterzeremoniell*, 2.1284–5.

37. Piccolomini, *Oratio ad Calixtum*, 923–8.

38. Piccolomini, *Oratio ad Calixtum*, 924.

39. Piccolomini, *Oratio ad Calixtum*, 926.

40. Pius says that ambassadors from the emperor, Spain, Hungary, Portugal, Bohemia, Burgundy, Austria, and Brandenburg came "pro veteri more adoraturi" (*Commentaries*, 2.24.1). The imperial legation "publice praestitit obedientiam; legatus vero Bohemiae regis in consistorio secreto idem egit" (2.24.2). See also Moroni, "Ubbidi-

enza," 23. Stenzig, *Botschafterzeremoniell*, 2.1285–6, lists obedience embassies to Pius from Venice, Milan, Florence, Avignon, Castile, Naples, Aragon, Portugal, Burgundy, Hungary, Bohemia, Brandenburg, Austria, Tirol, Saxony, and Frederick III. Frederick's orator this time was Johannes Hinderbach, the future bishop of Trent.

41. Pius II, *Commentaries*, 7.15.2.

42. Pius II, *Commentaries*, 7.13.5–9. At 7.13.7 Pius has Jouffroy say that in revoking the Pragmatic Sanction, Louis has "oboedientiam . . . Romanae Primaeque Sedi ac Pio pontifici tanquam Iesu Christi vicario veram et integram restituisse."

43. Pius II, *Commentaries*, 7.13.9.

44. Pius II, *Commentaries*, 7.13.9: "Everyone listened to [Pius] with profound attention, appearing to rouse themselves from the boredom they had suffered in listening to Arras."

45. Stenzig, *Botschafterzeremoniell*, 2.1287, lists Naples, Lucca, Siena, Mantua, Milan, Florence, Ferrara, Venice, the Knights of Rhodes, and Frederick III. There is also an oration for Savoy (see next note).

46. Cf. Ambrosius de Vignate, "Oratio ad pontificem maximum Paulum Secundum pro obedientia Sabaudiensium ducis," in Francesco Filelfo, *Epistolae* (Paris, 1503; repr., Paris, 1514).

47. Harold Acton, *The Pazzi Conspiracy: The Plot against the Medici* (London, 1979), 13.

48. Johannes de Aragonia, *Oratio ad Sixtum IV; Oratio ad Ferdinandum regem* (Rome: Johannes Philippus de Lignamine, [after December 19, 1471]) (ia00939800).

49. Aragonia, *Oratio ad Sixtum IV*, 2r–v.

50. See chapters two and three.

51. Bernardus Justinianus, *Oratio habita apud Sixtum IV contra Turcos* ([Rome: Georgius Lauer, after December 2, 1471]) (ij00606000), fol. 6r–v: "humilem devotamque obedientiam . . . deferamus."

52. Patricia H. Labalme, *Bernardo Giustiniani: A Venetian of the Quattrocento* (Rome, 1969).

53. Bernardus Justinianus, *Orationes et epistolae* (Venice: Bernardinus Benalius, [1492?]) (ij00611000).

54. Franco Martignone, "L'orazione di Ladislao Vetesy per l'obbedienza di Mattia d'Ungheria a Sisto IV," *Atti e memorie della Società Savonese di Storia Patria*, n.s., 25 (1989): 205–50; Stenzig, *Botschafterzeremoniell*, 2.1288–9.

55. Ladislaus Vetesius, *Oratio ad Sixtum IV pro praestanda oboedientia Mathiae Hungarorum Regis* ([Rome: Johannes Schurener, de Bopardia, after February 2, 1475]) (iv00266300); the reprint is iv00266400. Later reprinted again: ([Rome: Stephan Plannck, between 1481 and 1487]) (iv00266500).

56. Martin Lowry, *Nicholas Jenson and the Rise of Venetian Publishing in Renaissance Europe* (Oxford, 1991), 65, 100, and 115.

57. Aeneas Sylvius Piccolomini, *Oratio coram Calixto III de obedientia Friderici III* ([Rome: Stephan Plannck, 1488–90]) (ip00731000).

58. Determining the total is difficult, since Burchard lists more embassies than there are printed texts and does not mention every oration for which there is a printed text.

59. For the spatial aspects of diplomacy in Rome, see Fletcher, *Diplomacy in Renaissance Rome*, 122–44.

60. Stanisław Cynarski and Feliks Kiryk, "Jan z Targowisk," in *Polski Słownik Biograficzny*, ed. Mieczysław Horach and Paweł Jarosiński (Wrocław, 1962–64), 10.484. Thanks to Brian Krostenko for locating and translating this source.

61. Burchard, 2.434, 2.441. Traditionally, the pope sent a golden ornament in the shape of a rose or spray of roses each Laetare Sunday to a prince or state that had proved particularly loyal to the Church.

62. Burchard, 2.450 (English and French dispute) and 2.477 (French and Spanish). A similar dispute could have erupted in spring 1485, when ambassadors of Philip of Wittelsbach and Albert of Bavaria described each prince as both Elector of the Palatinate and Duke of Bavaria in their orations. Burchard does not mention any controversy at court; both orations were printed with the full set of competing titles.

63. The master of ceremonies under Julius II, Paris de Grassis, suggested that the orator touch on four topics in the course of his speech: joy at the news of the new pope's accession, praise of the new pope, the formal offer of obedience, and prayers and commendations for his successful reign; Stenzig, *Botschafterzeremoniell*, 2.1282.

64. Fliscus, *Oratio ad Innocentium VIII*, fols. 2v–3r.

65. Antonius Galeazius Bentivolus, *Oratio ad Alexandrum VI nomine Bononiensium habita* (Rome: Stephan Plannck, [not before September 1492]) (ib00331000), fol. 2v.

66. *Senensium obedientia publica* ([Rome: Stephan Plannck, after October 15, 1492]) (is00441000), fol. 1v.

67. Petrus Cara, *Oratio ad Alexandrum VI nomine Caroli II ducis Sabaudiae habita* ([Rome: Stephan Plannck, after May 21, 1493]) (ic00130000), fols. a4r–5r.

68. Setton, 2.392–4.

69. See, e.g., Petrus Cadratus, *Oratio ad Innocentium VIII* ([Rome: Bartholomaeus Guldinbeck, after February 11, 1485]) (ic00014000), fol. 2v: "Carolus VIII . . . te . . . verum Christi vicarium, beati Petri principis apostolorum verum successorem, verum papam, verum pontificem confitetur"; similarly, Franciscus Patritius, *Oratio ad Innocentium VIII* ([Rome: Stephan Plannck, after December 29, 1484]) (ip00153000), fol. 4r.

70. Guilelmus Caoursin, *Oratio ad Innocentium VIII* ([Rome: Stephan Plannck, after January 28, 1485]) (ic00107000), fol. 1r.

71. Fliscus, *Oratio ad Innocentium VIII*, fol. 5v.

72. Rogers, *Obedience of a King of Portugal*, 51–2.

73. Benvenuto da San Giorgio's oration to Alexander VI on behalf of Montferrat was delivered on Feburary 25, 1493. Stephan Plannck's edition of this text (is00128000) includes a prefatory letter by the author dated February 17. This could indicate that the author had not only prepared a copy of the text for readers but also delivered it to the printer before his ceremonial recitation on February 25. See *BMC* 6.779.

74. Petrus Cadratus, *Oratio ad Innocentium VIII habita* + Andreoccius de Shinucciis, *Oratio pro Senensibus ad Innocentium VIII* ([Rome: Stephan Plannck, after February 11, 1485]) (ic00015000).

75. Van der Laan, "Rudolph Agricola's Address," 434.

76. Cherubino Ghirardacci, *Historia di Bologna*, ed. Albano Sorbelli, *RIS*² 33.1.268; see also Alessandra Tugnoli, "Sala, Giovanni Gaspare da," in *DBI* 89.

77. Angelo (Fondi) da Vallombrosa, *Oratio elegantissima nomine serenissimi senatus apud Julium II ponitificem habita*, ed. Johannes Koechel ([Munich]: Johann Schobsser, 1505) (ia00707630).

78. Guilelmus Caoursin, *Ad Innocentium papam VIII oratio* ([Rome: Bartholomaeus Guldinbeck, after January 28, 1485]) (ic00106000); also ([Rome: Eucharius Silber, after January 28, 1485]) (ic00106500) and ([Rome: Stephan Plannck, after January 28, 1485]) (ic00107000).

79. Guilelmus Caoursin, *Rhodiorum Historia* (Ulm: Johann Reger, October 24, 1496) (ic00113000).

80. Caoursin, *Rhodiorum Historia*, sigs. h2v (woodcut) and h3r–h4r (oration).

81. Hieronymus Porcius, *Commentarius de creatione et coronatione Alexandri VI* (Rome: Eucharius Silber, September 18, 1493) (ip00940000).

82. Rafael Sabatini, *The Life of Cesare Borgia* (London, 1912): "court flattery could not be carried further than it was in this case by Hieronymus, an affected pedant, an empty-headed braggart, a fanatical papist."

83. Every one of the orations has a heading identifying the state that sponsored it; none of the headings identify the ambassador by name, but all credit Hieronymus Porcius with editing the text. See, e.g., "Sanctissimo ac beatissimo domino nostro Pape Senensium obedientia publica, per Hieronimum Porcium compilata," followed by "Alexandri Borgia sexti pontificis maximi responsio publica, per Hieronimum Porcium compilata."

84. Leonellus Chieregatus, *Oratio in funere Innocentii VIII etc.*, ed. Johannes de Velmede (Leipzig: Martin Landsberg, [ca. 1495]) (ic00453500). The printed title is *Orationes decoratissime Rome in ecclesia sancti Petri in funere Innocentii pape VIII coram cetu cardinalium prima, secunda coram papa moderno Alexandro sexto in prestanda obedientia habite.*

85. Michael Fernus, *Epistola de legationum adventu* (Rome: Eucharius Silber, [after May 23, 1493]) (if00104000), 5r.

86. Silvio A. Bedini, *The Pope's Elephant: An Elephant's Journey from Deep in India to the Heart of Rome* (New York, 1997). On the elephant's performance as a proffer of obedience, see Rogers, *Obedience of a King of Portugal*, 4–5.

87. See chapter three, note 72.

88. Caoursin was awarded appointments by Innocent and given presents on his return to Rhodes in 1485 (Setton 2.392–3); Scala received financial rewards from Innocent VIII and accolades on his return to Florence (Maxson, *Humanist World*, 173–5).

89. Broadsides of these texts were printed in Leipzig and Nuremberg: ii00100930, ii00109060, ii00114500, ii00116600.

90. Margaret Meserve, "The Papacy, Power, and Print: The Publication of Papal Decrees in the First Fifty Years of Printing," in *Print and Power*, ed. Nina Lamal (Leiden, 2021), 259–99.

91. Innocent's *In coena Domini* bull for 1486 was printed four times in the city, by Plannck and Silber (twice each: ii00106600, ii00107000, ii00107300, ii00107800); his bull for 1492 was printed once, by Plannck (ii000108000). Alexander's 1499 *In coena Domini* was published four times in Rome by Plannck, Silber, and Besicken (ia00371300, ia00371400, ia00371500, ia00372000) alongside four editions of another bull, issued on the same day, announcing the Jubilee of 1500.

92. Innocent VIII, *Bulla "Etsi ex iniuncto" condemnatoria libelli Conclusionum DCCCC Joannis Pici Mirandulani* ([Rome: Eucharius Silber, after August 4, 1487]) (ii00108900).

93. Innocent VIII, *Mandatum apostolicum et maledictio adversus Brugenses et Flamingos* ([Germany: n.p., after March 23, 1487]) (ii00136750).

94. Donald Weinstein and Valerie R. Hotchkiss, eds., *Girolamo Savonarola: Piety, Prophecy, and Politics in Renaissance Florence* (Dallas, 1994), esp. 65–71; Peter Amelung, "The Savonarola Collection of the Württembergische Landesbibliothek," in *Incunabula: Studies in Fifteenth-Century Printed Books Presented to Lotte Hellinga*, ed. Martin Davies (London, 1999), 549–57.

95. Alexander VI, Breve, May 12, 1497, "Cum saepe" [Italian] ([Florence: Francesco di Dino, after May 12, 1497]) (ia00369800). The letter appeared again in later editions with the text of the *processo* against Savonarola; this composite edition was printed twice in Florence, twice in Venice, and once in Rome.

96. Leonellus Chieregatus, *Sermo in publicatione confoederationis inter Innocentium VIII et Venetos* ([Rome: Bartholomaeus Guldinbeck, after February 2, 1487]) (ic00461000) and ([Rome: Eucharius Silber, after February 2, 1487]) (ic00462000); Chieregatus, *Sermo in publicatione confoederationis inter Alexandrum VI et Romanorum et Hispaniae reges, Venetorumque ac Mediolanensium duces*; Alexander VI, Bulla, April 6, 1495, "Quoniam pro communi" ([Rome: Stephan Plannck, after April 12, 1495]) (ic00458000) and ([Rome: Johann Besicken, after April 12, 1495]) (ic00459000); this text was reprinted in Milan, Pavia, Leipzig, and Strassburg.

97. Bartholomaeus Floridus, *Oratio confoederationis initae inter Alexandrum VI et Venetorum, Mediolani et Bari duces* ([Rome: Eucharius Silber, after April 25, 1493]) (if00228200).

98. Hadrianus Castellensis, *Oratio super foedere inter Alexandrum VI ac Romanorum, Hispaniae et Angliae reges, Venetorum Mediolanensiumque duces* ([Rome: Andreas Freitag, after July 31, 1496]) (ih00000800).

Chapter 8 • Brand Julius

1. "Julius Excluded from Heaven," in Desiderius Erasmus, *The Praise of Folly and Other Writings*, trans. Robert M. Adams (New York, 1989), 145. The attribution of this text to Erasmus, much debated, remains unresolved. For arguments in favor, see Silvana Seidel Menchi, ed., *Iulius exclusus e coelis*, in *Opera omnia Desiderii Erasmi*, I.8 (Leiden, 2013), 20–40.

2. "Julius Excluded from Heaven," 149–50.

3. Pastor, 6.185–607; Felix Gilbert, *The Pope, His Banker, and Venice* (Cambridge, MA, 1980); Christine Shaw, *Julius II: The Warrior Pope* (Oxford, 1993).

4. See, e.g., Julius II, *Breve ad reges duces et principes christianos in quo continentur potiores (licet plures sint alie) causae privationis cardinalium hereticorum scismatorumque* ([Rome:] Giacomo Mazzocchi, on or after October 25, 1511): "Nullus hoc breve imprimere audeat . . . sine licentia secretarii nostri domestici Sigismundi. / De mandato prefati magnifici domini Secretarii impressi ego Iacobus de Mazochis." Other editions invoke the authority of another secretary, Balthasar Tuerdus, and Antonio de Monte, an official of the Apostolic Camera.

5. These are enumerated and described in more detail in the appendix to Margaret Meserve, "The Papacy, Power, and Print: The Publication of Papal Decrees in the First Fifty Years of Printing," in *Print and Power*, ed. Nina Lamal (Leiden, 2021), 259–99.

6. Meserve, "Papacy," appendix, 1.1.

7. Meserve, "Papacy," appendix, 1.3.

8. Meserve, "Papacy," appendix, 1.4–7.

9. Meserve, "Papacy," appendix, 2.2–7.

10. Meserve, "Papacy," appendix, 2.1, 3.1–7.

11. Charles A. Stinger, *The Renaissance in Rome* (Bloomington, IN, 1985, 2nd ed. 1998); Giovanna Rotondi Terminiello and Giulio Nepi, eds., *Giulio II: Papa, politico, mecenate* (Genoa, 2005); F. Cantatore et al., eds., *Metafore di un pontificato: Giulio II (1503–1513)* (Rome, 2010); Nicholas Temple, *Renovatio urbis: Architecture, Urbanism and Ceremony in the Rome of Julius II* (Abingdon, 2011).

12. Massimo Rospocher, "Print and Political Propaganda under Pope Julius II," in *Authority in European Book Culture, 1400–1600*, ed. Pollie Bromilow (Farnham, UK, 2013), 97–119; Rospocher, *Il papa guerriero: Giulio II nello spazio pubblico europeo* (Bologna, 2015); Ottavia Niccoli, *Prophecy and People in Renaissance Italy*, trans. Lydia G. Cochrane (Princeton, NJ, 1990); Niccoli, *Rinascimento anticlericale: Infamia, propaganda e satira tra Quattro e Cinquecento* (Bari, 2005).

13. Nelson H. Minnich, "The Images of Julius II in the *Acta* of the Councils of Pisa-Milan-Asti-Lyons (1511–12) and Lateran V (1512–1517)," in Terminiello and Nepi, *Giulio II*, 79–90.

14. A broadside bull of Sixtus IV printed in Strassburg in 1480 (is00546700) includes the pope's coat of arms inset into the text. Alexander VI's 1498 Jubilee bull (ia00371200), printed by Eucharius Silber with a woodcut image of the Veronica flanked by Peter and Paul (fig. 5.17), is the only bull printed in Rome before 1501 that includes a woodcut, as far as I know.

15. Meserve, "Papacy," appendix, 1.1.a, features a simple title in the same type as the body of the text; some Roman editions of appendix, 1.2–3 (1508–9), feature a title in larger type and a small woodcut *stemma* at the start of the text; in appendix, 1.4, several editions of the bull against Alfonso d'Este, August 1510, feature full title pages featuring descriptive titles in large type, large woodcut *stemme*, and woodcut borders.

16. Shaw, *Julius II*, 127–35.

17. Shaw, *Julius II*, 148–9. This was the same Giovanni who went to Loreto in 1485 and then offered obedience to Innocent VIII. On his visit to Rome then, Giovanni stayed with Giuliano della Rovere at Santi Apostoli.

18. Shaw, *Julius II*, 149–61; Pastor, 6.277. The text of the bull is partially edited in *Annales ecclesiastici*, ed. Augustinus Theiner, 37 vols. (Paris, 1864–83), 30.458–60.

19. Meserve, "Papacy," appendix, 1.1.a.

20. Massimo Rospocher, "Guerre d'inchiostro e di parole al tempo di Cambrai," in *Dal Leone all'aquila: Communità, territori e cambi di regime nell'età di Massimiliano I*, ed. Marcello Bonazza and Silvana Seidel Menchi (Rovereto, 2012), 127–47, at 130.

21. Cecilia M. Ady, *The Bentivoglio of Bologna: A Study in Despotism* (Oxford, 1937), 197–9.

22. Meserve, "Papacy," appendix, 1.1.b; the date of this edition, printed by Johannes Beplin, is discussed below.

23. Pastor, 6.299–300. Ferrara, Mantua, and Savoy later joined the league as well (312). See also Giuseppe Gullino, ed., *L'Europa e la Serenissima: La svolta del 1509* (Venice, 2011).

24. Pastor, 6.300, 306–11.

25. Newberry Library: Wing Broadside ZP 5351.09. Thanks to Paul Gehl for calling this to my attention.

26. Pastor, 6.311–2; Sanudo, 8.187–205.

27. Meserve, "Papacy," appendix, 1.3.a.

28. Meserve, "Papacy," appendix, 1.3.b–n.

29. Sanudo, 8.203.

30. Sanudo, 8.161–2.

31. Sanudo, 8.187: "inteso questo, il papa fe' cavarle via, fulminando con grandissimo colora, etc."

32. *La rotta di Ghiaradadda: Agnadello, 14 maggio 1509: Studi, testi, e contributi per una storia della battaglia di Agnadello* (Treviso, 2009); Marco Meschini and Giancarlo Perego, *La battaglia di Agnadello: Ghiaradadda, 14 v. 1509* (Bergamo, 2009).

33. Silber also printed a ballad, *La miseranda rotta de' Veneziani*; Rospocher, *Il papa guerriero*, 227–56; Niccoli, *Prophecy and People*, 3–29.

34. Pastor, 6.316–7.

35. Pastor, 6.320, 444, 463; it is not clear whether this fictive letter was ever printed. Sanudo copied the text (9.567–70) in February 1510; Niccoli, *Rinascimento anticlericale*, 86–8.

36. Sanudo, 9.557: "Iesus Christus Mariae Virginis filius Iulio II vicario nostro indigno"; 9.560: "Datae ex Caelo nostro empireo, die 26 Decembris anno Nostrae Nativitatis in seculo 1509. / Ioannes Evangelista / de mandato subscripsi."

37. Pastor, 6.321–6.

38. Pastor, 6.326.

39. Pastor, 6.327–8.

40. Meserve, "Papacy," appendix, 1.4.a–e; Pastor, 6.328–9. Julius accused Alfonso of plotting with Cardinal d'Amboise (Louis XII's chief advisor, and brother of Louis d'Amboise) to depose Julius and make d'Amboise pope.

41. Pastor, 6.329.

42. Pastor, 6.330. This meeting eventually took place in Tours.

43. Pastor, 6.355–6.

44. Meserve, "Papacy," appendix, 1.8.a–b.

45. Pastor, 6.332.

46. Meserve, "Papacy," appendix, 1.5.a–e.

47. Meserve, "Papacy," appendix, 1.6.a–e.

48. Sanudo, 11.615: "La scomunicha, fata per il papa contra il gran maistro e altri francesi, ozi vidi vender su el ponte di Rialto, a stampa, latina e vulgar, un soldo l'una." I have not found an edition of this bull in translation. The Latin edition is not signed. The text could have been sent from Bologna to Rome to be printed, and printed copies could have been sent to Venice, but to accomplish this in a month, between October 14 and November 15, would have been quite a feat. Rospocher, "Guerre d'inchiostro," 130.

49. Pastor, 6.339.

50. Pastor, 6.339–46.

51. Theiner, *Annales ecclesiastici*, 30.559; Meserve, "Papacy," appendix, 1.9.a.

52. Theiner, *Annales ecclesiastici*, 30.539; Sanudo, 12.250–4; Pastor, 6.352–3.

53. Sanudo, 12.218–9.

54. Jean Lemaire de Belges, *Le Traité intitulé de la difference des scismes et des concilles de l'église* (Lyon, 1511). See also Jennifer Britnell, "The Antipapalism of Jean Lemaire de Belges' *Traité de la Difference des Schismes et des Conciles*," *Sixteenth Century Journal* 24 (1993): 783–800; and Margaret Meserve, "The Sophy: News of Shah Ismail Safavi in Renaissance Europe," *Journal of Early Modern History* 18 (2014): 1–30, at 21–3.

55. Pastor, 6.357–9; Michael A. Sherman, "Political Propaganda and Renaissance Culture: French Reactions to the League of Cambrai," *Sixteenth Century Journal* 8 (1977): 97–128; Jean-Claude Margolin, "Pamphlets gallicans et antipapistes (1510–1513): De la *Chasse du Cerf des Cerfs* de Gringoire au *Julius Exclusus* d'Erasme," *Moreana* 47 (2010): 123–47; Jennifer Britnell, *Le roi très chrétien contre le pape: Écrits antipapaux en français sous le règne de Louis XII* (Paris, 2011).

56. Niccoli, *Rinascimento anticlericale*, 79–91; Rospocher, *Il papa guerriero*, 259–92.

57. Sanudo, 12.218.

58. Pastor, 6.364–5.

59. Francesco Calvo called himself *impressore apostolico* in an edition of 1528; Antonio Blado signed himself *tipografo camerale* in 1535, a year after Calvo quit the city. See Francesco Barberi, "Blado, Antonio," *DBI* 10.753–7; Paolo Sachet, "Il Contratto tra Paolo Manuzio e la Camera Apostolica (2 Maggio 1561): La creazione della prima stamperia Vaticana privilegiata," *La Bibliofilia* 115 (2013): 245–62.

60. Besicken came to the city around 1493.

61. Fernanda Ascarelli, *Annali tipografici di Giacomo Mazzocchi* (Florence, 1961), esp. 15–9; Francesco Barberi, "Stefano Guillery e le su edizioni romane," in *Studi offerti a Roberto Ridolfi*, ed. Berta Maracchi Biagiarelli and Dennis E. Rhodes (Florence, 1973), 95–145, esp. 96–110; Tinto, 1–8; Martin Davies, "Besicken and Guillery," in *The Italian Book 1465–1800, Studies Presented to Dennis E. Rhodes*, ed. Denis V. Reidy (London, 1993), 35–54.

62. Massimo Miglio, *Saggi di stampa: Tipografi e cultura a Roma nel Quattrocento*, ed. Anna Modigliani (Rome, 2002); Concetta Bianca, "Le strade della 'sancta ars': La stampa e la curia a Roma nel XV secolo," in *La stampa romana nella Città dei Papi e in Europa*, ed. Cristina Dondi et al. (Vatican City, 2016). In 1512, Mazzocchi signed his editions "in vico Pellegrini," near Campo de' Fiori; Ascarelli, "Annali tipografici," 18. Silber seems to have worked out of Sweynheym and Pannartz's old premises in Palazzo Massimo.

63. See, e.g., *Monitorium contra Venetos* (Rome: Giacomo Mazzocchi, [1509]), fol. 12v: "Impressum . . . de mandato prelibati Sanct. D. N. domini Iulii divina providentia pape II."

64. Shaw, *Julius II*, 134–5.

65. For more on Guarini, see (of all places) Willard Fiske, *Chess in Iceland and Icelandic Literature with Historical Notes on Other Table-Games* (Florence, 1905), 211–3.

66. Rome, BNC, MS SS. Apostoli 4 (326), fols. 133r–138v. *Bulla, indulta, ac privilegia concessa civitati Forliviensi per sanctissimum dominum nostrum Iulium II pontificem maximum* (Forlì, 1507).

67. Julius II, *Bulla intimationis* (Rome: Jacopus Mazochius, July 31, 1511), sig. b4v. Mazzocchi's edition of the Italian translation of this bull has a more elaborate *rotula*

that includes a small image of the tiara and crossed keys. A Nuremberg reprint of the Latin edition includes an entirely woodcut *rotula*; R.I.IV.2116(3), sig. b2v.

68. AAV Arm. IV, t. 25, p. 269 (Bull of Paul III of 1542, with a woodcut *rotula* surrounded by a laurel wreath, divided into quadrants with Peter and Paul in the upper two, and "Paulus" and "PP III" below). Flavia Bruni, "In the Name of God: Governance, Public Order, and Theocracy in the Broadsheets of the Stamperia Camerale of Rome," in *Broadsheets: Single-Sheet Publishing in the First Age of Print*, ed. Andrew Pettegree (Leiden, 2017), 139–61.

69. Meserve, "Papacy," appendix, 1.3.a–c, 1.3.e–f.

70. Meserve, "Papacy," appendix, 1.3.i.

71. Meserve, "Papacy," appendix, 2.1.a–n.

72. *Bulla Julii pape ii. edita contra Johannem Bentivolum in civitate Bononiensi libertatem ecclesiasticam occupantem* (Rome: Johannes Besicken, November 12, 1506). At least one copy (BAV R.I.IV.1414 (int. 9)) has a variant title that omits the last three words.

73. See, e.g., Silber's edition of Julius II, *Bulla prima annatarum*, after July 28, 1505 (Tinto, 30); *Bulla super electione Pontificis futura*, after January 14, 1506 (Tinto, 33); *Bulla confirmatoria bulle Sixti Quarti de testando*, after December 20, 1507 (Tinto, 42).

74. Meserve, "Papacy," appendix, 1.1.b. The edition is signed "Impressum Rome per Ioannem Beplin de Argentina ad instantiam magistri Ioannis Carminate de Lodi Anno salutis M.D.VI. Die vero . xii . mensis Novembris," but there are good reasons to doubt this information. Beplin is recorded as active in Rome in 1506–17, but this is the only edition attributed to him for the year 1506 and in fact is the only Beplin book to bear a date before 1511. Giovanni Battista Carminate di Lodi is recorded as a partner in several Beplin colophons, but these all date to 1511 or later as well. The criblé woodcut Beplin used is more commonly found in editions of Marcellus Silber (Tinto lists fifteen Marcellus Silber editions in which it appears), all from 1510 or later, with the exception of one undated edition of another 1506 bull that was formally reissued by the chancery in 1510. All this suggests that this edition dates to 1510 or 1511 as well: Beplin would have reprinted Besicken's 1506 edition, copying not only the text but also the original printing date from Besicken's colophon. Many Julian bulls were published in multiple editions, some of them issued years after their original publication.

75. Tinto, 43.

76. *Littere apostolice institutionis Collegii scriptorum Archivii Romane Curie* (Tinto, 41). Tinto's "stemma di Giulio II no. 1" is also reproduced in Tinto, tav. XII.

77. Tinto, 46.

78. Manuel I's *Epistola ad Julium II* of June 12, 1508 (Tinto, 47), and Nicoletto Dati's *Oratio in die Circumcisionis*, October 6, 1508 (Tinto, 50).

79. Tinto, 57, 65, 100.

80. *Bulla . . . super privatione*, August 9, 1510; Meserve, "Papacy," appendix, 1.4.b.

81. *Bulla censurarum . . . et interdicti generalis*, October 9, 1510; Meserve, "Papacy," appendix 1.5.a–b.

82. *Bulla declarationis incursus censurarum . . .* , October 14, 1510; Meserve, "Papacy," appendix, 1.6.a–b.

83. *Bulla intimationis . . .* , July 18, 1511, in two different editions of this announcement; Meserve, "Papacy," appendix, 2.1.d and f.

84. *Bulla monitorii . . .* , July 28, 1511; Tinto, 104, Meserve, "Papacy," appendix 2.2.b.

85. Tinto, 105.

86. *Breve ad reges duces et principes christianos . . .* , October 24, 1511; Meserve, "Papacy," appendix, 2.5.

87. Tinto 130, 137 (postmortem).

88. Barberi, "Stefano Guillery"; Davies, "Besicken and Guillery." For his woodcut borders and other ornaments in the *stilo francese*, see Lamberto Donati, "Stampe quattrocentine di Stefano Guillireto," in *Essays in Honour of Victor Scholderer*, ed. Dennis Rhodes (Mainz, 1970), 144–58.

89. Guillery's edition is listed in Frank Isaac, *An Index to the Early Printed Books in the British Museum. Part II: MDI–MDXX* (London, 1938), no. 12150; Silber's is Tinto, 64.

90. Alfred W. Pollard, *Last Words on the History of the Title-Page* (London, 1891); Theodore Low De Vinne, *The Practice of Typography: A Treatise on Title-Pages* (New York, 1902); Alastair Fowler, *The Mind of the Book: Pictorial Title Pages* (Oxford, 2017); Francesco Barberi, *Il Frontispizio nel libro italiano del Quattrocento e del Cinquecento* (Milan, 1968).

91. Tinto, stemma no. 2, reproduced on tav. XII.

92. The cut identified by Tinto as stemma no. 3 appears much less frequently in Silber's production.

93. I identify this edition as Guillery's based on the woodcut initial *I* on fol. 2r, which also appears in the *Bulla intimationis* (1511) signed by Guillery, which features the smaller woodcut of the Julian *stemma*. If Guillery kept the *I* in his possession from 1510 to 1511, then its appearance here would indicate that this edition is his as well.

94. Tinto's cornice no. VIII and no. IX.

95. Meserve, "Papacy," appendix, 2.1.j, 2.2.d, 2.5.d, 2.7.d.

96. Meserve, "Papacy," appendix, 2.1.j (var), 2.2.e.

97. Meserve, "Papacy," appendix, 1.4.d, 1.6.d.

98. *Ablas Büchlein der Stationes der Stat Rom unnd der Kirchen mit irem Ablas durch das gantz Jahr. Babst Julius der Zehendt* (Nuremberg, 1515).

99. Meserve, "Papacy," appendix, 1.1.

100. Meserve, "Papacy," appendix, 2.6.

101. Meserve, "Papacy," appendix, 1.5.

102. Meserve, "Papacy," appendix, 1.6.

103. Meserve, "Papacy," appendix, 2.7.

104. Meserve, "Papacy," appendix, 3.2.

105. Meserve, "Papacy," appendix, 2.5: "Breve ad reges duces et principes christianos in quo continentur potiores (*licet plures sint alie*) causae privationis cardinalium hereticorum scismatorumque" (emphasis mine).

106. See note 25 above; see also Munich, BSB, Einblatt. XI,96. A German translation of the letter, also a broadside, was printed in Munich, BSB, Einblatt. VI,22.

107. I am grateful to Orsolya Mednyánszky, who identified the source of this image for me. See the catalogue notes to New York, Pierpont Morgan Library, MS M.272.

108. The first known edition was printed in Bologna in 1515 (*Ioachini abbatis Vaticinia circa apostolicos viros et ecclesiam Romanam* [Bologna: Girolamo de' Benedetti, 1515]), with the image of the pope, the bird, and the stars at sig. c1v.

109. Meserve, "Papacy," appendix, 2.1, 3.1–7.

110. Nelson H. Minnich, "The First Printed Editions of the Modern Councils: From Konstanz to Lateran V (1499–1526)," *Annali dell'Istituto storico italo-germanico in Trento/Jahrbuch des italienisch-deutschen historischen Instituts in Trient* 29 (2003): 447–68.

111. Nelson H. Minnich, "*Rite convocare ac congregare procedereque*: The Struggle between the Councils of Pisa-Milan-Asti-Lyons and Lateran V," in *Councils of the Catholic Reformation* (Aldershot, 2008), study IX.

112. Meserve, "Papacy," appendix, 1.8.

113. *Littere clare memorie Ludovici XI Francorum regis Christianissimi super abrogatione pragmatice sanctionis in quarta sessione sacroscancti Lateranensis concilii publice lecte et recitate* ([Rome: Marcellus Silber, 1512]): USTC 838702.

114. Meserve, "Papacy," appendix, 2.1.a and reprints, fol. 8v: "Constitutio Concilii Constantiensis de auctoritate et potestate sacrorum generalium conciliorum temporibusque et modis eadem convocandi et celebrandi."

115. *Donatio Constantini*, published in Rome in three different editions between 1504 and 1513: USTC 826855, 955616, 801952.

116. Tinto, cornice no. VIII.

117. Tinto, cornice no. IX.

118. Angelo Fondi da Vallombrosa, *Epistole Angeli Anachorite Vallisumbrose pro Christiana unitate servanda* ([Rome: Marcellus Silber, after October 25, 1511]), fol. 1v.

119. J. H. Burns, "Angelo da Vallombrossa and the Pisan Schism," in *The Church, the Councils, and Reform: The Legacy of the Fifteenth Century*, ed. Gerald Christianson et al. (Washington, DC, 2008), 194–211.

120. Burns, "Angelo da Vallombrossa," 197–8.

121. Burns, "Angelo da Vallombrossa," 199–206.

122. E.g., "A tergo autem supradictarum litterarum Apostolicarum infrascripta verba erant . . . " and "A tergo vero erant scripta hec verba, videlicet . . . "

123. Randolph C. Head, *Making Archives in Early Modern Europe: Proof, Information, and Political Record-Keeping, 1400–1700* (Cambridge, 2019), 52–3.

124. Nelson H. Minnich, "The Official Edition (1521) of the Fifth Lateran Council," in *Councils of the Catholic Reformation* (Aldershot, 2008), study III.

125. *Bulla concilii in decima sessione super materia montis pietatis lecta per r. p. dominum Bertrandum episcopum Adriensem, oratorem ducis Ferrarie in Romana curia* ([Rome: Marcellus Silber, 1515]). For other examples of "lecta" in the titles of Leonine bulls printed by Marcellus Silber, see Tinto, 215, 224, 225: *Bulla undecime sessionis reformationis predicatorum divini verbi, lecta per R. p. d. Ioannem episcopum Revaliensem illustrissimorum principum dominorum Ioachim principis electoris ac Alberti magni magistri Militie ordinis Theotonicorum, marchionum Brandenburgensium oratorem*.

126. *Bulla concilii in decima sessione super materia montis pietatis lecta . . .* ([Rome: Marcellus Silber, 1515]), sig. a4v. The same acclamations are recorded in Tinto, 198 and 199 (bulls also read out at the tenth session).

127. Reproduced in Rospocher, *Il papa guerriero*, fig. 7.

128. Leo X, *Bulla super treugis* [*sic*] *et induciis quinquennalibus inter principes Christianos* ([Rome: Etienne Guillery, after March 14, 1518]) (Barberi, "Stefano Guillery," 142).

129. Leo X, *Bulla super impressione librorum, lecta in decima sessione sacrosanctae Lateranensis concilii, per r. p. d. episcopum Nanatensem* ([Rome: Marcellus Silber, 1515]).

See Rudolf Hirsch, *Printing, Selling, and Reading, 1450–1550* (Wiesbaden, 1967), 90; Hirsch, *"Bulla super impressione librorum*, 1515," in *The Printed Word: Its Impact and Diffusion* (London, 1978), study XIV; Pastor, 8.397–8.

130. Leo X, *Monitorium penale contra Franciscum Mariam ducem Urbini* (Rome: [Marcellus Silber, after March 1, 1516]): Tinto, 214. Leo issued another *monitorium* against Thomas de Foix, governor of Milan, in the summer of 1521, printed by Silber in the same format: *Monitorium penale contra Thomam de Foix in ducatu Mediolani gubernatoris locumtenentem* ([Rome: Marcellus Silber, after July 27, 1521]): Tinto, 271.

131. Leo X, *Bulla contra errores Martini Lutheri et sequacium* (Rome: Giacomo Mazzocchi, 1520) (Ascarelli, "Annali tipografici," 138). Reprinted at least eleven times in Paris, Leipzig, Erfurt, Cologne, Rostock, Strassburg, Antwerp, and Landshut: see USTC entries under the title *Bulla contra errores*.

132. Sylvester Prierias, *In presumptuosas Martini Lutheri conclusiones de potestate dialogus* ([Rome: Marcellus Silber, after June 1518]) and *Replica ad fratrem Martinum Lutherum* ([Rome: Marcellus Silber, end of 1518]) (Tinto, 236 and 244).

133. Tommaso Radini Tedeschi, *In Martinum Lutherum oratio* (Rome: Giacomo Mazzocchi, August 1520) (Ascarelli, "Annali tipografici," 139); Giovanni Antonio Modesti, *Oratio ad Carolum caesarem contra Martinum Lutherum* (Rome: Giacomo Mazzocchi, October 2, 1520) (Ascarelli, "Annali tipografici," 140).

134. Charles V, *Contra Martinum Lutherum et eius libros, doctrinam, sectatores, receptatores, ac nonnullos famosos libellos atque chalcographos edictum* (Rome: Giacomo Mazzocchi, July 6, 1521) (Ascarelli, "Annali tipografici," 145).

135. *Determinatio theologice facultatis Parisiensis super doctrina Lutheriana hactenus per eam visa* (Rome: Etienne Guillery, after April 15, 1521); Henry VIII, *Assertio septem sacramentorum adversus Martinum Lutherum* (Rome: Etienne Guillery, [1521]) (Barberi, "Stefano Guillery," 142).

136. Tommaso da Vio (Caietanus), *De divina institutione pontificatus Romani pontificis super totam ecclesiam a Christo in Petro* (Rome: Marcellus Silber, March 22, 1521) (Tinto, 265).

137. The cardinal forbade his flock from purchasing, reading, owning, sharing, or talking about Luther's books; Hirsch, *Printing, Selling, and Reading*, 94.

Conclusion

1. "Roma invasa da manifesti contro Papa Bergoglio," *Il Tempo*, February 4, 2017. The text of the poster read, "A Francè, ha commissariato Congregazioni, rimosso sacerdoti, decapitato l'Ordine di Malta e i Francescani dell'Immacolata, ignorato cardinali . . . Ma 'ndo sta la tua misericordia?"

2. E.g., notices at *Il Fatto Quotidiano* ("Roma: Manifesti anonimi contestano Papa Francesco," February 4, 2017); *Il Messagero* ("Roma, manifesti contro Papa Francesco: spunta l'ombra dei conservatori," February 4, 2017); *Corriere della Sera* ("Roma, manifesti contro il Papa: 'Ma n'do sta la tua misericordia?,'" February 4, 2017); and *Il Post* ("I manifesti contro il papa a Roma," February 5, 2017). Most English-language reports derive from "Anonymous Posters Criticizing Pope Appear in Rome," Reuters, February 4, 2017.

3. Comment by Mauro Mattetti, 9:00 a.m., February 5, 2017, at *iltempo.it*: "È TORNATO !!!!!!!! È tornato Pasquino !!!!!! Ne avevamo proprio bisogno."

4. Antonio Spadaro, Facebook, February 5, 2017. This did not prevent another critic of Francis from suggesting that the pope himself, or his allies, had, "like a new Nero," ordered the poster campaign as a way of generating sympathy for himself.

5. See http://www.vatican.va/offices/papal_docs_list.html.